chemical bonding
clarified through
quantum mechanics

GEORGE C. PIMENTEL
University of California,
Berkeley

RICHARD D. SPRATLEY
University of
British Columbia

chemical bonding clarified through quantum mechanics

HOLDEN-DAY, INC.
San Francisco
Düsseldorf Johannesburg London Mexico
Panama São Paulo Singapore Sydney Toronto

Published by Holden-Day, Inc.
500 Sansome Street, San Francisco, California

Library of Congress Catalog Card Number 71-75914

Printed in the United States of America

234567890 CO 807987654

preface

The hydrogen molecule ion H_2^+ elucidates chemical bonding in the same way that the hydrogen atom clarifies atomic structure. When electron repulsions and the Pauli Principle are considered, the atomic orbitals of the hydrogen atom explain that marvelous array, the Periodic Table—its periodic repetitions and its trends within rows. With these same factors considered, electron repulsions and the Pauli Principle, the molecular orbitals of the hydrogen molecule ion give a parallel coherence to the many-faceted subject of chemical bonding.

With the quantum mechanical view of the atom, we can exploit this unifying approach to the central concept of chemistry, the chemical bond. Mathematical apparatus is not needed, so a valid conceptual portrayal can be conveyed to a student with general interest. We address this monograph to such a group—we feel it will be valuable in a general introductory chemistry course at the college and university level. We hope it will also interest students of chemical bonding at other levels and on both sides of the lectern.

The material presented in this monograph has been molded by the authors' classroom experience at Berkeley over the past five years, first in the accelerated and then the general freshman chemistry courses. Because the subject is abstract, it is lodged in the first half of the third quarter—rather late in the course. This placement permits a concluding return to descriptive chemistry, now with opportunity to exercise the principles of bonding and to illustrate their power.

We have the usual debts to our colleagues, whose encouragement and critiques were invaluable. Professor R. E. Powell, in particular, deserves thanks for the many times he lent his encyclopedic knowledge and critical judgment to the development of some ideas here presented. A number of outstanding graduate students have contributed too—as teaching assistants who cared a lot about teaching—in particular, Martin Ackermann, John N. Cooper, Jerome V. V. Kasper, Malcolm Nicol, Joanne Moore Herr, Mark Rochkind, and James Shirk. Our students, too, have helped us more than they could know through their conscientious efforts as our materials were still taking form. Finally, we warmly thank Miss Susan Arbuckle for her unimpeachable stenography, her endless patience and her unquenchable good cheer.

G. C. Pimentel
R. D. Spratley

contents

Chemistry is primarily concerned with the making and breaking of chemical bonds. Consequently rules of combination, systematics of reactivity, and, ultimately, theories of chemical bonding, have occupied central positions in the activity of chemists and in their scientific literature.

The beginnings

Rudimentary bonding concepts actually predated the atomic theory by many centuries. The postulate of four "elements," earth, air, fire, and water, contains the germ of the compositional idea. Furthermore, the theme of chemical bonding can be seen in early writings about the "modes of combination" expressed in the three principles called sulfur, mercury, and salt. Consider these words by Paracelsus, written in about 1525.*

"As to the manner in which God created the world, take the following account. He originally reduced it to one body, while the elements were developing. This body He made up of three ingredients, Mercury, Sulphur, and Salt, so that these three should constitute one body. Of these three are composed all the things which are, or are produced, in the four elements [earth, air, fire, and water]. These three have in themselves the force and the power of all perishable things. In them lie hidden the mineral, day, night, heat, cold, the stone, the fruit, and everything else, even while not yet formed."

That this early scientist was groping for principles of constitution is evident in his writings about how these three "principles" act together to make up the "element" air.

prologue

"Mercury, Sulphur, and Salt are so prepared as the element of air that they constitute the air, and make up that element. Originally the sky is nothing but while Sulphur coagulated with the spirit of Salt and clarified by Mercury, and the hardness of this element is in this pellicle and shell thus formed from it. Then, secondly, from the three primal parts it is changed into two—one part being air and the other chaos—in the following way. The Sulphur resolves itself by the spirit of Salt and in the liquor of Mercury, which of itself is a liquid distributed from

* H. M. Leicester and H. S. Klukstein, *Source Book in Chemistry,* p. 19, McGraw-Hill, N.Y., 1952.

heaven to earth, and is the albumen of the heaven, and the mid space. It is clear, a chaos, subtle, and diaphanous. All density, dryness, and all its subtle nature are resolved, nor is it any longer the same as it was before. Such is the air."

Such efforts to clarify chemical constitution can be seen, in retrospect, to have been abortive in absence of the atomic theory. Until the existence of atoms was recognized, chemistry was destined to remain a collection of recipes and empirical prescriptions, only encumbered by the embryonic and metaphysical theories of the time.

The atomic hypothesis: cornerstone of chemistry

Nevertheless, chemical analysis became such a well developed and powerful technique that it furnished the first compelling evidence that matter is particulate. Thus, in the first half of the nineteenth century, the accumulating knowledge of chemical composition led Dalton to propose the existence of atoms. This proposal liberated and stimulated views of chemical bonding whose usefulness can still be seen today.

Bonding rules—called "valence rules"—began to evolve during the second half of the nineteenth century. These empirical rules were given systematic foundation when the importance of the periodic table as a regularizing guide became evident. A rich harvest of progress followed—the preparation, characterization and practical use of new compounds not found in nature came to be a day-to-day activity. This success, in turn, confronted the valence rules with an enormous and growing volume of new descriptive chemical facts. To cope with them, the valence rules became more and more ornate. Without a basic unifying theory, chemists resorted to classification according to "bond type" and, as the first half of the twentieth century drew to its close, chemists manipulated with dexterity a complex catalogue of covalent bonds, ionic bonds, metallic bonds, coordinate bonds, charge-transfer bonds, dative bonds, chelate bonds, bridge bonds, one-electron bonds, and hydrogen bonds. Chemical bonding was attributed to electron sharing, to "exchange" forces, to magnetic spin, "overlap," available space, and lowered kinetic energy. All of it was kept in some sort of workable order with the aid of the periodic table. This powerful ordering device came to be engraved in the mind of every working chemist, and it gave him at least an intuitive basis for predicting what chemical composition and which type of chemical bond to anticipate as new preparations were attempted. The lack of a "first principles" approach drew the scorn of theoretical physicists, but no apology was either forthcoming or needed. Chemists had prepared almost a million compounds by the year 1950 and the number of new compounds was growing at the rate of two or three hundred per day.

The dawn of quantum mechanics

We now realize that the evolution of a single, unifying theory of chemical bonding had to await the development of quantum mechanics. Yet it was a giant step forward when, at the turn of the century, Rutherford determined that the atom consisted of negatively charged electrons moving around a small, massive, positively charged nucleus. This atomic structure suggested that chemical bonds might be associated with electron-sharing. Electron counting became a popular game, for stable compounds seemed to be connected to the specially stable electron populations of the inert gases. G. N. Lewis observed that a large part of chemical bonding could be explained in terms of atoms striving for these inert gas populations through the sharing of pairs of electrons. A variety of useful schemes developed —the octet rule, the electron-dot diagram, the pigeonhole representations, and most recently, the Linnett quartets. These schemes accelerated the juggernaut of chemical progress.

Then the quantum mechanical picture of the atom began to emerge. Physicists found they now possessed a description of the atom that could accurately predict atomic properties whenever the mathematics could be solved. By 1930 the scene was set for a true understanding of chemical bonding. It was a while coming, however, because of mathematical difficulties. Calculations for the simplest molecules H_2 and H_2^+ proved that quantum mechanics did contain the explanation of chemical bonding. Urged on by a few bold individuals, like Linus Pauling and Robert Mulliken, chemists began incorporating quantum mechanical ideas into their bonding theories, even though quantitative calculations were quite out of reach. Then, as ever is so, each human problem is finally overtaken by a solution. The electronic computers appeared and they multiplied many many fold our computational ability. Now even polyatomic molecules with many electrons can be handled; our quantum mechanical understanding of bonding is firm.

So now we will pause to examine what quantum mechanics tells us about chemical bonds. We will review the chemist's valence rules in the light of these "first principles." But first we will examine the quantum mechanical view of the atom. Here can be seen most clearly the qualitative differences and likenesses between physical behavior on the scale of atomic dimensions and that of the macroscopic world we sense directly around us. These differences and likenesses furnish the setting needed to understand what bonding is all about.

one quantum mechanics and the hydrogen atom

A successful theory is a useful paraphrase of observational knowledge. When a time-tested conflict develops between a theory and new observations, the theory always takes second place. First the theory will be patched and tinkered to deal with the new facts. Even if the conflict persists, the deficient theory will continue in use, with reservations, until a new theory is found that does encompass and usefully paraphrase both the previous knowledge and the new facts. Then the old theory will fall into disuse and be cheerfully ignored by scientists bent upon further expansion of our scientific frontiers.

The development of quantum mechanics displays this process beautifully. We'll touch a few highlights of this story, but we'll pay more attention to the logical than to the chronological relationships.

1-1 An intuitive basis for quantum theory

Toward the close of the nineteenth century, the two brightest areas in physics were separable and not in conflict. The pre-1900 classical laws of mechanics accurately described the motions of particulate bodies from pebbles to planets. In a similar way, electromagnetic theory was stunningly successful in explaining electrical and magnetic phenomena, including the behavior of light. Both theories were, of course, based solidly in experimental knowledge.

Within the next three decades, the foundations of both areas were shaken by crucial failures. In retrospect, one experiment can be seen to be the experimental key that pointed inexorably toward quantum mechanics, though it is not usually presented in that way. This experiment was the Michelson–Morley search for an "electromagnetic ether" to explain the propagation of light.

(a) THE ELECTROMAGNETIC ETHER THAT WASN'T THERE

All knowledge of how acoustic waves are propagated through gas, liquid, or solid media, led quite naturally to the assumption that light, a wave phenomenon, must have some sort of propagating medium. Nineteenth century physicists called this hypothetical medium an "ether," a term that causes doubt about their chemical familiarity. A. Michelson, an American physicist, set out in 1881 to verify this confident model with an experiment that would measure sensitively any motion of the earth through the expected "ether." His negative result* challenged our most fundamental descriptions of the spatial and temporal relationships between objects and events. Then two decades passed before a significant attempt to adjust to Michelson's surprising result was published by a Dutch scientist named H. A. Lorentz. He showed how the description of positions and times of events had to be changed if light is propagated in a vacuum at a fixed velocity. Thus, the stage was set for a young Austrian scientist named Albert Einstein to make a revolutionary advance.

(b) $E = mc^2$

In 1905, a year after Lorentz's work, Einstein reopened another subject with a previous history—relativity. He considered the implications of these new views of space and time on the description of dynamic events (events that involve changes in positions, times, velocities, momenta, and energies) in a moving laboratory. The new model led to the then startling result that energy and mass are merely different manifestations of a single entity. Every energy transfer is accompanied by a mass transfer and every mass, whether moving or not, represents a store of energy. Mass is usually measured with certain devices and is expressed in grams. Energy is measured with other probes and is expressed in ergs. The proportionality constant between these units is c^2, the square of the velocity of light:

$$E = mc^2 \tag{1-1}$$

The appearance of c in this relationship is a logical consequence of the Lorentz modification of our views of space and time, as required by the Michelson–Morley experiment. Good Dr. Einstein showed this to be a natural result of our use of light as a communication link when measuring energy relations between moving masses.

* Six years later, in 1887, Michelson and a colleague from chemistry, E. W. Morely, verified the experimental result with improved accuracy. Nevertheless, a quarter of a century later, some reputable scientists were still repeating the experiment with the conviction that the result might not actually be negative. In fact, the experiment has been considered worthy of repetition and confirmation even within the last few years using lasers and the most precise measuring devices available.

(c) ENERGY HAS MASS: MASS IS ENERGY

There were some immediately obvious implications that didn't sit well with scientists of the time. The physicists saw their precious Law of Conservation of Energy thus married to the chemists' Law of Conservation of Mass. Neither law was correct as it had been applied earlier. The two laws became one and the same statement, since energy and mass are the same thing. Mass is conserved only if its energy forms are taken into account, and vice versa.

Quite as significant and more germane to quantum theory are the implications of $E = mc^2$ on the relationship between light and mass. Light, then recognized as a wave phenomenon, was already known to be a form of energy. Absorption of light can be measured with traditional energy-measuring devices, such as calorimeters, thermometers, thermocouples, and the like. Now light must also have mass and, presumably, all the properties of mass. Mass must also do its part and possess the properties of energy. Neither classical mechanics nor electromagnetic theory could provide these links.

(d) $E = mc^2$ AND THE IMPLICATION OF THE ATOMIC THEORY

Light is an energy form. Energy must have the properties of mass. What can we learn about light from what we know about mass? Let's see.

Suppose a gram of table salt is ground so finely that it blows in the wind. Now examine it under the very best microscope you can find. Even under the highest magnification, you'll see only smaller and smaller crystals of the same stuff, sodium chloride. Matter is continuous on a macroscopic scale. All our evidence that matter is made up of tiny, discrete particles called atoms is indirect. Nevertheless, we are firmly convinced of the atomic theory because this model is so widely useful in explaining the properties of matter. *Matter looks continuous but its properties are explained with a particulate model.*

Light emitted by a hot radiating source can be passed through a prism to produce a spectrum. The color spectrum we see is continuous—apparently every color is represented. There is a smooth variation of intensity that peaks at a wavelength fixed by the source temperature. If the source is moved away from the prism, the intensities of all colors decrease together uniformly until the whole spectrum gradually becomes too weak to see. Light energy, like mass, appears continuous.

Now we can ask about the implications of the identity of energy and mass. If many properties of matter are best described with a particulate model, should not the same be true for light energy? This is exactly what Max Planck decided (without the benefit of Fig. 1-1 or $E = mc^2$) as he struggled with the problem of explaining the intensity–wavelength relationship described above. This was the first property of light that displayed real inadequacy of the wave description; Planck found it necessary to postulate the existence of little light

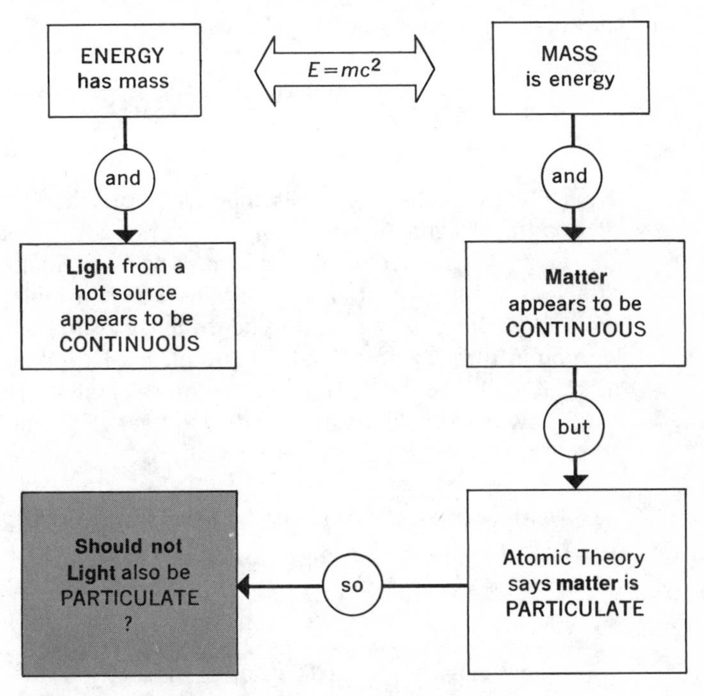

Figure 1-1 Energy has mass: Mass is energy. The beginnings.

"packets" or "quanta" to explain the radiation from a hot source. This was the beginning of the *quantum theory of light* and, in retrospect,* we see this as the first fulfillment of the parallelism implied by $E = mc^2$.

(e) HOW BIG IS A QUANTUM?

This particulate model of light quantitatively connects the energy of a light packet to its frequency or wavelength. Light waves are usually described in terms of either the frequency (the number of electromagnetic waves per second, designated by ν, the Greek letter "nu"), or the wavelength (designated λ,

* Planck was not aided at the time by $E = mc^2$; in fact, he proposed his view about six years before Einstein published his results on relativity. Planck was wholly concerned with the failure of electromagnetic theory to explain the experimental observations of the intensity–color relationship.

the Greek letter "lambda"). To convert to the equivalent energy in ergs, Planck found that he needed a conversion factor of $6.6 \cdot 10^{-27}$ ergs seconds. He called this factor h:

$$E = h\nu \qquad \text{and,} \qquad \text{since } \nu = c/\lambda$$

$$E = h(c/\lambda) \tag{1-2}$$

and,

$$\text{ergs} = (\text{erg sec})\left(\frac{\text{cycles}}{\text{sec}}\right) = (\text{erg sec})\,\frac{(\text{cm/sec})}{(\text{cm/cycle})}$$

Thus Planck's constant h tells the "size" of these "particles" or quanta of light, just as Avogadro's number N tells us the size atoms must have. Everyone senses how small an atom is—after all, a big gulp of water contains about a mole, $6 \cdot 10^{+23}$ molecules! Suppose we tried to heat this 18 gram gulp of water with an infrared lamp to make a cup of tea. How many quanta of light would we need? Here's how the calculation goes:

(i) How many calories are needed to heat 18 grams of water from 0°C to 100°C?

$$\begin{aligned}\text{Heat} &= (\text{mass})(\text{heat capacity})(\text{temperature change})\\ &= (18 \text{ grams})\left(\frac{1.0 \text{ calorie}}{\text{gram deg}}\right)(100^\circ) = 1800 \text{ calories}\end{aligned}$$

(ii) How many ergs is 1800 calories of heat?

$$\begin{aligned}\text{Ergs} &= (\text{calories})(\text{conversion factor, ergs to calories})\\ &= (1800 \text{ calories})(4.2 \cdot 10^{7} \text{ ergs/calorie})\\ &= 7.6 \cdot 10^{10} \text{ ergs}\end{aligned}$$

(iii) How many quanta of infrared light of wavelength 10 microns (frequency $3 \cdot 10^{13}$ cycles/sec) are needed to give $7.6 \cdot 10^{10}$ ergs?

$$\begin{aligned}&\text{Number of quanta}\\ &\quad = (\text{ergs needed})/(\text{ergs per quantum})\\ &\quad = (7.6 \cdot 10^{10} \text{ ergs})/h\nu\\ &\quad = \frac{(7.6 \cdot 10^{10} \text{ ergs})}{(6.6 \cdot 10^{-27} \text{ erg sec})(3.0 \cdot 10^{13} \text{ cycles/sec})}\end{aligned}$$

$$\text{Number of quanta} = 3.8 \cdot 10^{23} \text{ quanta}$$

$$\text{Number of moles of quanta} = \frac{3.8 \cdot 10^{23}}{6.0 \cdot 10^{23}} = 0.63 \text{ moles}$$

We see that a mole of water can be warmed from 0°C to 100°C with about $\frac{2}{3}$ of a mole of infrared light quanta. We conclude that quanta or packets of light energy are indeed small compared with our macroscopic measuring devices—they are appropriate in magnitude to the molecular scale of things.

(f) MASS AND MOMENTUM OF A LIGHT QUANTUM

It is an easy arithmetical task to combine (1-1) and (1-2) to calculate the mass associated with a single quantum of light energy with frequency ν (nu).

In general

$$E = mc^2$$

and, for light,

$$E = h\nu$$

hence

$$h\nu = m_\nu c^2$$

or

$$m_\nu = \frac{h\nu}{c^2} \tag{1-3}$$

Momentum p is just mass times velocity. Expression (1-3) gives the mass of a quantum. Since light is propagated with velocity c, we can calculate the momentum p_ν of a quantum.

$$p_\nu = m_\nu c = \frac{h\nu}{c^2} c$$

$$p_\nu = \frac{h\nu}{c} \tag{1-4}$$

and since $\lambda = c/\nu$,

$$p_\nu = \frac{h}{\lambda} \tag{1-5}$$

Experimental verification of this momentum is provided by the scattering of light. In 1922, A. H. Compton found that X rays, when scattered by electrons, suffer a momentum change, just as would two colliding billiard balls. He measured the resulting frequency shift as a function of the scattering angle. The shift was accurately explained by the laws of conservation of energy and momentum using the particulate model of the light quantum.

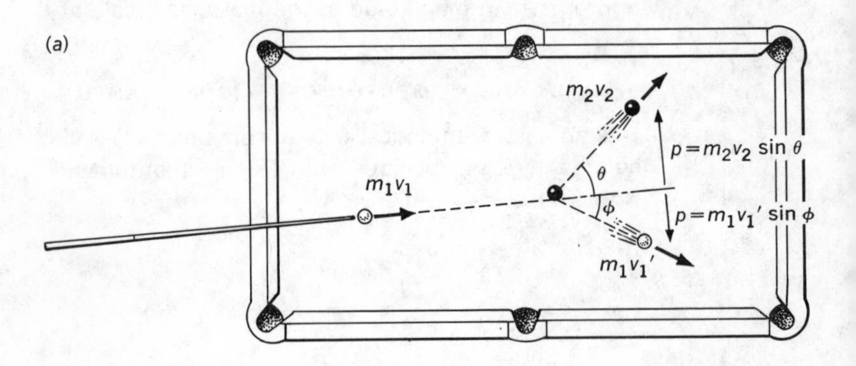

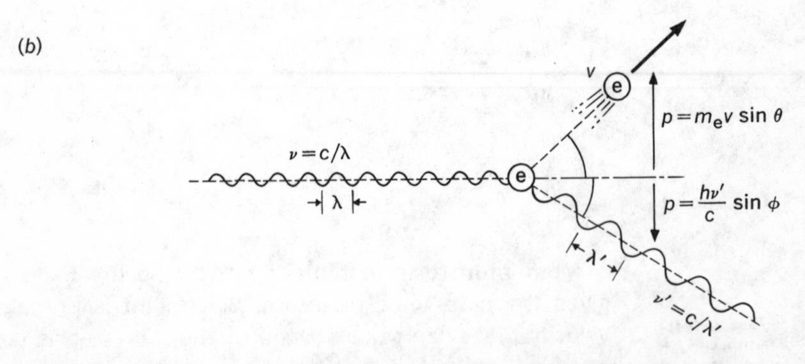

Figure 1-2 Light ball in the corner pocket. (a) The momenta p of the two balls after collision are correctly predicted by the Law of Conservation of Momentum. (b) Optical billiards. The frequency of the light wave, hence the momentum of the photon, is changed after collision with an electron. The same Law of Conservation of Momentum applies to photons as to billiard balls, reinforcing our belief in the particle properties of light.

(g) $E=mc^2$ AND THE IMPLICATION OF THE WAVE NATURE OF LIGHT

This parallelism of energy and mass is, of course, a two-way street. If light must (and does) have the particulate properties of mass, is it not necessary to expect mass to possess properties that would normally be attributed to light? In particular, we are used to describing light as an electromagnetic wave. It has a frequency of ν cycles per second, and a corresponding wavelength λ. What "wavelength" should be ascribed to a particle to make the parallelism of Figure 1-1 complete?

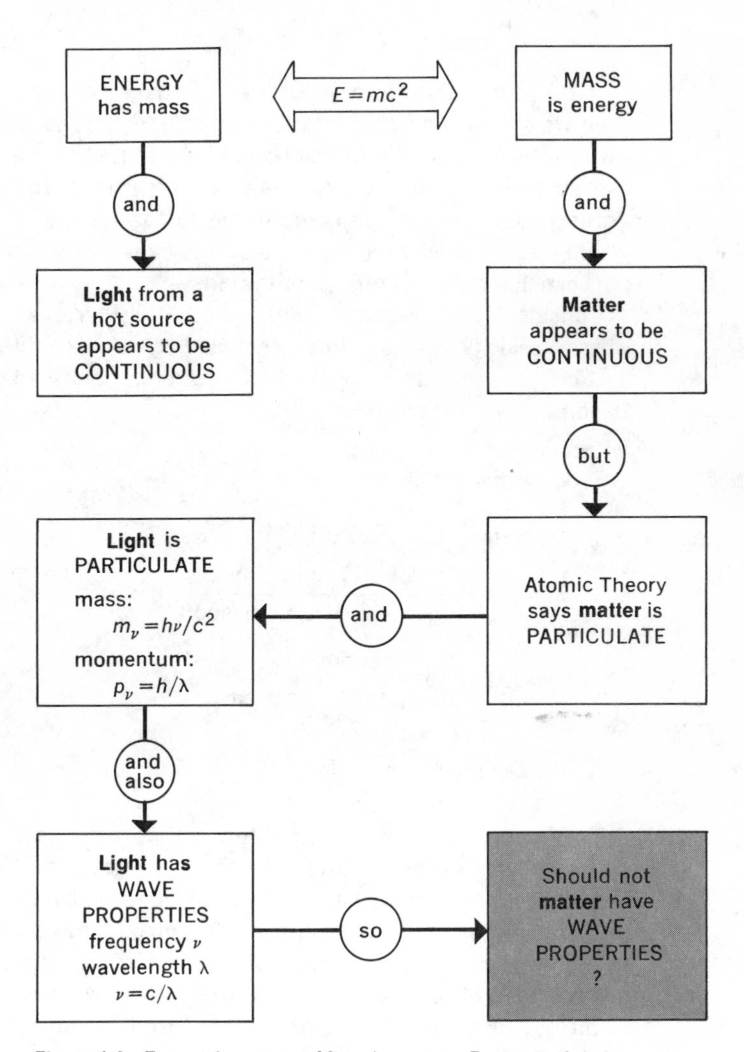

Figure 1-3 Energy has mass: Mass is energy. Part way there.

Equations (1-4) and (1-5) provide the link we need. We know how to express the momentum of a mass particle, $p_m = mv$, and expression (1-5) shows how a light wave connects momentum and wavelength. Why not assume the same relationship for mass? Since $p_\nu = h/\lambda$, perhaps

$$p_m = \frac{h}{\lambda_m}$$

or

$$\lambda_m = \frac{h}{p_m} = \frac{h}{mv} \qquad (1\text{-}6)$$

If this connection is a valid one, as suggested by $E = mc^2$, then we should find characteristic wave phenomena associated with a moving particle. Diffraction of light as it strikes a grating-like surface is one such example. To display diffraction, the grating spacing must be comparable to the wavelength. What grating spacing is needed to test the wave properties of a particle? Expression (1-6) gives the answer.

Consider an electron (mass $= 9.1 \cdot 10^{-28}$ grams) moving with a velocity of 0.01 times the velocity of light ($0.01 \cdot c = 3 \cdot 10^{-8}$ cm/sec is the velocity of an electron accelerated through 26 volts).

$$p_m = mv = (9.1 \cdot 10^{-28})(3 \cdot 10^{8})$$
$$= 27.3 \cdot 10^{-20} \text{ gm cm/sec}$$
$$\lambda = \frac{h}{p_m} = \frac{6.6 \cdot 10^{-27} \text{ erg sec}}{27.3 \cdot 10^{-20} \text{ gm cm/sec}}$$
$$= 0.24 \cdot 10^{-7} \frac{\text{gm cm}^2}{\text{sec}} \frac{\text{sec}}{\text{gm cm}}$$

So,

$$\lambda = 2.4 \cdot 10^{-8} \text{ cm} = 2.4 \text{ Å}$$

To demonstrate diffraction, then, all we need is a surface with regular grooves cut into it at intervals of about 2.4 Å. Such a cutting job would be beyond our very best milling machines but, fortunately, nature abounds with crystals that have atoms regularly aligned with spacings of this order. So electrons that are bounced off the face of a crystal should show the same sort of diffraction patterns as light reflected from a grating! So they do, as Davisson and Germer showed at the Bell Laboratories in 1927! This result confirmed the reality of the wave properties of a moving particle, as expressed in (1-5) and as proposed in 1925 by a French scientist, Louis de Broglie.

This demonstration of the diffraction of "particle waves" completes our parallelism between energy and matter. The relationships qualitatively portrayed in Figure 1-5 finally took

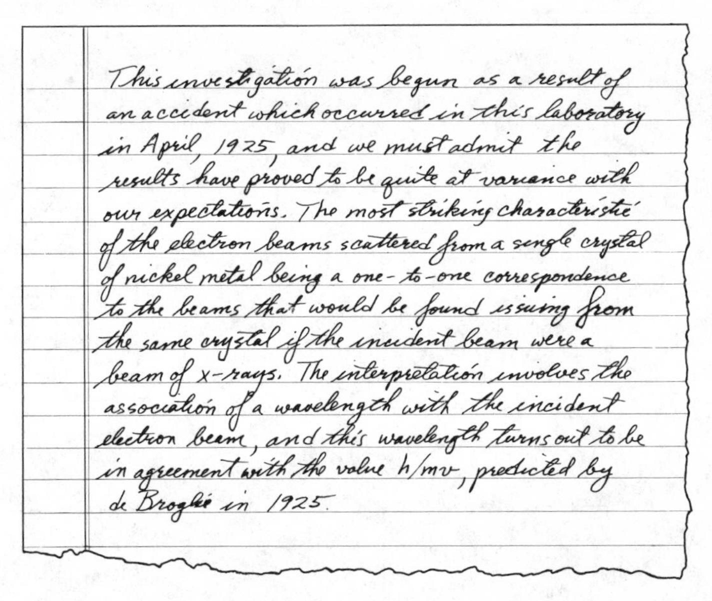

This investigation was begun as a result of
an accident which occurred in this laboratory
in April, 1925, and we must admit the
results have proved to be quite at variance with
our expectations. The most striking characteristic
of the electron beams scattered from a single crystal
of nickel metal being a one-to-one correspondence
to the beams that would be found issuing from
the same crystal if the incident beam were a
beam of x-rays. The interpretation involves the
association of a wavelength with the incident
electron beam, and this wavelength turns out to be
in agreement with the value h/mv, predicted by
de Broglie in 1925.

Figure 1-4 Electrons behave like waves. This quotation, taken from the 1927 paper of Davisson and Germer, illustrates a frequent route to scientific advance. Some unexpected results when followed up lead to important discoveries. In this case, a laboratory mishap led to the conversion of some polycrystalline nickel into a crystalline form. Diffraction patterns similar to those observed with X radiation were observed. De Broglie's predictions were in this way given their first experimental verification.

form in the first three decades of this astonishing twentieth century. The stage was set for quantum mechanics.

1-2 Quantum mechanics

To understand atomic structure, we begin with the simplest atom, the hydrogen atom. We would like to describe the dynamics of motion of an electron moving in the vicinity of a proton. The "dynamics of motion" means the electron energy, momentum, trajectory, and the changes in these quantities as the hydrogen atom makes its way through life. Ultimately we might hope that this description would tell us why two H atoms bond to form a molecule H_2 but three H atoms do not form H_3. These were questions without answer in the pre-$E = mc^2$ classical physics. The chemists had to "go it alone" with no mathematical model of chemical bonding to support the burgeoning knowledge of chemical changes. Energy relationships were hung upon the sturdy frame of thermodynamics; chemical bonding was an empirical and mysterious art hunting for its guiding rationale. Again one particular experimental result

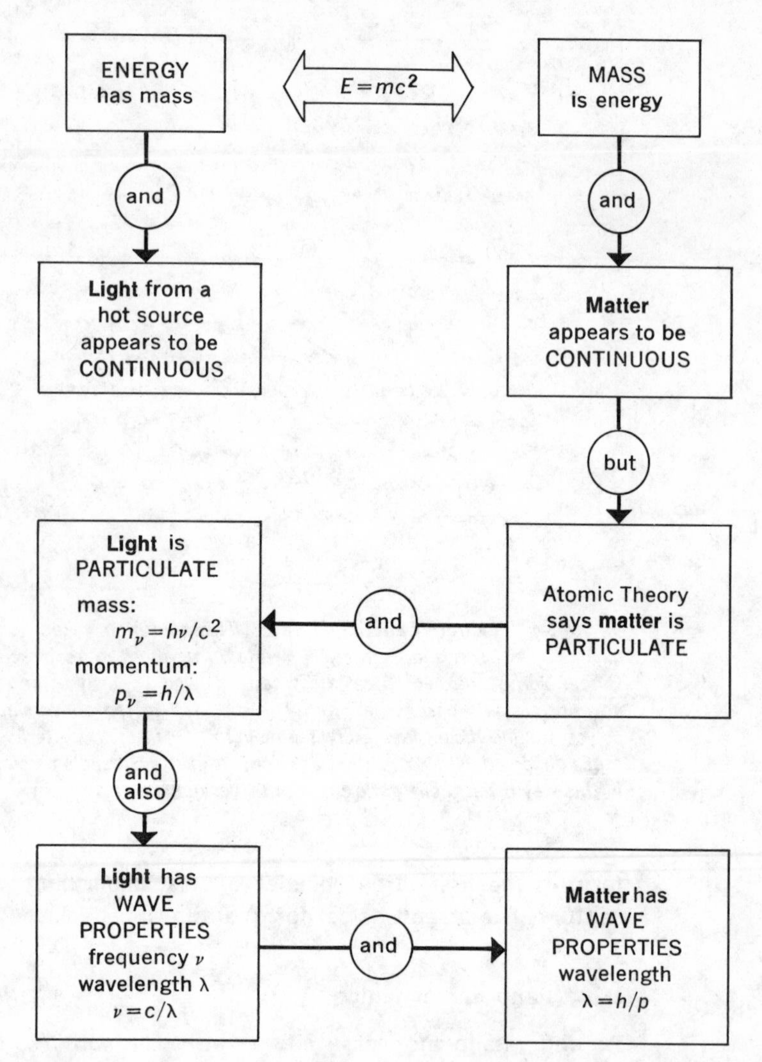

Figure 1-5 Energy has mass: Mass is energy. The story is complete.

provided the key to this mystery, the spectrum of the hydrogen atom.

(a) THE LINE SPECTRUM OF A HYDROGEN ATOM

Hydrogen atoms are produced when an electric discharge is struck through gaseous hydrogen. Their presence is revealed in the light emitted; a fact recognized long before there was any explanation for the color spectrum of this light. Figure 1-6 shows the hydrogen atom spectrum. Unlike the radiation emitted from hot sources, the electric discharge glow emits only special colors in a "line" spectrum. Interpreted in terms

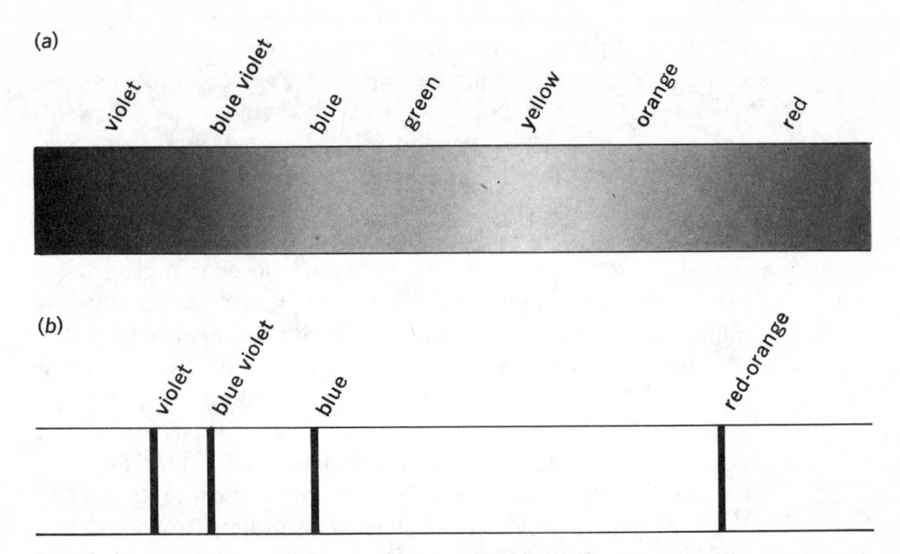

Figure 1-6 The visible region of the spectrum (a) and the line spectrum of the hydrogen atom in the visible region (b).

of expression (1-2), $E = h\nu$, *only selected energies can be emitted by the hydrogen atom in the form of light quanta.* As research study of this phenomenon progressed, it developed that the photon energy of $15.35 \cdot 10^{-11}$ ergs was observed (which is 235.2 kcal per mole of quanta) and also photon energy $19.36 \cdot 10^{-11}$ ergs (278.8 kcal/mole), but no quanta of energy in between. To heighten the interest, photons of energy exactly equal to the difference $(19.36 - 15.35) \cdot 10^{-11} = 4.01 \cdot 10^{-11}$ ergs (43.6 kcal/mole) were also observed.

These observations were quite unexplainable within classical physics. Consider the classical, planetary model of the atom, the view popular just after the turn of the century. The electron was pictured to be orbiting around the nucleus with the centrifugal force just balanced by the coulombic attraction. Such an atom should be able to absorb or release any energy whatsoever, the change merely altering the orbital radius. Scientists of the day were forced to consider either an unpalatable modification of this model or its complete abandonment. Rarely is a theory abandoned as long as no alternative is available. In fact, this planetary model, despite its obvious inadequacy, was not abandoned until, at last, the quantum mechanical model was developed.

It seemed unlikely to propose that the atom could contain any energy (as a planetary model dictates) and yet absorb or release energy only in special, metered amounts. The alternative was to postulate that the atom could hold only special energies. This would automatically imply that only special energies ($E_1 - E_2$) could be released as the atom changed from one of its "allowed" energies E_1 to another of its "allowed" energies E_2. What was needed was a rationale upon which to hang this then unlikely proposal.

In 1913 Niels Bohr furnished a possible rationale. He discovered that the hydrogen atom spectrum could be explained if in Nature's hydrogen atom the electron moved in those special planetary orbits which limit the angular momentum to integer multiples of $h/2\pi$. This rather arbitrary proposal was made acceptable by its tie-in with other phenomena (the radiation from a hot surface, through h), plus the opportunity to save, in part, the planetary model of the atom. It is an interesting commentary on the nature of scientific activity that this proposal by Bohr opened the way to quantum mechanics despite the fact that it proved to be incorrect in almost every detail. Accepting, as everyone now does, the quantum mechanical view of the atom, the following failures of the Bohr planetary atom can be catalogued:

—The electron does *not* have a planetary trajectory.
—The integer-related momenta proved to be incorrect, as shown in Table 1-1.

Table 1-1 Comparison of the Momenta of Hydrogen Atom States: Bohr's Planetary Model and the Quantum Mechanical Atom

Hydrogen Atom State	Planetary Atom $p = n(h/2\pi)$ $n =$ integer	Q. M. Atom $p = \sqrt{l(l+1)}\,(h/2\pi)$ $l =$ integer
1s	$1(h/2\pi)$	0
2s	$2(h/2\pi)$	0
2p	$2(h/2\pi)$	$1.41(h/2\pi)$
3s	$3(h/2\pi)$	0
3p	$3(h/2\pi)$	$1.41(h/2\pi)$
3d	$3(h/2\pi)$	$2.45(h/2\pi)$
4s	$4(h/2\pi)$	0
4p	$4(h/2\pi)$	$1.41(h/2\pi)$
4d	$4(h/2\pi)$	$2.45(h/2\pi)$
4f	$4(h/2\pi)$	$3.46(h/2\pi)$

—The momentum criterion used for the one-electron hydrogen atom failed to explain the observed energy levels of *any* atom with two or more electrons.

—The model proved to give no clue whatsoever to the origin of chemical bonding.

—The model provided no basis for understanding *why* quantization occurs or *why* an orbiting electron would fail to radiate its energy as electrodynamics dictates; both properties were imposed without justification.

Despite this dismal record, Bohr's courage in recognizing the need for a departure from classical physics won for him a place in history. Still today, the "allowed" energy states of an atom are called "stationary states," as named by Bohr. These stationary states were and still are characterized by "quantum numbers," integers that account for the marvelous pattern in the hydrogen atom spectrum (see Fig. 1-7). It took a few

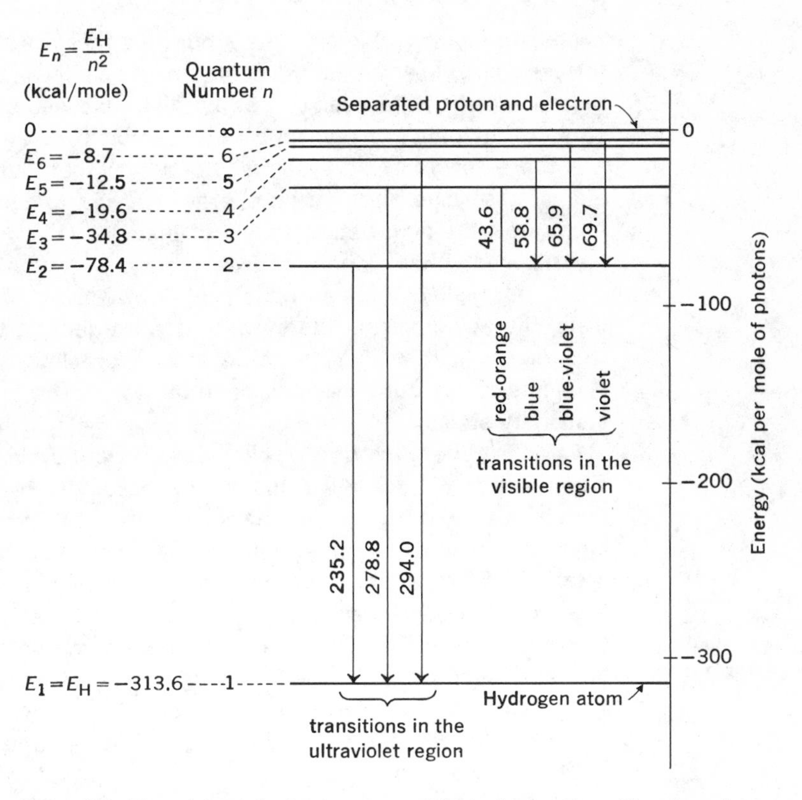

Figure 1-7 The origin of the line spectrum of the hydrogen atom. The allowed electronic energies of the stationary states are shown by horizontal lines. The energies of these states are related by a simple formula involving the quantum number n. Some of the many possible transitions are shown by vertical arrows. The energy of the emitted light is just the difference between the energies of the two stationary states involved. The visible lines of Figure 1-6 result from transitions to the $n = 2$ level.

decades for the limitations of the Bohr atom to be admitted, but Bohr's contribution was the crucial break from the physics of the macroscopic world.

In 1926, things finally fell into place when a German scientist, Erwin Schroedinger, recognized the connection between Bohr's stationary states and de Broglie's wave properties of the electron. This step in the development of quantum mechanics must be considered as one of the most important ever made in science—to be compared with the contributions of Galileo, Newton, and Maxwell. To understand this connection, we will consider the "wave properties" of two other physical systems, a vibrating string and a vibrating drum. Their tones are in harmony with the music of the atom!

(b) THE LINE SPECTRUM OF A GUITAR STRING

There are innumerable systems around us that can accept or release energy in any amount: an automobile can be accelerated to any velocity; a baseball can be bunted, smashed out of the park, or anything in between; a mass can be lifted any height. Are there macroscopic systems around us that behave like atoms, absorbing or releasing energy in special amounts? Yes, there are—the guitar string is one of them.

A plucked guitar string produces a musical sound of definite pitch and quality. If the string is held down against one of the frets, thereby changing the string's effective length, the pitch is changed. In this familiar and aesthetic example, we find the elements of quantum mechanics and the essence of the atomic stationary states.

Obviously the pitch of a plucked string is determined by its vibrational frequencies, which set up acoustic disturbances in the air. As the string vibrates back and forth, its movement at any point along its length and its frequency of oscillation characterize the vibration. If we examine this movement, we find two spots at which the string cannot move—at the ends. The string is fixed at the bridge, and again along the neck by one of the frets (according to the fingering). These restraints on the guitar string's movement are called *boundary conditions.* The *boundaries* in this case, are the ends of the string, and the *conditions* are that the string displacement at the ends always be zero. We shall see that *because of the boundary conditions, only special vibrations can occur: the vibrations are quantized.*

Figure 1-8 shows some of the ways in which the string can vibrate. The simplest vibration (i) is called the fundamental and it provides the principal tone, or pitch. The next simplest vibration (ii) distorts the string like a sine wave. As the string vibrates

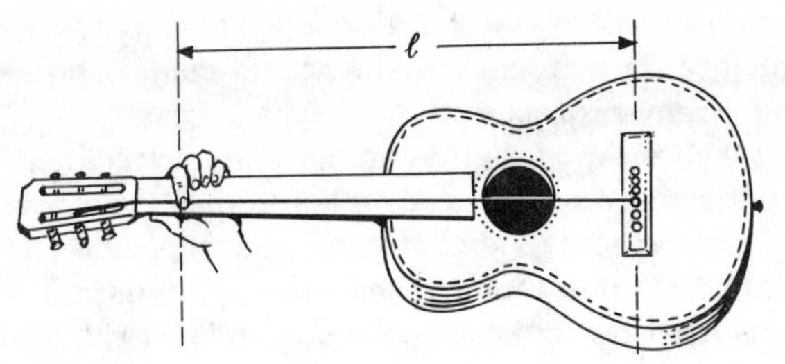

(a) allowed vibrations

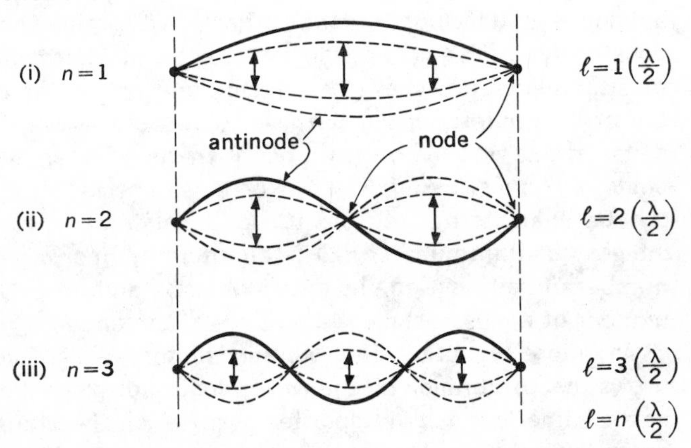

(b) forbidden vibrations

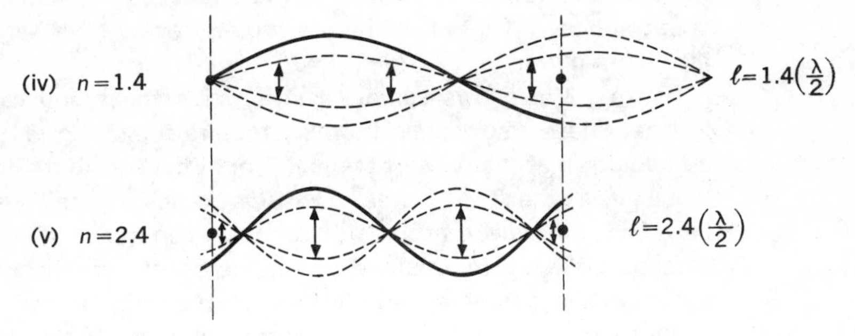

(c) the line spectrum

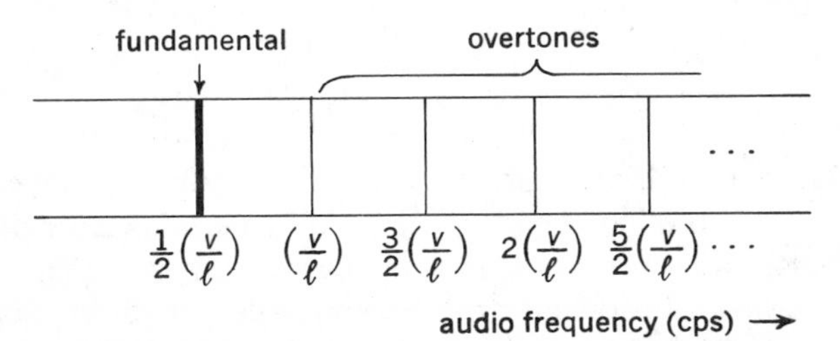

Figure 1-8 The allowed (a) and unallowed (b) vibrations of a guitar string and its spectrum (c). The wavelength of the string is related to its length by an integer number n: $\lambda = (2/n)\ell$.

The audio frequency is given by: ν audio $= v/\lambda$, in which v is the velocity of sound.

back and forth, every point on the string passes periodically through a zero displacement. At the center, however, there never is any displacement. This spot is called a "node." The spots where the string has maximum displacement are called "antinodes." We see that vibration (i) has two nodes, those at the ends forced by the boundary conditions, and one antinode. Vibration (ii) has three nodes, the two ends and one in the middle, and two antinodes. This vibration divides the length ℓ into one wavelength. Vibration (iii) has four nodes and three antinodes, and includes one and a half wavelengths. The vibrations of type (ii), (iii), and higher, are called "harmonics" in musical theory, and they provide the "quality" or "timbre" of the guitar's notes. For our purposes, however, we see that the guitar string vibrates only in special frequencies and, hence, emits only special sounds. Each of these special frequencies can be characterized by the fundamental frequency and an integer quantum number n. The quantum number n gives the number of half-wavelengths in the vibration and $(n + 1)$ is the number of nodes (including the nodes at the ends).

One more aspect of these vibrations interests us. Figure 1-8 shows that in vibration (ii) the two antinodes displace the string at the same *time* but in opposite sense. When the string is up at the left antinode, the string is down on the right. Later in the vibration the opposite is true. The two antinodes are said to be "out-of-phase." There is a change in phase every time we pass through a node.

Figure 1-8(*b*) shows some ways in which the string *cannot* vibrate. These displacements would require the string to move up and down at the base of the guitar or at the fret, but at both these places the string is held. *Thus, the boundary conditions are the origin of the quantization.* With boundary conditions, the wave motion of the vibrating string has a line spectrum whose frequencies are characterized by integer quantum numbers, a nodal pattern, and phase relationships. All of these characteristics appear in the quantum mechanical description of the atom.

(c) THE LINE SPECTRUM OF A DRUM

The same concepts and language of the guitar string carry over to the vibrations of a drum. Most of us think of a drum as a musical instrument with little personality, but it isn't so. By analyzing the motions of a drumhead, you will see that it too has both pitch and quality.

Around the circumference of the drum the leather of the drumhead is tightly held so it cannot move. This is the boundary

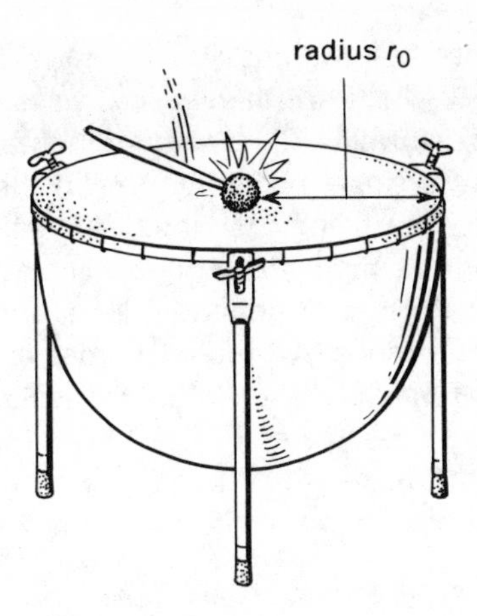

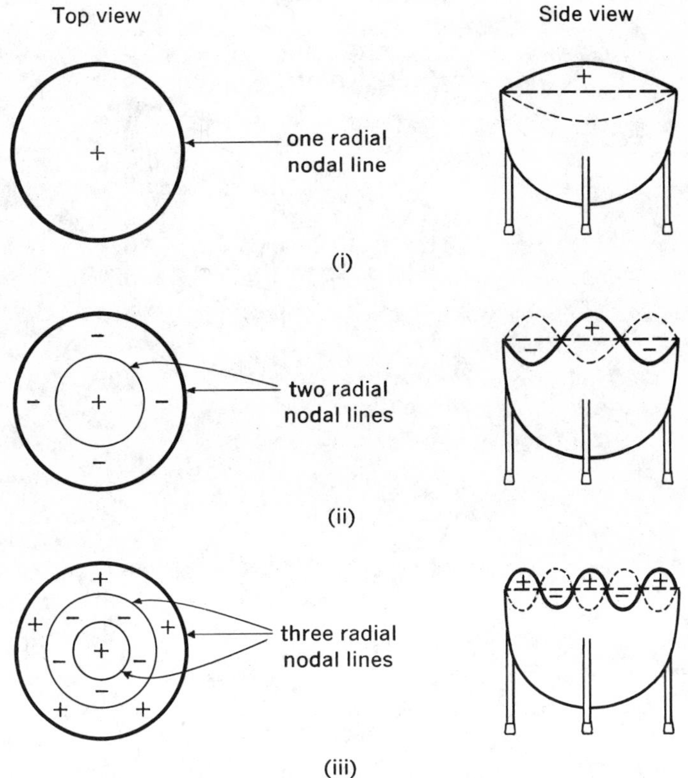

Figure 1-9 Radial vibrations of a drum: radial nodes.

condition of a drum. Thus, the two-dimensional vibrator has a nodal *line* just as the one-dimensional vibrator (the string) has a nodal *point*. This boundary condition limits the ways in which the drumhead can vibrate and gives the drum a specific set of vibrational frequencies.

Suppose the drum is struck just in the center, as shown in Figure 1-9. There are then special ways in which the leather can move. In vibration (i), the fundamental, there is a single anti-node and a single nodal line. In (ii) there are two nodal lines, the outer one being the circle defined by the drum edge, and the second being a circular nodal line at an intermediate radius. There are also two antinodal lines and the entire movement maintains circular symmetry. Across each nodal line there is a change of phase: if the leather is displaced upward on one side of the line (designated in Fig. 1-9 by a plus sign), it is displaced downward on the other side (shown by a negative sign).

The drum vibrates differently, however, if it is struck off

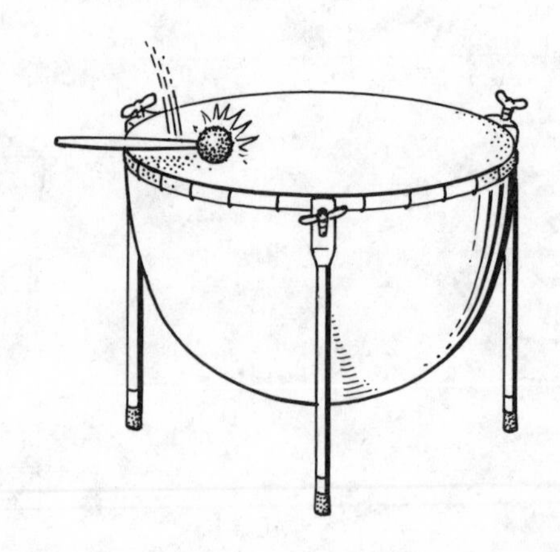

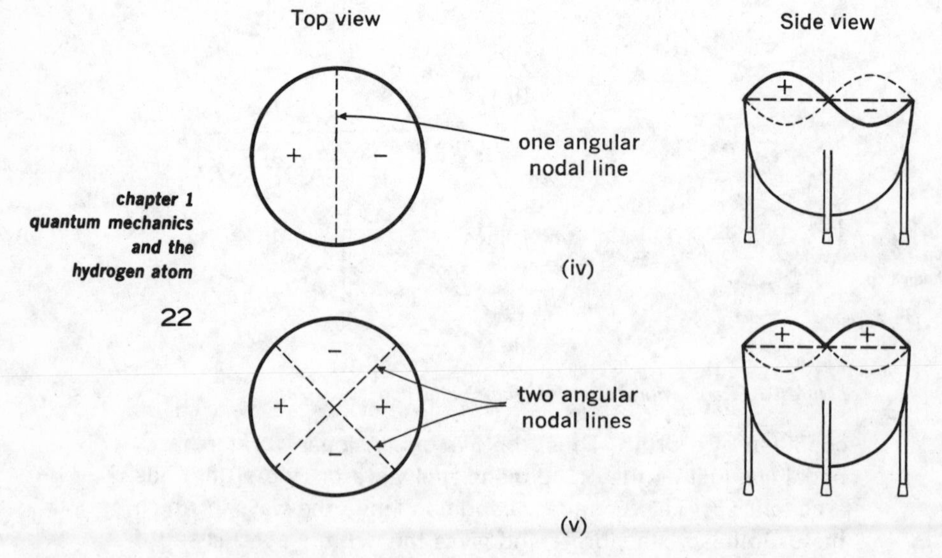

Figure 1-10 Angular vibrations of a drum: angular nodes.

center. This displaces the leather skin but without the circular symmetry. Again it turns out, though, that the vibrations are quantized and that they have nodal lines, now extending across the drum, as shown in Figure 1-10. In vibration (iv) there is one nodal line across the drum dividing it into two halves. In (v) the vibration divides the head into quadrants, two going up while the other two are displaced downwards. In such motions, there is a disturbance of the circular symmetry by the nodal lines across the drum. If we imagined a 0 to 360° scale around the drum's edge, the displacement would vary with angle (except at the edge where the boundary conditions still hold). Hence, the nodes shown in Figure 1-10 are called *angular nodes.*

(d) MUSIC AND THE ELECTRON

What we have learned is that musical instruments have line spectra like those of an atom. The wave-like vibrations of a guitar string or a drumhead are quantized, apparently like the dynamic motion of an electron in an atom. These musical motions can be described in mathematical detail by a characteristic mathematical equation, the "wave equation."

Along comes Erwin Schroedinger, a mathematical physicist who could write down and solve the wave equation for a vibrator on a napkin during lunch. He heard tell that a hydrogen atom has a line spectrum (like a vibrating string) and that an electron can be diffracted like a wave (the de Broglie prediction). Two and two equals four, said Dr. Schroedinger, and the line spectrum of the hydrogen atom shows that its equation of motion must be a wave-type equation with boundary conditions that fix the possible energy values. This bold decision was the real birth of quantum mechanics.

The classical equation of motion did not fit this description. It can be simply stated in terms of the sum of kinetic and potential energy, as follows:

$$\text{kinetic energy} + \text{potential energy} = \text{total energy}$$

or, in conventional symbols,

$$T \quad + \quad V \quad = \quad E \qquad (1\text{-}7)$$

The kinetic energy T is just $\frac{1}{2}mv^2$. The potential energy is the electrostatic energy of an electron attracted to a proton at some distance r, $-e^2/r$. Hence (1-7) can be rewritten as

$$\tfrac{1}{2}mv^2 - \frac{e^2}{r} = E \qquad (1\text{-}8)$$

To get this equation into momentum form, as Niels Bohr preferred it, we substitute $p = mv$.

$$\frac{1}{2}\frac{p^2}{m} - \frac{e^2}{r} = E \tag{1-9}$$

That's no wave equation, thought Erwin. He saw how to make it into one, though. Schroedinger replaced the classical momentum p by a mathematical "operator" (which is merely an instruction to carry out a mathematical operation upon some mathematical quantity). He selected, of course, an operator that converted (1-9) into a wave equation. This required the invention of a new function, called the *wave function* and symbolized ψ (psi). This wave function was inserted to give the operator something to operate on. Now the "Schroedinger equation" looks like this:

$$\left(\frac{1}{2}\frac{p^2}{m} - \frac{e^2}{r}\right)\psi = E\psi \tag{1-10}$$

and we must remember that in this new "wave mechanics," p is no longer simply m times v, it is an instruction to "operate" upon the function ψ.* It developed that ψ contains our knowledge of the electron in the atom.

Now we need some boundary conditions—they give us the line spectrum. Physical reality must be the guide here. If the function ψ contains our knowledge of the whereabouts of the electron in the hydrogen atom, the boundary conditions must be built on limiting conditions that make sense for the problem. One of these boundary conditions can be likened to the edge of the drumhead, where the displacement must always be zero. The edge of the atomic drum is at infinity—if the electron reached an infinite separation from the proton, it would no longer be attached. So we place the mathematical restriction on ψ that it always be zero at infinity. This and two other physically reasonable boundary conditions† give equation (1-10) a line spectrum, and it turns out to be exactly the hydrogen atom spectrum.

This achievement was only the beginning of a stunning series of successes for the new quantum mechanics. Without modifi-

*Some typical mathematical operators that might be applied to a function ψ are "find the slope" of ψ and "find the curvature of ψ." The last one is the one that Schroedinger substituted for p^2 to obtain a wave-type equation.

†These other two conditions are a bit more mathematical. In essence, they guarantee that ψ remains sensibly finite and does not hop about, changing in a discontinuous fashion. These, too, are based in the physical reality that prevents electrons from doing things at infinite velocity.

cation, it explains the energy levels of many-electron atoms. More important, it has proved to be in quantitative accord with the properties of molecules: calculations of bond energies, bond lengths, molecular vibration frequencies, and energy levels agree with experiment as accurately as calculational approximations and experimental uncertainty permit test. Only a few years after Schroedinger proposed his equation, it was found to be applicable to the simplest molecules, H_2 and H_2^+, but mathematical difficulties stood in the way of similar tests for more complicated molecules. Approximations were needed, but gradually it became clear that *all of chemistry is explained in quantum mechanics.* As computers have come onto the scene, calculations can now be valuable for molecules with as many electrons as carbon monoxide, methane, water, and ammonia. In principle, all chemical changes could be predicted if the mathematical obstacles were not so insurmountable. Nevertheless, with the advent of quantum mechanics, chemical bonding stepped out of empiricism. Now we have a firm theoretical framework to guide us in understanding and predicting chemical phenomena.

(e) WHAT DOES THE WAVE FUNCTION TELL US?

The solution of the wave equation gives us immediately two quantities of interest, E, the "allowed" energy levels of the atom or molecule, and, for each allowed energy, a function ψ of the electron's position coordinates. What information is contained in ψ?

We have two models with which to relate. The energy conservation equation (1-7), $T + V = E$, relates to the classical description of the motion of the particles in the system. The Schroedinger equation (1-10) is a wave-type equation and it relates to the vibratory motion of a string or a drumhead. In classical mechanics, particles move in trajectories, whereas wave motion is described by a displacement function that gives the nodal pattern, the amplitudes of displacement, and the phases. Examination of ψ reveals greater similarity to the latter, the wave model. The function ψ has nodal surfaces and phase changes across these nodal surfaces. When two wave functions interact, they interact either constructively or destructively, as waves do, depending on whether they interact in-phase or out-of-phase. On the other hand, there is no information whatsoever in ψ about the trajectory of an electron. In fact, on the microscopic scale, the *concept* of electron trajectory in an atom or molecule has lost significance. Instead, we find that the function ψ, or rather, its square ψ^2, gives only the probability of finding the electron within a given volume. Where ψ^2 is high, the probability of finding the electron is high. Where ψ^2 is low, the electron is rarely found. On a nodal surface the value of ψ is zero, so the electron is never found there.

The information contained in ψ^2 can be likened to the information contained in the holes in a dart board. The pattern of holes shows the result of many individual dart throws—it also would guide a gambler wagering on the next throw. There is a high probability that the dart will land within a circle containing many dart holes and a low probability for a circle of equal area but containing few dart holes. The density of dart holes gives probability information. If we were to express mathematically the density of dart holes as a function of position on a dart board, we would have a function exactly analogous to ψ^2.

To take full advantage of the easily interpreted dart board analogy, we might make a similar graph for the wave function of a hydrogen atom. For any set of coordinates x,y,z, the function ψ has a particular numerical value. We could make a graph in which dots are shown with density (dots per unit volume) proportional to the square of this numerical value. Figure 1-11 shows such a plot for a 1s wave function, this being the ψ that corresponds to the lowest allowed energy state of hydrogen. Of course, the electron in the hydrogen atom occupies space in three dimensions around the proton, so Figure 1-11 shows only a cross-sectional cut through the atom. Nevertheless, the picture gives an immediate qualitative picture of where the electron is most likely to be found—somewhere reasonably near the nucleus.

This is, of course, only one way to represent the probability picture of how the electron moves in the vicinity of the proton.

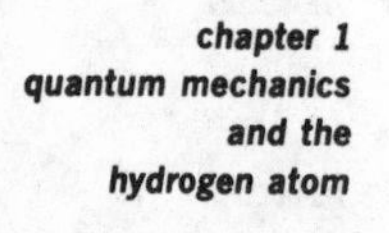

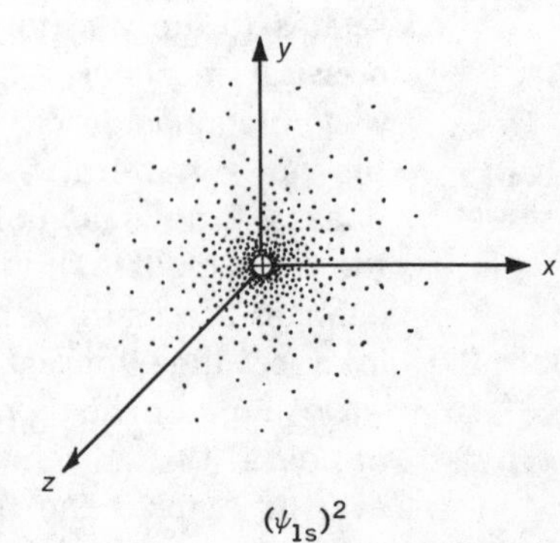

Figure 1-11 The dart board and a hydrogen 1s function: probability density plots.

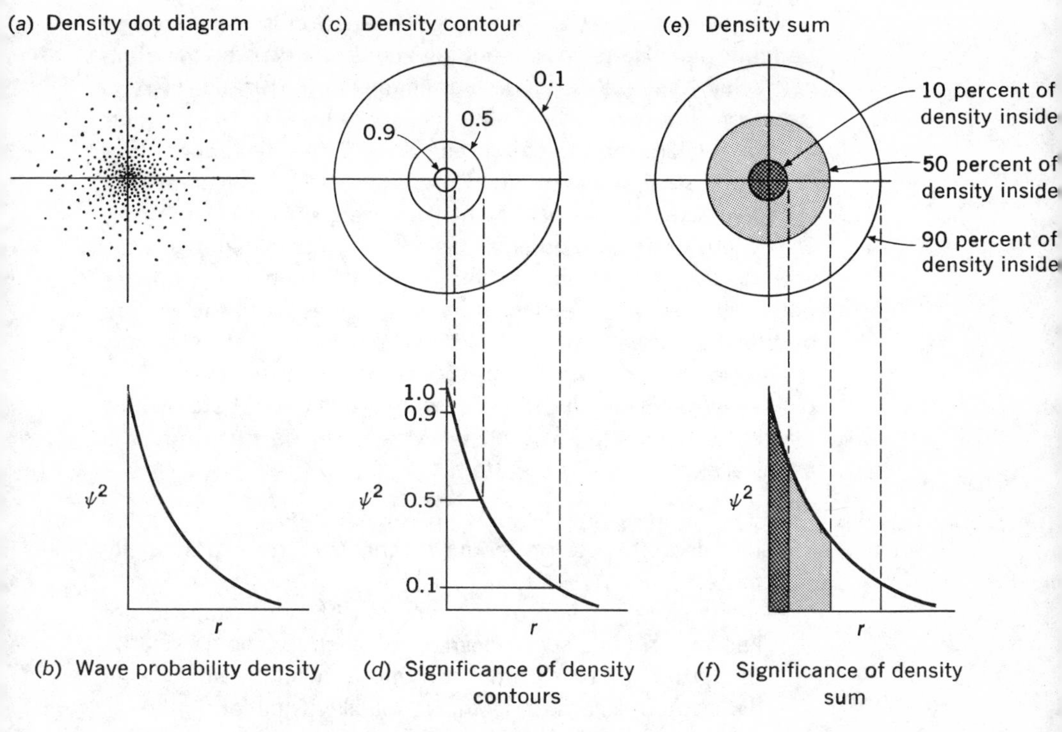

Figure 1-12 Representations in two dimensions of the probability distribution for a hydrogen 1s wave function.

Figure 1-12 shows three additional ways. Figure 1-12(*b*) shows a graph of the dot density (ψ^2) along some line radiating from the proton. Along the horizontal axis we have plotted the value of *r*, the distance from the nucleus, and along the vertical axis we plot ψ^2.

Figure 1-12(*c*) shows density contours on a cross-sectional cut through the probablity dot diagram of Figure 1-12(*a*). Figure 1-12(*d*) shows the significance of the density contours: they present a contour map of where the electron spends its time. Such a map has the advantage that it displays qualitatively the shape of the distribution (in two dimensions) and it displays quantitatively how rapidly the probability drops off far from the nucleus. This representation is much used because it is easy to calculate.

Figure 1-12(*e*) is probably the most informative representation. The innermost contour line is that line that *encloses* 10 percent of the probability in the smallest possible volume. The next contour shows the line that encloses 50 percent in the smallest possible volume. Finally, the third contour shows the 90 percent enclosure line. These contours are most useful to the gambler, whether betting on where the dart will fall or where the electron will be. They not only show where the electron is most likely to be found but also tell us something about how "large" the atom is. Plainly, the size is not well defined—*there is no boundary to the atom*. One could arbitrarily take its size to be the 0.5 contour in Figure 1-12(c) or one could take the 0.5 density-sum contour in 1-12(e). Some workers might prefer the 0.9 density-sum contour in 1-12(e) on the argument that the electron spends only 10 percent of its time outside this 0.9 line, a small enough fraction to be unimportant. At this moment we needn't choose between these alternatives. Our main interest is in the picture of the atom given by quantum mechanics. It differs greatly from the classical view, but since this quantum mechanical picture agrees with all the experimental facts about atoms and molecules, it surely must be preferred. In the quantum mechanical description of the atom,

—the electron trajectory is completely unspecified;
—the electron position is known only through a probability pattern;
—though the position of the electron at any instant is not known, it is at some *point;* the electron should *not* be thought of as being atomized into a "cloud" with charge distributed according to the probability function;*
—there are a variety of ways to designate the probability pattern that shows where the electron is likely to be found;
—this probability pattern shows that there are no boundaries to the atom—it extends to infinity.

(f) THE UNCERTAINTY PRINCIPLE

Before proceeding to examine other properties of the quantum mechanical atom, we should look briefly at the Uncertainty Principle—because many people are uncertain about the meaning of this principle. About the same time that Schroedinger was framing the wave picture of the atom, a German scientist named Heisenberg discovered a limitation on our possible knowledge of the atom. He realized that a consequence of the particulate nature of light is that light disturbs the object viewed. If a photon can be represented as a bullet with momentum appropriate to its frequency, then as this photon bullet ricochets off of an object, there will be some recoil momentum transferred to the object. This is the same momentum transfer that causes a change in the photon frequency, as pictured in Figure 1-2.

*Such a hypothetical electron cloud has a potential energy associated with the repulsion of its parts that does not appear in the potential function used in the Schroedinger equation. Nevertheless, this "electron-cloud" model has significant use merely because it has some computational simplicities.

Because of the low electron mass, this recoil effect is not small. Furthermore, the higher the photon energy, the larger will be its momentum [see expression (1-4)] and, hence, the greater the recoil effect. Unfortunately, the high-energy photons give the sharpest probes for fixing the electron's position.

These facts present a dilemma if we try to measure simultaneously *both* the position and the momentum of an electron. To pinpoint the electron's position accurately, we need a high-energy photon, but the high-energy photon disturbs the electron's momentum. If we use a low-energy photon to permit an accurate momentum measurement, the photon is too "soft" to locate the electron precisely. We can either optimize conditions to measure position or to measure momentum, but we cannot measure *both* simultaneously to any accuracy we wish.

This qualitative statement of the Uncertainty Principle is sufficient for our needs. It says that we cannot precisely locate the electron and at the same time measure its momentum. However, it places no limit on how closely we can locate the electron. It *is* somewhere and we can find its position to any degree of certainty we wish. The Uncertainty Principle only tells us that the closer we determine position, at a given instant, the less we will be able to learn about its momentum at that same instant.

1-3 The hydrogen atom

With the assurance that quantum mechanics explains the properties of atoms and molecules, it behooves us to adapt our thinking to this model. The hydrogen atom is our most informative example because an exact solution of its Schroedinger equation is possible. No such exact solution is possible for any atomic (or molecular) system with two or more electrons, although approximations can be made that allow us to come very close to the correct solutions. These many-electron atoms will be considered in Chapter Two when we have the benefit of a clear understanding of the hydrogen atom.

(a) QUANTUM NUMBERS

We have learned that only particular energies can be held by a hydrogen atom. These particular energies, called energy levels, and the electron probability pattern are given by the Schroedinger equation. Both the energy and the electron probability distribution depend upon integral numbers that are closely related to the integral numbers that identify the nodal properties of a vibrating string or drum. For a string, there is displacement in only one dimension (the y direction), and we need only one number to describe its permitted motions. The two-dimensional vibrations of the drum, on the other hand, need two numbers, one of which describes the radial nodal properties and the other of which describes the angular nodal properties.

The Schroedinger wave equation for the hydrogen atom describes the electron wave in *three dimensions.* It is entirely appropriate, then, that *three* integral numbers are necessary to describe fully each energy state of the hydrogen atom. These numbers are called quantum numbers. Each set of quantum numbers fixes a possible energy of the atom and also the probability pattern that describes what we know of the electron's location. Figures 1-11 and 1-12 show the atom in its lowest energy state. More energetic energy levels correspond to more complex spatial distributions. These spatial distributions are counterparts to the orbital trajectories that describe the classical motions of planets in a solar system. If we picture a solar system that could be shrunk as much as desired, by the time the sun had shrunk to the mass of a proton, the orbital trajectory would have become the quantum mechanical probability distribution expressed by ψ^2. Because of this correspondence, scientists refer to a probability pattern as an "orbital." We must remember, though, that the word orbital now refers to a picture like that in Figure 1-11 and the trajectory meaning is lost.

The three quantum numbers that define a particular probability distribution (a particular orbital) are symbolized n, ℓ and m. The first of these, n, is the most important, so it is called the *principal* quantum number. This number n can have any integral value, 1, 2, 3, 4, ... That integral value determines the energy of the hydrogen atom according to the formula

$$E_n = -R \cdot \frac{Z^2}{n^2} \qquad (1\text{-}11)$$

where R = a constant = $313.6 \dfrac{\text{kcal}}{\text{mole of H atoms}}$

Z = charge on the nucleus (+1 for a hydrogen atom)

n = principal quantum number = 1, 2, 3, 4, . . .

The negative sign in (1-11) indicates that the energy is lowered relative to the reference state, a proton and an electron separated from each other at an infinite distance.

The principal quantum number also fixes the nodal characteristics of an orbital. While a vibrating string has nodal *points* (see Fig. 1-8) and a vibrating drum has nodal *lines* (see Fig. 1-9), a hydrogen atom has nodal *surfaces*. At such a surface the wave function ψ changes phase, like the up-down phase change at a node in the guitar string. At this nodal surface, ψ^2 is zero; the electron is not found at this location. The number of nodal surfaces is equal to n, one of which is the boundary condition nodal surface at r = infinity.

The spatial extent of the orbital is determined by two of the quantum numbers, n and ℓ. For example, quantum mechanics tells us the average distance $\bar{r}$ of the electron from the nucleus:

$$\bar{r} = a\frac{n^2}{Z}\left[\frac{3}{2} - \frac{\ell(\ell + 1)}{2n^2}\right] \tag{1-12}$$

where a = a constant = 0.529 Å
Z = charge on the nucleus (+1 for a hydrogen atom)
n = principal quantum number, and
ℓ = angular quantum number

The second quantum number ℓ can have integral values beginning at zero but not exceeding $(n - 1)$. Thus, if n is 3, ℓ can have any of the values 0, 1, or 2 but no others. If n is 1, ℓ can have no value other than 0.

The quantity ℓ also relates to the nodal pattern of the orbital.

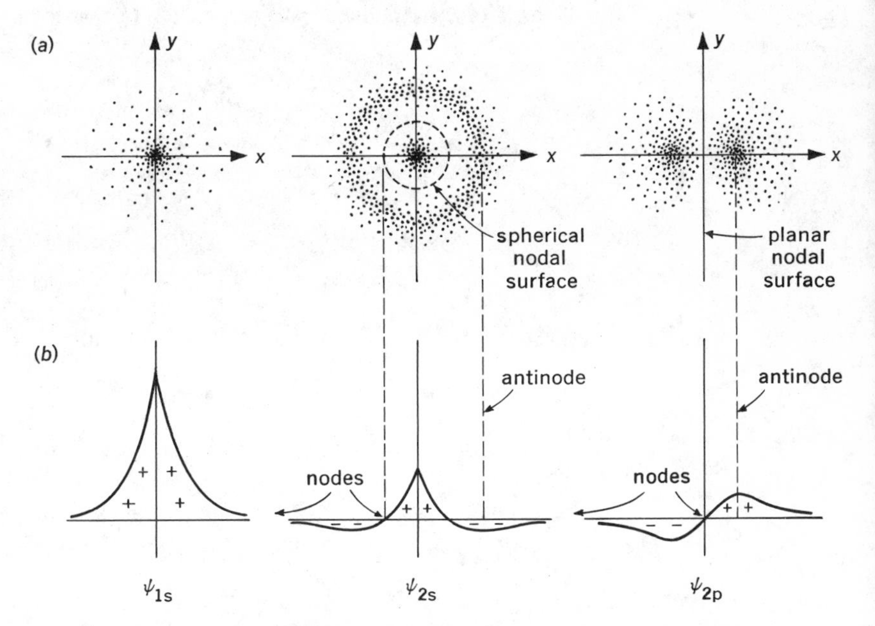

Figure 1-13 Nodal surfaces for the hydrogen atom. (a) Electron probability density. (b) Electron wave functions (plus and minus signs designate phase). Notice that the 2p ($n = 2$, $\ell = 1$) function changes phase at the nucleus. The two lobes of the p functions are out of phase, like a guitar string vibrating in the first overtone.

There are ℓ nodal surfaces with angular dependence. Since there is a total of n nodal surfaces, there must be $(n - \ell)$ nodal surfaces without angular dependence, that is, spherically symmetric. For example, if $n = 2$, then ℓ can be either 0 or 1. If $\ell = 0$, there are no nodal surfaces with angular dependence; the two radial nodal surfaces are both spherical in shape. One of these is the surface at r = infinity. If $\ell = 1$, there is one angularly dependent nodal surface, a plane. The second nodal surface $(n - \ell = 2 - 1 = 1)$ is a sphere, again the nodal surface at infinity. Figure 1-13 shows cross-sectional views of the probability distributions corresponding to $n = 2$. The change of phase at the nodal surface is designated by plus and minus signs. These signs have significance related to the up-down displacements in wave phenomena (as portrayed in Figs. 1-9 and 1-10) and do not signify electric charge.

The $n = 2$, $\ell = 1$ orbital is shown in Figure 1-13 to be directed along the x axis, and the yz plane is its nodal surface. This brings to mind the possibility that this orbital could be otherwise oriented. There are three dimensions, why should the x axis be preferred? Indeed, it is not. There are three orbitals with $n = 2$, $\ell = 1$, and these can be considered to be directed along the three axes, as shown in Figure 1-14. These three orbitals

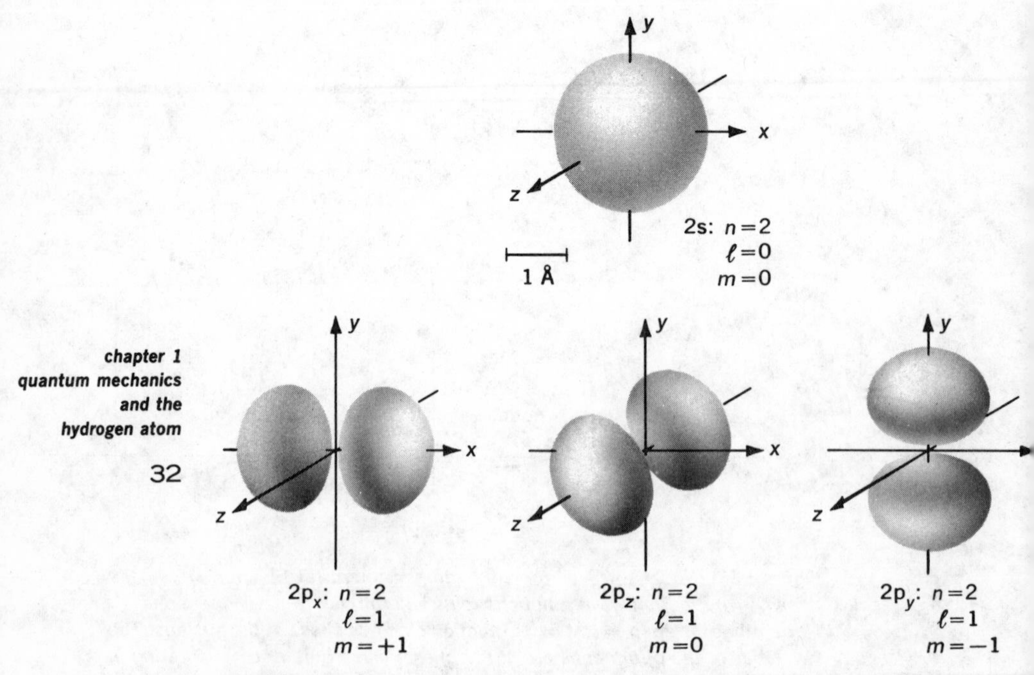

Figure 1-14 The $n = 2$ orbitals for the hydrogen atom: orientations and nodal surfaces.

are identified by the third quantum number m. This number is called the *magnetic quantum number* and it can take integral values either positive or negative in sign but not exceeding ℓ:

$$m = \ell, \ell - 1, \ell - 2 \cdots 1, 0, -1, -2 \cdots -\ell \qquad (1\text{-}13)$$

Again, using the example of $n = 2$, we now find the following set of orbitals, all with the same energy,

$$E = -R\frac{Z}{n^2} = -R\frac{1}{2^2} = -\tfrac{1}{4}R$$

			E	$\bar{r}$
$n = 2$	$\ell = 0$	$m = 0$	$-\frac{1}{4}R$	6a
$n = 2$	$\ell = 1$	$m = +1$	$-\frac{1}{4}R$	5a
$n = 2$	$\ell = 1$	$m = 0$	$-\frac{1}{4}R$	5a
$n = 2$	$\ell = 1$	$m = -1$	$-\frac{1}{4}R$	5a

(1-14)

These four orbitals have already been pictured in Figure 1-14. We note that with $n = 1$ there is only one orbital ($n = 1, \ell = 0, m = 0$). With $n = 2$ there are four orbitals (1-14) and four is equal to n^2. This is so for any choice of n; there are n^2 orbitals with that value of the principal quantum number.

(b) THE HYDROGEN ATOM ENERGY LEVEL DIAGRAM

In summary, the three quantum numbers, n, ℓ and m, fix the possible energy states of the hydrogen atom and the corresponding probability distributions or orbitals. The energy and the number of orbitals with a particular energy are fixed by the value of n. The nodal patterns depend as well upon ℓ and m.

Table 1-2 shows how the number of states rises as n rises. It also shows the popular notation used to designate each orbital. Instead of indicating three numbers n, ℓ and m, an orbital is designated first by a number, the value of n, and then by a letter, s, p, d, or f, to indicate the value of ℓ. Thus the orbital $n = 2$ and $\ell = 0$ is called the 2s orbital. We shall see that all orbitals with $\ell = 0$ have spherical symmetry, so the symbol s can be considered to stand for "spherical."* An orbital with $n = 2$ and $\ell = 1$ is called a 2p orbital. There are three of these, corresponding to the three values of m ($+1, 0, -1$). Since the three 2p orbitals can be pictured to be oriented at right angles

* Actually the letters s, p, d, and f were chosen before the hydrogen atom problem was solved and the letters were descriptive of characteristic spectral behaviors. Hence the original abbreviations no longer have value.

Table 1-2 The Hydrogen Atom Quantum Numbers

n	$E = -\frac{313.6}{n^2}$ kcal/mole	$\ell = 0, 1, 2, 3, \ldots, n-1$ s, p, d, f, ...	$m = \ell, \ell - 1, \ldots,$ $1, 0, -1, \ldots \ell$	Number of orbitals n^2
1	$-\frac{313.6}{1^2} = -313.6$	0(1s)	0	1
2	$-\frac{313.6}{2^2} = -78.4$	0(2s) 1(2p)	0 +1, 0, −1	1 + 3 = 4
3	$-\frac{313.6}{3^2} = -34.8$	0(3s) 1(3p) 2(3d)	0 +1, 0, −1 +2, +1, 0, −1, −2	1 + 3 + 5 = 9
4	$-\frac{313.6}{4^2} = -19.6$	0(4s) 1(4p) 2(4d) 3(4f)	0 +1, 0, −1 +2, +1, 0, −1, −2 +3, +2, +1, 0, −1, −2, −3	1 + 3 + 5 + 7 = 16

to each other, the symbol p can be related to the descriptive word "perpendicular" and the three p orbitals can be designated p_x, p_y and p_z.

The information in Table 1-2 can be represented pictorially through an energy level diagram, as in Figure 1-15. Horizontally, we represent an orbital as a pigeonhole which might be occupied by an electron. The orbital is placed on a vertical energy scale to indicate how much the energy is lowered as an electron is captured into that orbital by a proton.

Such a diagram simplifies discussion of the spectra of atoms. Picture an electron in a 2p orbital of hydrogen. This atom has an energy lower than that of a separated electron and proton by 78.4 kcal/mole. If, now, this atom were to release energy and drop to a lower state, quantum mechanics and experiment tell us there is only one such state, the 1s orbital. This energy level, the lowest, is 313.6 kcal/mole below our zero reference energy. Hence the atom can make the transition 2p → 1s only if it can dispose of the energy difference

$$(-78.4) - (-313.6) = 235.2 \text{ kcal/mole}$$

This is, of course, one of the energy changes actually observed in the hydrogen atom line spectrum. *All of the lines in the hydro-*

gen atom are accounted for by energy differences like the two shown in Figure 1-15.

(c) OTHER ONE-ELECTRON ATOMS

The ion He^+ consists of a nucleus with charge $+2$ and a single electron. Its Schroedinger equation is, mathematically speaking, identical to that of a hydrogen atom with only one change, the nuclear charge Z is doubled. Equations (1-11) and (1-12) apply, provided we make $Z = 2$ instead of $Z = 1$. The energy level scheme looks identical to that in Figure 1-15 except that every energy is multiplied by Z^2, or 4. To remove the electron from a hydrogen atom requires 313.6 kcal. To remove the electron from He^+ requires $(313.6)(4) = 1254.4$ kcal. Thus we see

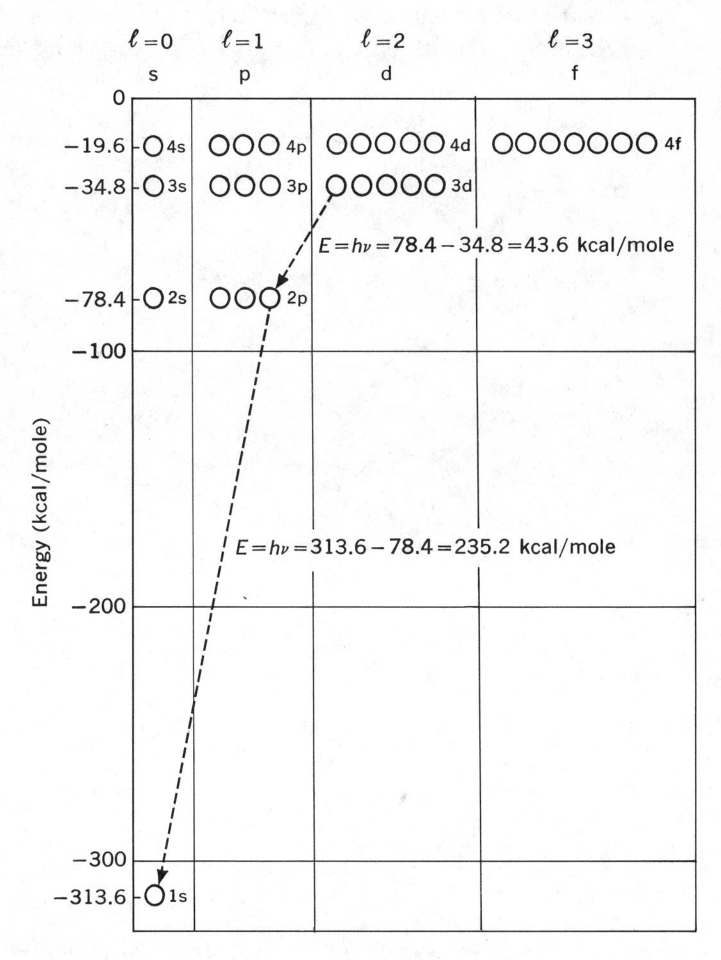

Figure 1-15 Hydrogen atom energy level diagram.

that the higher positive charge on the nucleus holds the electron more tightly—an entirely reasonable result.

It is also reasonable to expect that the +2 nuclear charge will pull the electron in closer. Equation (1-12) shows this is also true. Whereas the 1s state of the H atom has an average radius of $\frac{3}{2} \cdot 0.529 = 0.79$ Å, the He^+ ion has an average radius of 0.39 Å, half as large.

On the other hand, the orbital distributions are the same for He^+ as for H (see Fig. 1-16), except smaller in all dimensions by the $1/Z = \frac{1}{2}$ factor.

In a similar way, the energy levels and orbital characteristics of the doubly charged lithium ion, Li^{+2}, are like those of the hydrogen atom except for the effect of nuclear charge. A Li^{+2} ion consists of a nucleus with $Z = 3$ and a single electron. Its ionization energy is $(313.6)(3^2) = 2822.4$ kcal and the spatial extent of its orbitals is one-third that of H.

Figure 1-17 contrasts the energy level diagrams predicted

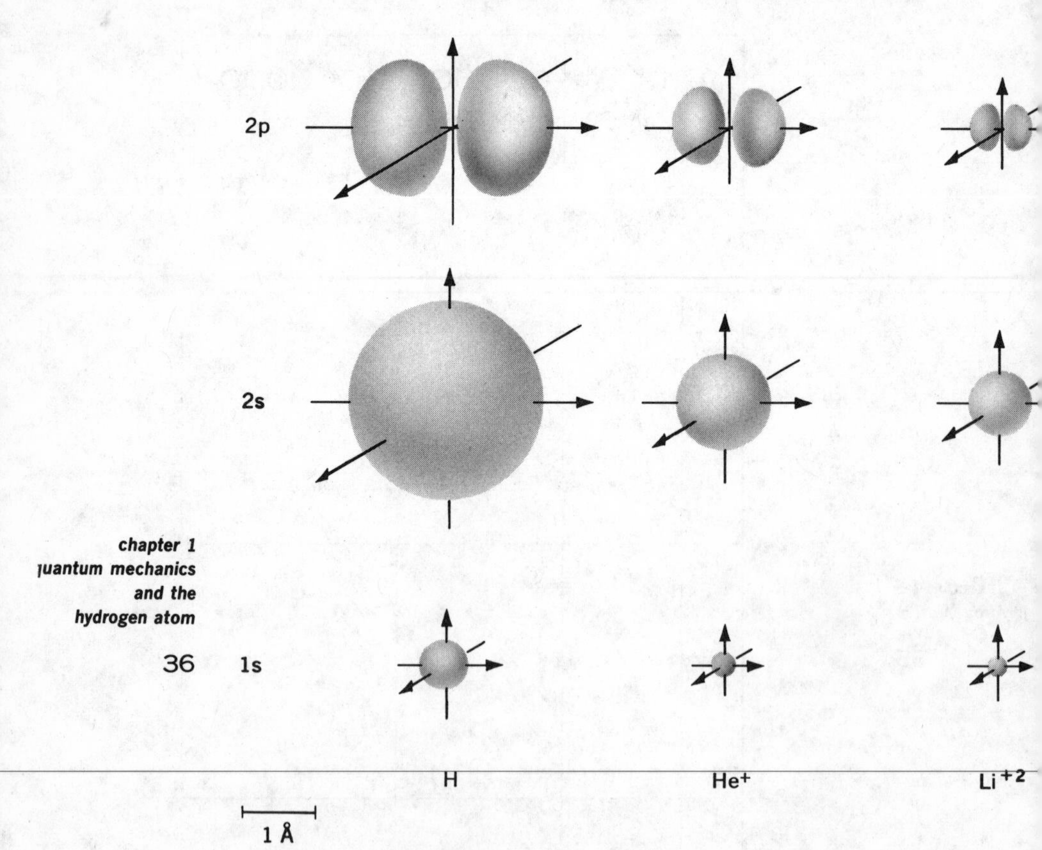

Figure 1-16 Relative sizes of orbitals of one-electron atoms and ions, H, He^+ and Li^{+2}. Boundary surfaces represent the 99 percent density sum.

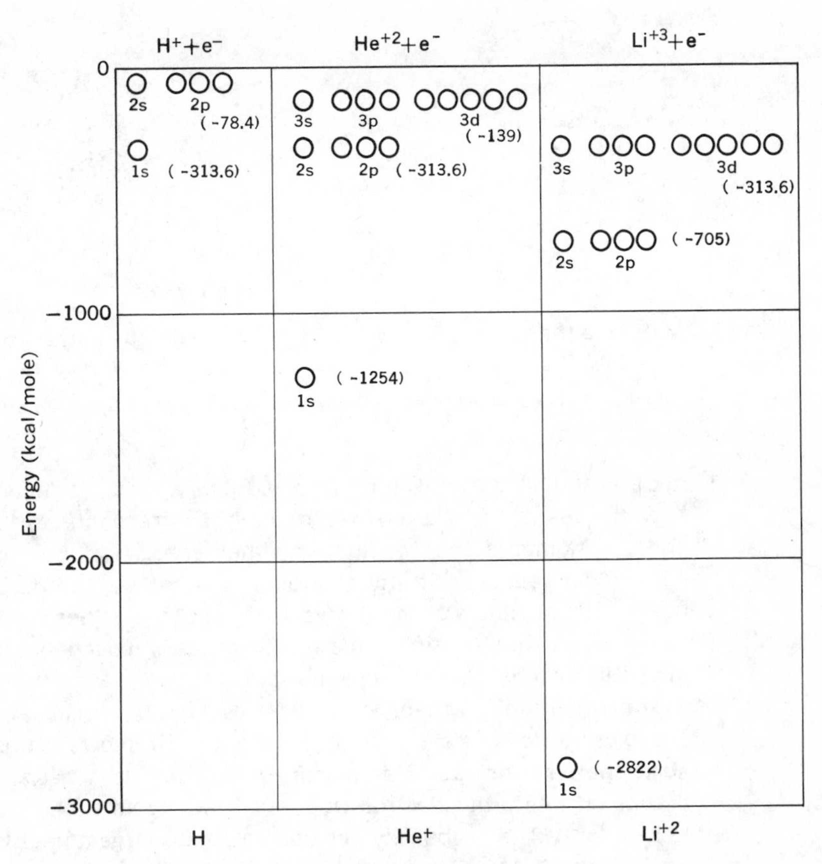

Figure 1-17 Energy level diagrams of one-electron atoms and ions.

for these three one-electron atoms, H, He^+ and Li^{+2}. These energy level diagrams are in perfect accord with the experimental spectra emitted by these species in glow discharge experiments. Quantum mechanics scores again!

Qualitatively we see two important results. First, as nuclear charge increases, electrons are held more tightly. Second, as nuclear charge increases, the electrons are pulled closer—the atom gets smaller.

(d) CONTRAST TO THE PLANETARY ATOM

Figure 1-18 pictures the planetary and the quantum mechanical views of the atom. The planetary model is simple. The electron endlessly circles about its proton sun, always at a fixed radius (in the 1s state). It doesn't radiate, despite instructions to do so by classical electromagnetic theory, because Niels Bohr said it had better not. Otherwise the electron would spiral into the

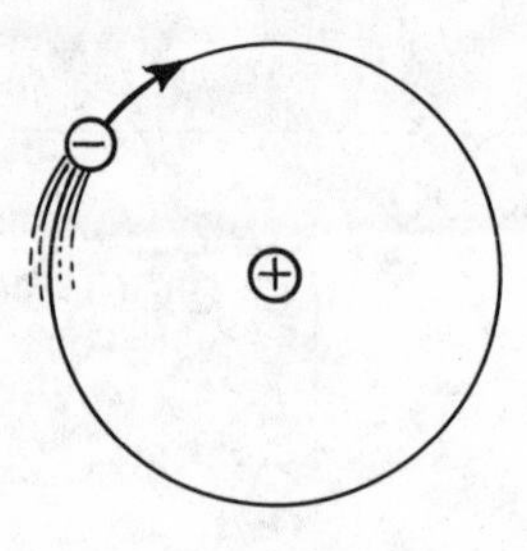

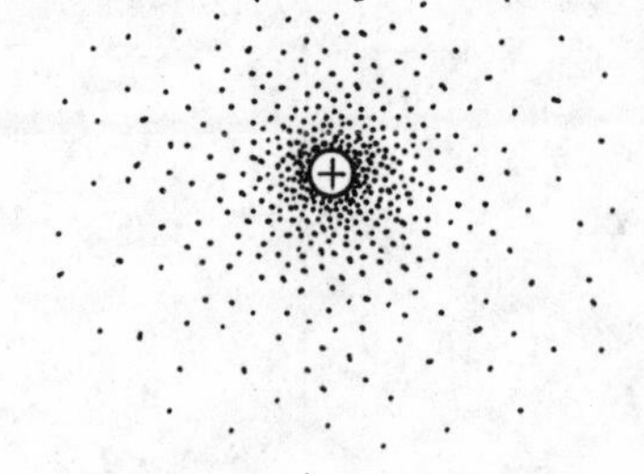

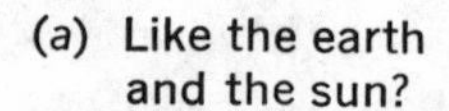

(*a*) Like the earth and the sun?

(*b*) Like the pattern of holes on a dart board?

Figure 1-18 The planetary (a) and quantum mechanical (b) hydrogen atom.

nucleus and all matter would collapse. It has a particular energy because one of the integer-fixed momenta corresponds to that energy. Momenta that would give other energies, lower and higher, were again arbitrarily forbidden by Bohr's authoritarian model. The planetary concept gave no clue as to why these other energies are not possible; quantization was imposed on the model to force it to agree with the facts.

The quantum mechanical view is less readily assimilated. The picture doesn't even show the electron; it shows a probability pattern instead. This pattern only gives us a basis for betting on where the electron would be if we did an experiment to look for it. The probability per unit volume is largest near the nucleus and it dwindles off as we look farther away. There is no electron trajectory shown at all. That was lost in the mathematics when the kinetic energy term was modified to convert the energy conservation law into a wave-type equation. The electron seems to move about—but the model includes no trajectory! This is a new way to have to think!

Indeed it is; we must be reminded again and again of the guitar string and put out of our minds the solar system. But why? Why use the complicated, abstract model instead of the simple, planetary one? The answer lies in science's implacable test of usefulness. The probability view fits many different kinds of observations. The planetary model fits only one kind: the energy levels of the one-electron atoms. That is not enough when a better model is at hand.

A striking difference between the two pictures in Figure 1-18 is found in the electron–proton distance. The planetary electron never comes closer to the nucleus than dictated by its fixed orbital radius. In contrast, in Figure 1-18(*b*) the probability density near the nucleus is greater than anywhere else. Are there any observations that test these two conflicting conclusions? Yes, there are several, three of which we will mention here.

(i) *Electron capture by the nucleus.* A few nuclei are unstable with respect to a change in which an electron is captured into the nucleus, decreasing its positive charge by one unit. This process requires the electron to be extremely close to the nucleus part of the time—in accord with the quantum mechanical view.

(ii) *Nuclear magnetic splittings.* Most nuclei, including protons, are tiny magnets. In a strong magnetic field, these nuclear magnets can be lined up in a quantized way. The magnets can then be reoriented by the absorption of radiation (light) whose frequency just matches the energy levels of the magnets in the field. However, the electron is itself a tiny magnet, so it, too, communicates with the nuclear magnets. Certain aspects of the nuclear-magnet energy level diagrams require that the electron magnet come very close to a proton magnet part of the time. In fact, the quantum mechanical atom correctly tells how much.

(iii) *Spectral splittings in the H atom spectrum.* When the hydrogen atom spectral lines are examined with the most powerful optical equipment, each line is found to have two components. The minute frequency difference is explained with the aid of the quantum mechanical view of the H atom. The splitting turns out to be due to the part-time proximity of the electron and the nucleus, again in accord with Figure 1-18(*b*).

With this reassurance, let's return to the troubles of the planetary atom. First, Bohr had to impose fixed energy levels (through his momentum restriction) on an unreceptive model. But Schroedinger's guitar-string model is perfectly tuned to the quantization of energy. A wave phenomenon with boundary conditions *naturally* leads to special energy levels. Instead of appearing as a strange constraint, special energies are an intrinsic part of a wave description.

The second headache of the planetary model is that radiation and loss of energy is implied by the movement of an electron charge in a curved trajectory. Quantum mechanics simply does not have this difficulty since the trajectory concept is gone. The wave description of the electron in an atom does not describe a trajectory, so it cannot be accused of implying energy loss through radiation. Furthermore, it cannot be said that there is a trajectory but quantum mechanics doesn't give it. The change of the classical equation of motion to the quantum mechanical one specifically altered the kinetic energy term. We can paraphrase this change into an instruction—"Stop

Figure 1-19 The Planetary Atom: R.I.P.

thinking about trajectories—we can't fit nature with a trajectory-type equation of motion." And why should we complain? Without a trajectory, the electron need not radiate its energy and we know that it does not.

But the most compelling nails are yet to be driven into the planetary model's coffin. This simple picture never explained *any* many-electron atom, beginning with helium, nor did it provide even a tiny step forward in explaining chemical bonding. Quantum mechanics deals successfully with both. The many-electron atoms are considered in Chapter Two, and the rest of this monograph explores chemical bonding, all illuminated with the aid of our quantum mechanical light.

two many-electron atoms

The hydrogen atom is simple because it contains only one electron—there is but one electrostatic attraction contributing to the potential energy. This simplicity is lost even in the next element, the helium atom. With two electrons, this atom has a potential energy made up of three contributions: each electron is attracted to the nucleus and the electrons are repelled by each other. Immediately the mathematical difficulty increases to the point at which laborious approximation methods are required. The situation is magnified many-fold if we delve further into the periodic table. Just consider the oxygen atom which, with its eight electrons, has eight electron–nucleus attractive terms and *twenty-eight* separate electron–electron repulsive terms. No wonder it takes a digital computer to keep track of such calculations!

Nature is kind to us here. Astonishingly, the energy levels of the many-electron atoms are obviously related to those of the hydrogen atom. The connection is so clear that chemists use the hydrogen atom quantum number designations for atoms with many electrons. With this head start on the energy level diagram, disturbances due to electron–electron repulsions can be identified and understood.

2-1 The energy levels of many-electron atoms

Atoms in an electrical discharge emit light, and the colors seen reveal that atom's energy level scheme. Like the hydrogen atom, many-electron atoms also emit line spectra—only particular energies are observed. A quantum mechanical model is needed. However, no atom with two or more electrons has energy level spacings as simply related as the integer-connected levels of the H atom. The rather complicated frequency patterns are understandable, though, with the aid of the H atom quantum numbers when electron–electron repulsions are considered. How this works out can be seen by looking at the energy level diagrams of the next two elements, helium and lithium.

(a) HELIUM ATOM

A naive way to approach the two-electron, helium atom would be to assume that somehow we could turn off the electron–electron repulsion. Then each electron would move around the nucleus oblivious to the presence of the other. Each electron would occupy the 1s orbital of the He^+ ion, the lowest energy level in Figure 1-17. Each electron would be bound by 1254 kcal, as is the one electron of He^+. Each electron would occupy an orbital with an average radius of 0.39 Å.

With this start, let's turn up the knob on the electron–electron repulsion. How will the movement of the electrons and their energy levels change? First, their repulsion will tend to keep them apart. If one electron is on the north side of the nucleus, the other will tend to be on the south side. *Their motions will tend to correlate.* Second, the energy levels will change because the electrons repel each other. How much? Well, we can make a very rough guess in a simple way. Suppose the two electrons were each to occupy a 1s orbital (average radius 0.39 Å) and to stay, by mutual agreement, on opposite sides of the nucleus. They would then be, on the average, two times 0.39 Å apart. The repulsion energy of two electrons 0.78 Å apart is just 420 kcal. This crude estimate tells us that if two electrons occupy the He 1s orbital, it will be easier to remove one of them because of the electron–electron repulsion. How much easier? Well, something like 420 kcal easier. The ionization energy might be near 800 kcal.

In fact, the ionization energy of a neutral helium atom is not 1254 kcal (as it would be if there were no electron–electron repulsion) but 567 kcal. As our naive model predicted, electron repulsions are quite substantial—in fact, our naive model underestimated their effect by 60 percent.

Figure 2-1 shows a part of the experimentally determined energy level diagram of He. Look at the 2s and 2p energy levels. These pigeonholes are placed at the energies appropriate to the states in which one of the electrons has been lifted from the 1s orbital and placed in the 2s or 2p orbital. The first feature that catches the eye is that the 2s and 2p orbitals are quite close in energy to the 2s and 2p orbitals of a simple hydrogen atom. In these excited states, the effect of the electron–electron repulsion is to give the 2s or 2p electron the silly idea it is in a hydrogen atom! Instead of showing the ionization energy appropriate to an electron moving near a helium nucleus, the 2s or 2p electron sees the combination of the helium +2 charge surrounded by the tightly held 1s negative electron. It myopically decides that the nuclear charge is $+2 - 1 \cong +1$. The inner 1s electron effectively "shields" half of the nuclear charge.

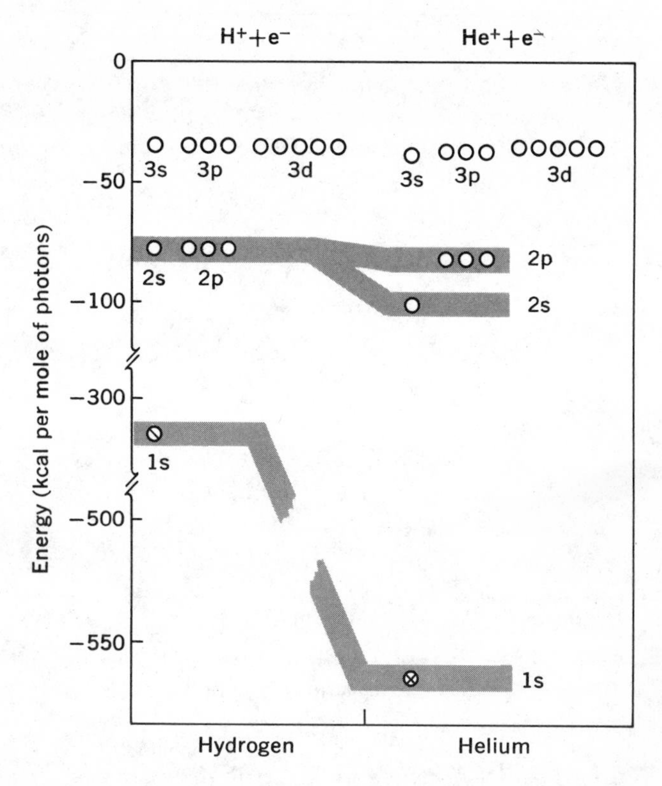

Figure 2-1 Energy level diagram for hydrogen and helium atoms. (Notice breaks in the vertical scale.)

The other feature of the 2s and 2p orbitals that needs mention is that the 2s and 2p energies are no longer identical. The 2s orbital lies 20 kcal lower than the 2p level. Thus the 2s and 2p orbitals are still close in energy but no longer identical. This splitting of energy levels with the same principal quantum number is always present in many-electron atoms, not only between the s and p orbitals, but also between p and d orbitals. The latter splitting has enormous significance, since it reshapes the periodic table.

(b) LITHIUM ATOM

If we returned to our naive model, the "turned-off" electron repulsion model, three electrons might try to crowd into the lithium 1s orbital. Each would be bound with an energy of 2822 kcal (see Fig. 1-17) and would have an average radius of 0.26 Å. There would, of course, be very large electron–electron repulsions with three electrons in the same neighborhood. This situation need not be pursued, however, for all of our

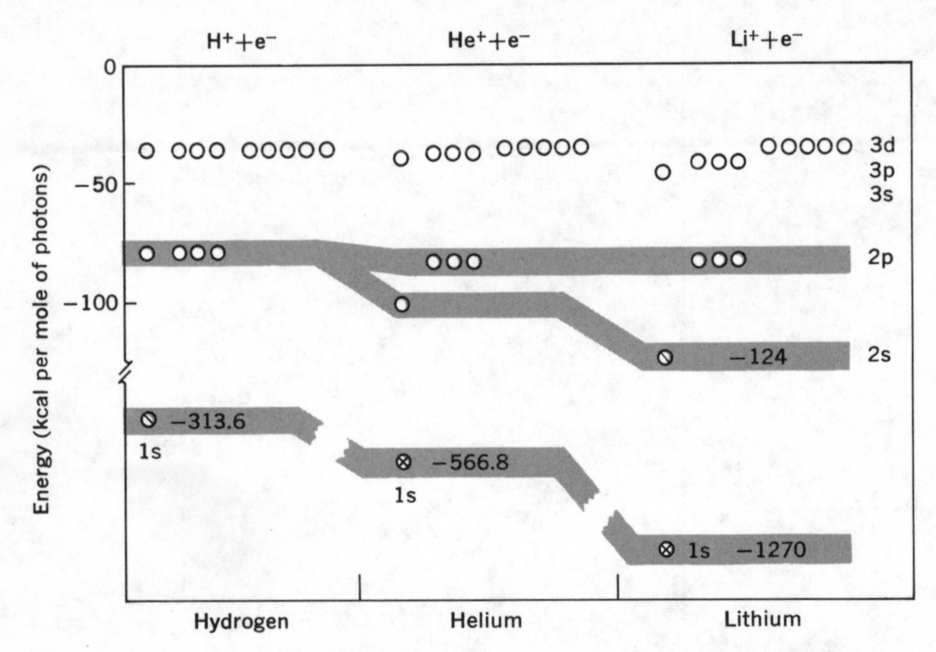

Figure 2-2 Energy level diagram for hydrogen, helium and lithium atoms. (Notice break in the vertical scale.)

experimental evidence indicates that *no orbital ever accommodates more than two electrons.* This is an empirical result whose explanation has *not* been found to lie in electron–electron electrostatic repulsion. It is called the *Pauli Principle* and it is the key to the structure of the periodic table. We will see its impact immediately when we apply this rule to the neutral lithium atom.

According to the Pauli Principle, if two electrons already occupy the 1s orbital, a third cannot be accommodated, so it must enter some higher energy state. The lowest one available is the 2s state. We will designate this state by a notation that lists the orbital occupancies of the electrons: $Li(1s^2 2s)$. The notation indicates that two electrons are in the 1s orbital ($1s^2$) and one electron is in the 2s orbital ($1s^2 2s$). How tightly is the third electron bound in this case? Figure 2-2 shows that the ($1s^2 2s$) state has an ionization energy of 124 kcal/mole.

$$Li(1s^2 2s) \rightarrow Li^+(1s^2) + e^- \qquad E_{ion'n} = 124 \text{ kcal/mole} \quad (2\text{-}1)$$

In contrast, the next higher state of the neutral lithium atom, $Li(1s^2 2p)$ has an ionization energy of only 81 kcal/mole.

$$Li(1s^2 2p) \rightarrow Li^+(1s^2) + e^- \qquad E_{ion'n} = 81 \text{ kcal/mole} \quad (2\text{-}2)$$

Just as in the helium atom, the 2s and 2p states no longer have the same energies. *Electron repulsion splits energy levels of the same principle quantum number.*

It is informative to investigate in an approximate way the significance of these ionization energies. Suppose we were to construct a one-electron atom with ionization energy of a 2s electron equal to 124 kcal/mole. Equation (1-11) tells us the magnitude of the nuclear charge Z^* we would need.

$$E = -313.6 \frac{Z^{*2}}{n^2} \tag{2-3}$$

so,

$$E = -124 \text{ kcal} = -313.6 \cdot \frac{Z^{*2}}{2^2}$$

$$Z^* = 1.26$$

Furthermore, with this effective nuclear charge, the average orbital radius, according to equation (1-12), would be

$$\bar{r} = 0.529 \cdot \frac{n^2}{Z^*} \left\{ \frac{3}{2} - \frac{\ell(\ell+1)}{2n^2} \right\} \tag{2-4}$$

$$\bar{r} = 0.529 \frac{2^2}{1.26} \left\{ \frac{3}{2} - 0 \right\} = 2.52 \text{ Å}$$

We see the same effect that appeared in the excited helium atom He(1s2s). The outermost electron acts as though the nuclear charge is much smaller than its actual magnitude. For lithium, the 2s electron "feels" the +3 nucleus through the "shield" provided by the two tightly held 1s electrons. The 2s electron feels an effective nuclear charge of 1.26.

We can make the same argument for the lithium state Li($1s^2 2p$). The 2p electron is more easily removed—its ionization energy is only 81 kcal. Calculating effective nuclear charge, this time we obtain $Z^* = 1.02$. The average radius of such an electron is calculated as follows:

$$\bar{r} = 0.529 \cdot \frac{2^2}{1.02} \left\{ \frac{3}{2} - \frac{1 \cdot (2)}{2 \cdot 2^2} \right\}$$

$$\bar{r} = 2.59 \text{ Å}$$

The 2p electron feels a lower effective nuclear charge than the 2s electron, despite the fact that the average radius seems to be about the same. The explanation of this difference is evident in the orbital pictures in Figure 1-13. The 2s orbital penetrates right down to the nucleus, whereas the 2p orbital has a nodal plane at the nucleus. Thus, the 2s electron has the opportunity to "penetrate" the electron shield provided by the two 1s electrons in the lithium atom. Evidently the 2p electron does this much less effectively. Because of its nodal pattern, the probability of finding the electron at the nucleus is zero; hence we find the 2p electron barely penetrates the 1s electron shield —it feels effectively only the difference between the $+3$ charge of the nucleus and the -2 charge of the two 1s electrons.

This penetration effect explains the 2s–2p splitting effect seen in the energy levels of both helium and lithium. As remarked earlier, this same effect persists and is even larger in the higher states, such as 3s, 3p and 3d. In fact, we shall see that the combination of the Pauli Principle (two electrons per orbital) and the splitting of s, p, and d orbitals enables us to understand the regularities of the entire periodic table.

(c) ON TO NEON

We now have the guidelines needed to look at the trends across an entire row of the periodic table from lithium to neon. Figure 2-3 shows the important orbitals, the 2s and 2p orbitals, for these eight atoms. For each atom the 1s orbital is shown to be fully occupied (two electrons only, according to the Pauli Principle!) but its energy is shown on a compressed scale. Also, the higher orbitals 3s, 3p, 3d, 4s, . . . are not shown; they are vacant and need not concern us now.

The orbitals that *are* shown suffice to hold all the electrons in excess of the two in the tightly bound 1s orbital. The orbital occupancy, as we proceed across the periodic table, is simple to understand, given the Pauli Principle. Moving from lithium to beryllium, the second electron can be accommodated by the 2s orbital, so the 2p orbitals need not be used. The ionization energy is, of course, higher than that of lithium, because beryllium has a higher nuclear charge. Proceeding to boron, however, the fifth electron is denied occupancy either in the 1s orbital or the 2s orbital. Consequently the lowest energy state for boron is $B(1s^2 2s^2 2p)$, with one electron in the higher-energy 2p orbital. Hence, despite the higher nuclear charge of boron, its ionization energy is lower than that of beryllium. The reason is obvious from Figure 2-3—the nuclear charge effect is more than offset by the fact that the most easily removed electron is in a 2p orbital instead of a 2s orbital.

Proceeding now to carbon, the six electrons will be distributed as follows. Two electrons occupy the 1s orbital, two occupy the 2s orbital, and the remaining two electrons make themselves

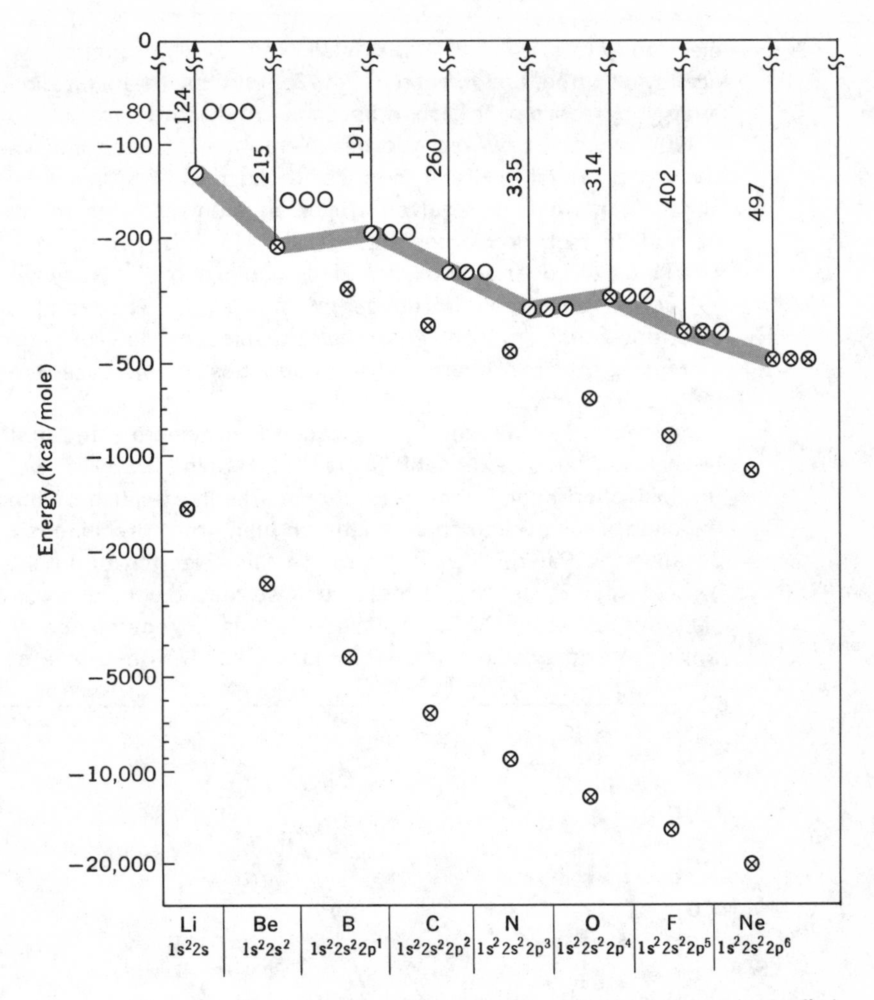

Figure 2-3 Energy levels, orbital occupancy and ionization energies for first-row elements. (Notice scale compression with rising energy.)

comfortable among the three 2p orbitals as they wish. Two of them *could* occupy the $2p_x$ orbital, as far as Pauli is concerned. However, that would place them in the same region of space, concentrated along the x axis, resulting in high electron–electron repulsion. A better situation results if one electron goes into the $2p_x$ orbital and the second enters either the $2p_y$ or $2p_z$ orbital. That keeps the two electrons in different neighborhoods and reduces electron repulsion. So the lowest energy state of carbon atom is $C(1s^22s^22p_x2p_y)$. Its ionization energy exceeds that of boron because the nuclear charge of carbon is higher than that of boron, and for both atoms a 2p electron must be removed.

Nitrogen is more of the same. By the Pauli Principle, three

electrons are left for the 2p orbitals and, to keep apart, they occupy all three, one in each of $2p_x$, $2p_y$ and $2p_z$. The ionization energy increases over carbon because of nuclear charge.

With oxygen, however, the last electron no longer finds a vacant 2p orbital in which to lodge. It must enter a 2p orbital already containing an electron. Hence electron repulsion shows up and, in fact, dominates the increase in the nuclear charge. The ionization energy of oxygen is less than that of nitrogen.

Fluorine and neon introduce no new ideas. Their added electrons must, in turn, enter half-occupied orbitals but the increasing nuclear charge effect continues to increase the ionization energy.

We see, then, the trend of ionization energy across the first two rows of the periodic table in Figure 2-4. Going from H to He, the ionization energy increases very markedly because of the doubling of the nuclear charge. Lithium then drops precipitously because the Pauli Principle forces the third electron up into a high-energy, 2s orbital. Across the first row, the next seven electrons all enter 2s–2p orbitals. There is a general rise in ionization energy due to nuclear charge, but with a couple of

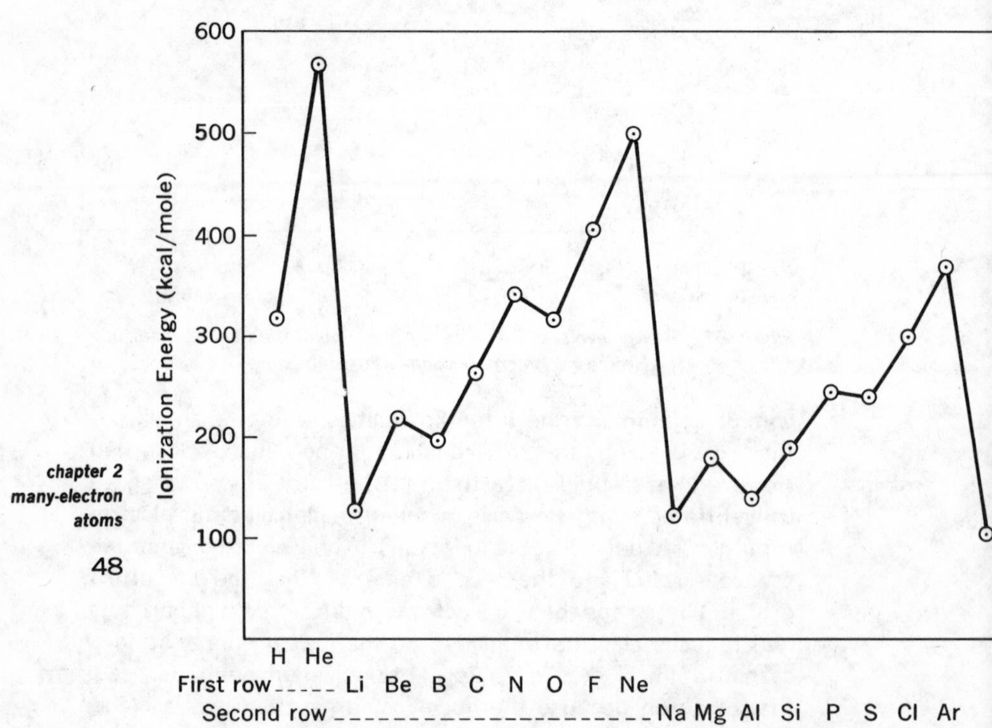

Figure 2-4 Trends in ionization energies in the first and second rows of the periodic table.

minor jogs due, first to the 2s–2p energy difference, and then to the first double occupancy of the 2p orbitals with the higher electron repulsion.

Figure 2-4 goes on to show that going from neon to sodium results in another precipitous drop, just like that of lithium. The reason is obvious. Once again, the filling of the 2s and 2p orbitals forces the next electron up into a higher-energy orbital, the 3s orbital.

(d) EFFECTIVE NUCLEAR CHARGE ACROSS THE FIRST ROW

It is informative to use our one-electron approximation to examine the effective nuclear charge felt by the most easily removed electron and the effective orbital radius for that electron. For each element we can calculate Z^*, the nuclear charge that would give rise to the observed ionization energy for a 2s or 2p electron ($n = 2$), using our modified form of (1-11).

$$E = -E_{\text{ion'n}} = -313.6 \frac{Z^{*2}}{n^2} \tag{2-5}$$

Then with this Z^* we can calculate the average radius that one-electron orbitals would have, using the modified form of (1-12), with $\ell = 0$ for the 2s orbital and $\ell = 1$ for the 2p.

$$\bar{r} = \frac{0.529 \cdot n^2}{Z^*} \left\{ \frac{3}{2} - \frac{\ell(\ell + 1)}{2n^2} \right\} \tag{2-6}$$

The results are summarized in Table 2-1. We see that the

Table 2-1 Effective Nuclear Charge and Orbital Radius in the One-Electron Approximation

	Z^*	$\bar{r}$(Å)
H	1.00	0.76
He	1.34	0.59
Li	1.26	2.52
Be	1.66	1.92
B	1.56	1.70
C	1.82	1.45
N	2.07	1.28
O	2.00	1.32
F	2.26	1.17
Ne	2.52	1.05
Na	1.84	3.88
Mg	2.25	3.18

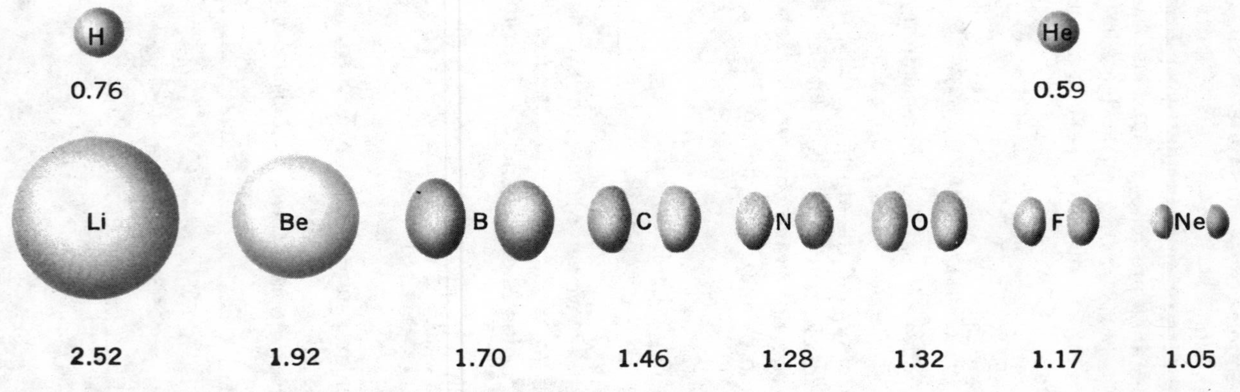

Figure 2-5 Average radii of atoms: one-electron, effective-charge model.

extent to which the most easily removed electron feels the nuclear charge increases across a row of the periodic table. This increasing attraction causes the orbital to shrink—the average radius decreases (see Fig. 2-5). Much of chemistry is explainable in terms of the two trends displayed in Table 2-1: the rise in ionization energy and decrease in orbital size across a row of the periodic table.

(e) ELECTRON SPIN AND THE PAULI PRINCIPLE

Careful measurements show that the electron has magnetic as well as electrostatic properties! If a stream of electrons is passed through a strongly varying magnetic field, it is split into two streams. The amount of the deflection indicates that each electron acts like a small magnet. One of the electron streams bends in a fashion appropriate to being attracted by the imposed field and the other appropriate to an equal and opposite repulsion. The evidence is that each tiny magnet is obliged (by nature) to make a yes-or-no choice—it must be aligned either completely *with* the imposed field or *against* it. *The magnetic interaction is quantized and there are only two possible states.*

From where does this electron magnetic field come? On a macroscopic scale, a magnetic field is generated when electric charge moves in a circular path. An instinctive interpretation, then, is that there must be some way in which the charge within an electron moves in a circular path. *Perhaps the electron is spinning!*

This explanation is the one now accepted. This simple model permits a calculation of the magnetic field that would result from a given spin angular momentum. However, such a calculation gives a result that is double the observed value. To interpret this, we must recall that, as the electron spins, the distribution of the mass will determine the angular momentum and the distribution of charge will determine the magnetic field. Now the discrepancy can be given an intuitive meaning. If the mass and charge are not distributed identically, the classical calculation will predict the wrong relationship between angular momentum and magnetic field—as it does.

We needn't dwell on this dilemma. It suffices to say that the electron spin (or magnetic field) is said to be able to take on only two values, $+\frac{1}{2}$ and $-\frac{1}{2}$. The $+$ and $-$ signs denote the empirical observation that is depicted in Figure 2-6; the electron must be *for* or *against* the magnetic field, nothing in between. The $\frac{1}{2}$ factor is the momentum/field discrepancy factor between the simple calculation and the observed result. The importance of all this to chemistry is that magnetic measurements show a

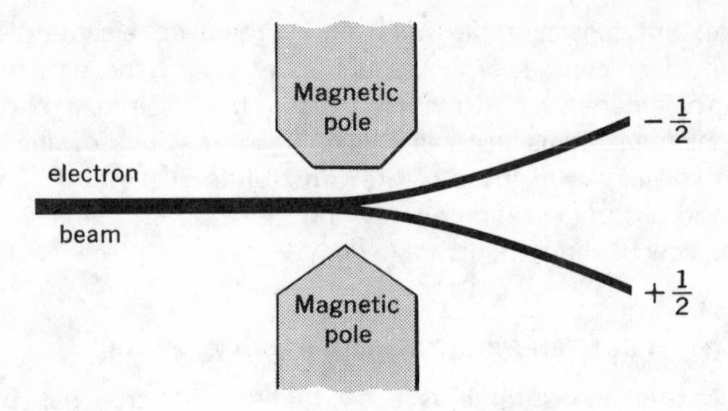

Figure 2-6 Electron spin revealed in an uneven magnetic field produced by pole faces of different shapes.

relation between these spins and the Pauli Principle. Such magnetic experiments coupled with spectroscopic studies of orbital occupancy show that the Pauli Principle can now be stated in a more sophisticated form: Only two electrons can occupy a single orbital and *those two electrons must have opposite spins.*

This idea is simply represented by the slant of the slash across a pigeonhole representation of an orbital. When one electron is placed in the orbital, shown by a slash from left to right, ⦸ a second can be added, shown by a slash from right to left, and the resulting symbol ⊗ designates a filled orbital. Alternate representations use arrows, pointed up and down.

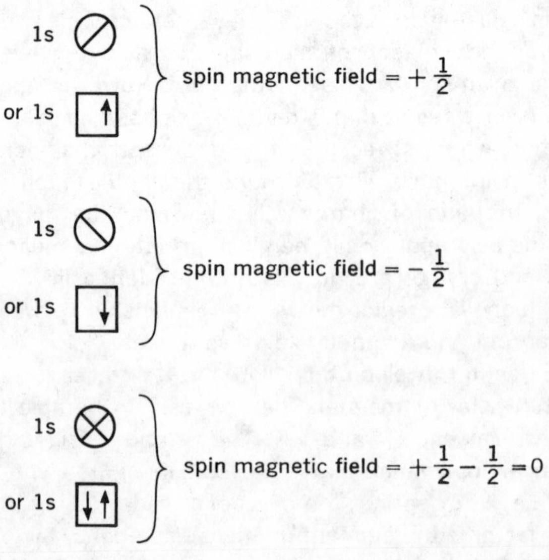

1s ⊘
or 1s ↑ — spin magnetic field $= +\frac{1}{2}$

1s ⦸
or 1s ↓ — spin magnetic field $= -\frac{1}{2}$

1s ⊗
or 1s ↓↑ — spin magnetic field $= +\frac{1}{2} - \frac{1}{2} = 0$

Figure 2-7 Representations of electron spin in filled and partially filled orbitals.

It is immaterial which pictoral representation is used (see Fig. 2-7).

Unfortunately, the origin of this rule of nature is essentially as empirical as the Pauli Principle itself. We observe that the orbital occupancy and the magnetic properties of atoms (and molecules) are all consistent with this rule: two electrons, at most, are allowed per orbital and they must have opposite spin. The explanation is apparently *not* found in the magnetic attraction that occurs between two adjacent magnets oppositely oriented and the repulsion that results from parallel orientations. The explanations now offered delve into relativistic effects, and these sophisticated arguments shed no light on chemistry that isn't already brought there by the Pauli Principle itself.

2-2 Ionization energy and valence electrons

There is a strong connection between the ease of removal of an electron from an atom and the chemistry of that atom. The electrons that are most easily removed are most likely to become involved in chemical bonding. Let's look at the successive ionization energies of the atoms lithium through neon—the first row of the periodic table.

(a) LITHIUM AND BERYLLIUM

It is entirely reasonable that it should be more difficult to remove a second electron from an atom than the first. The first is pulled away from a neutral atom, whereas the second (and successive ones) must be removed from a positively charged ion. Even with this expectation, there are some crucially important variations from atom to atom. Compare, for example, the ionization of gaseous lithium and beryllium atoms.

$$Li \rightarrow Li^{+} + e^{-} \qquad E_1 = 124 \quad \text{kcal/mole}$$

$$Li^{+} \rightarrow Li^{+2} + e^{-} \qquad E_2 = 1744 \quad \text{"}$$

$$Li^{+2} \rightarrow Li^{+3} + e^{-} \qquad E_3 = 2824 \quad \text{"}$$

$$Be \rightarrow Be^{+} + e^{-} \qquad E_1 = 215 \quad \text{"}$$

$$Be^{+} \rightarrow Be^{+2} + e^{-} \qquad E_2 = 420 \quad \text{"}$$

$$Be^{+2} \rightarrow Be^{+3} + e^{-} \qquad E_3 = 3548 \quad \text{"}$$

$$Be^{+3} \rightarrow Be^{+4} + e^{-} \qquad E_4 = 5020 \quad \text{"}$$

It is more difficult to remove one electron from beryllium than from lithium—as noted in Figure 2-3. The nuclear charge on the beryllium nucleus is larger than that of lithium, so this should be so. The surprise is that it is *easier* to remove a second electron from beryllium (E_2 = 420 kcal) than from lithium (E_2 = 1744 kcal). The situation then reverses again for the third electron!

There is, however, nothing mysterious about these ionization energies when the orbital occupancies are taken into account. Figure 2-8 shows the ionization processes in terms of energy level diagrams. The huge energy jump for lithium between E_1 and E_2 is simply determined by the need to pull the second electron from the tightly held 1s orbital. In beryllium, the second electron must occupy the 2s orbital, so E_2 for beryllium exceeds E_1 only because of the extra electron repulsion in the neutral beryllium atom with its two 2s electrons. The actual nuclear charge is the same and so is the orbital. The first electron leaves from a neutral atom in which four electrons repel each other. The second electron leaves from a positively charged atom in which only three electrons repel each other.

However, E_3 is enormous for beryllium, since now a 1s electron must be removed. The magnitude of E_3 is enhanced, and E_4 still more, by the increasing positive charge of the ion (de-

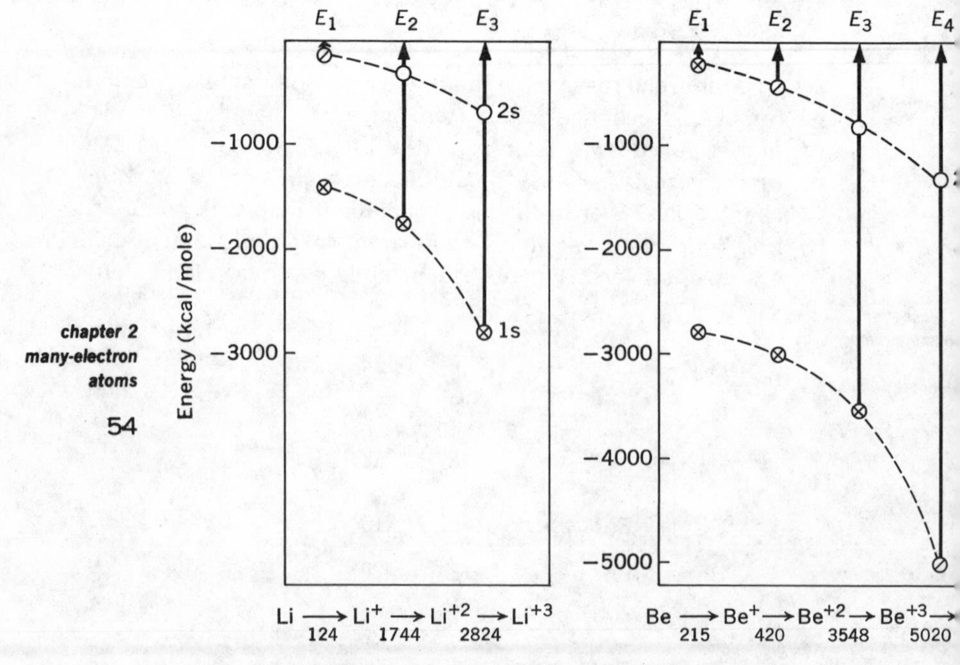

Figure 2-8 Orbital occupancy and ionization energy for Li and Be.

creased electron repulsion with constant nuclear charge), but the reason for the large jump between E_2 and E_3 is due to the need to reach down to a 1s electron for E_3.

It is informative to apply the one-electron, effective-charge model to the second and third most easily removed electrons of Li and Be. From the values of E_2, the one-electron values of Z^* and $\bar{r}$ are contrasted in Table 2-2 with the corresponding

Table 2-2 Effective Nuclear Charge and Orbital Radii Contrasted for First and Second Ionizations of Li and Be

	Electron Removed	Z^*		$\bar{r}$**(Å)**
Li	2s	1.26		2.52
			1.10	
Li^+	1s	2.36		0.34
			0.64	
Li^{+2}	1s	3.00		0.26
Be	2s	1.66		1.92
			0.65	
Be^+	2s	2.31		1.37
			1.05	
Be^{+2}	1s	3.36		0.24
			0.64	
Be^{+3}	1s	4.00		0.20

values for the first electron of each atom. There are two interesting points made in this table. First, note again the very much smaller average radius associated with the 1s electrons, both in Li^+ and Be^{+2}. This accounts for their effective shielding of the nuclear charge insofar as the 2s electrons are concerned. The other contrast is the value of Z^* for the first 2s and the second 2s electron ionizations of beryllium. Here we see that the second 2s electron "feels" an effective nuclear charge $2.31 - 1.66 = 0.65$ higher than the first. Then the third electron, a 1s electron, feels an effective nuclear charge $3.36 - 2.31 = 1.05$ higher than the second. In turn, the fourth electron, again a 1s electron, feels an effective nuclear charge $4.00 - 3.36 = 0.64$ higher than the third. We see that as successive electrons are removed, the effective nuclear charge rises by about two thirds of a proton charge if the electrons have the same principal quantum number. If the principal quantum number changes, then Z^* rises by a full proton charge. *Inner electrons shield the nucleus more fully than do electrons in the same orbital.*

(b) THE FIRST ROW OF THE PERIODIC TABLE

The factors at work in Li and Be are visible in the successive ionization energies of all the atoms in the first row of the periodic table. Table 2-3 and Figure 2-9 show these atoms and their ionization energies related according to orbital occupancy.

It will suffice to examine the successive ionizations of fluorine to see the significance of the table. Of course, the neutral fluorine atom orbital occupancy is $F(1s^2 2s^2 2p_x^2 2p_y^2 2p_z)$. Hence the first two electrons are removed from doubly occupied 2p orbitals, requiring 402 and 807 kcal/mole, respectively. Then the

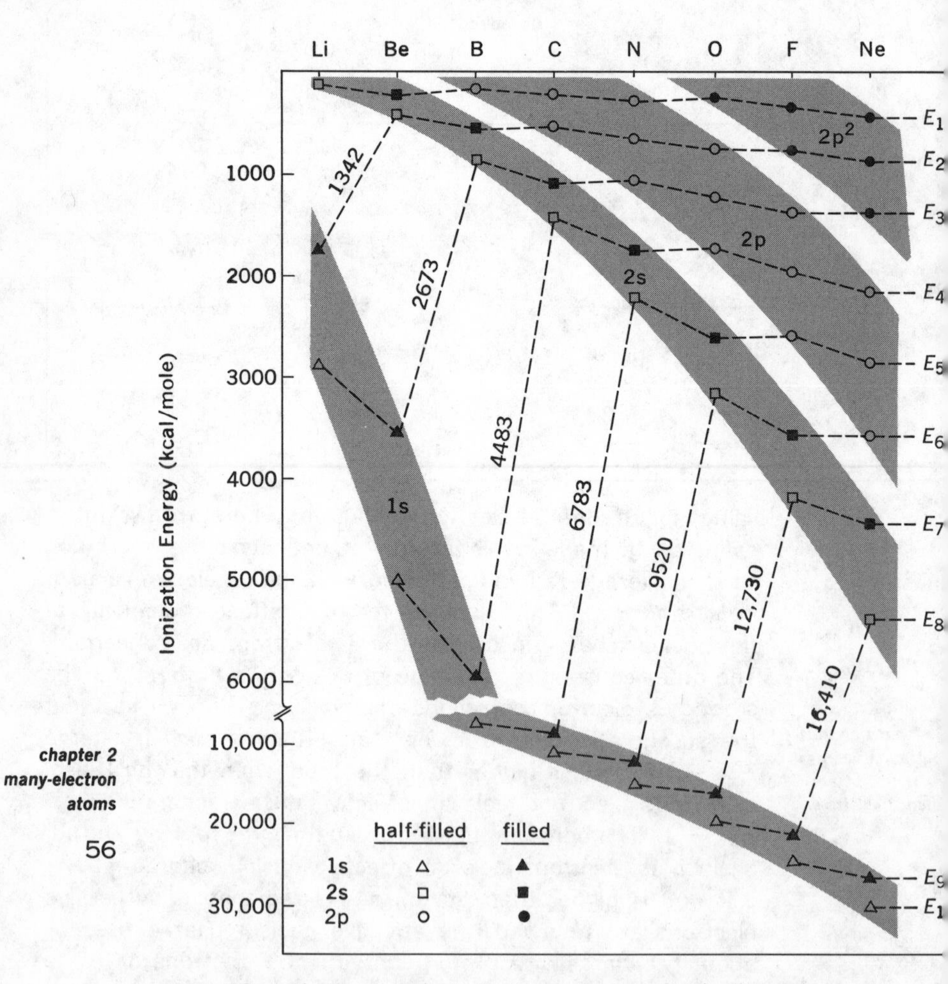

Figure 2-9 Ionization energies and valence electrons for the first-row atoms. The four shaded bands refer to removal of an electron from the 1s orbital (△), the valence 2s orbital (□), and a valence 2p orbital (○). Open symbols refer to singly occupied orbitals and solid symbols to doubly occupied orbitals. (Notice the large energy gap between the valence orbitals and the 1s orbitals.)

Table 2-3 Ionization Energies of the First-Row Elements (kcal/mole)

	Li	Be	B	C	N	O	F	Ne
E_1	124	215	191	260	335	314	402	497
E_2	1744*	402	579	562	683	809	807	945
E_3	2824	3548	875	1102	1098	1271	1445	1463
E_4		5020	5980	1487	1785	1782	2010	2235
E_5			7845	9040	2257	2622	2634	2916
E_6				11,200	12,700	3180	3620	3620
E_7					15,300	17,000	4270	4470
E_8						20,000	21,900	5490
E_9							25,400	27,400
E_{10}								31,100

* A sharp jump in ionization energies occurs on removal of the first 1s electron (shaded portion).

third electron, removed from a half-occupied orbital, requires 1445 kcal/mole, 638 kcal more than the second electron. The next big jump comes between E_5 and E_6, since $E_5 = 2634$ kcal removes the last 2p electron and $E_6 = 3620$ kcal removes one of the 2s electrons at the cost of an extra 1000 kcal.

The biggest point made by Figure 2-9 and Table 2-3 is the huge gap between the energy needed to remove the second 2s electron and the first 1s electron. For fluorine, these energies are 4270 and 21,900 kcal/mole, respectively. Plainly the first seven electrons, requiring from 402 to 4270 kcal for removal, are qualitatively different from the last two electrons, requiring 21,900 and 25,400 kcal. We see a gap like this for every atom. These energy differences account for the chemistry of each atom and the trend across a row of the periodic table. For the second row, the 2s and 2p orbitals are comparable in energy—their electron occupancy determines the chemistry of the elements lithium to neon. Hence these orbitals are given a special name.

—*All the uppermost occupied or partially occupied orbitals of comparable energy are called the* **valence orbitals.**

—*Electrons that occupy the valence orbitals are called* **valence electrons.**

We see that lithium has one valence electron and has valence orbital capacity for an additional seven electrons. Carbon has four valence electrons and valence orbital capacity for four additional electrons. Fluorine is a sort of mirror image of lithium—it has seven valence electrons and valence orbital capacity for only one more. These are the factors we'll use to explain chemical bonding—the availability of valence electrons and valence orbital vacancies determine the number and strength of the chemical bonds that an atom can form.

(c) ELECTRON AFFINITY

It is found experimentally that some gaseous, negative ions are energetically stable, though extremely reactive. The halogen atoms are the most thoroughly studied; for example, the energy required to remove an electron from a negative fluorine ion is 81.0 kcal/mole.

$$F^-(g) \rightarrow F(g) + e^-(g) \qquad \Delta H = +81 \text{ kcal/mole}$$

Plainly this energy is analogous to an ionization energy, and it might logically be denoted E_0. However, such quantities are usually written in terms of the reverse process, the capture of an electron by a neutral atom. The energy released is called the electron affinity. It is usually given as a positive number, an arbitrary but widely used exception to normal thermodynamic practice (scientists are fickle, too!).

$$F(g) + e^- \rightarrow F^-(g) \qquad \Delta H = -81 \text{ kcal/mole}$$
$$\text{electron affinity } E = +81 \text{ kcal/mole}$$

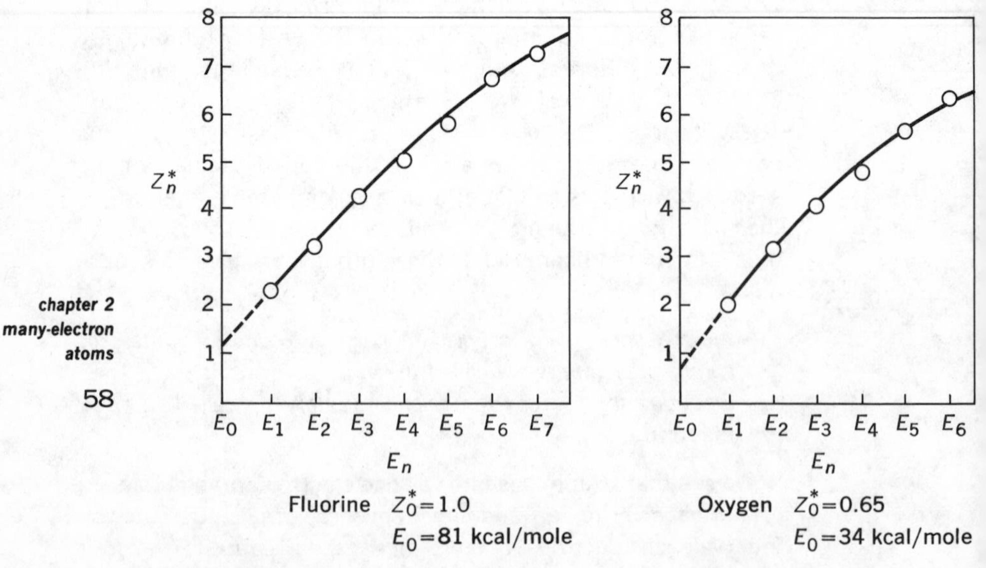

Figure 2-10 Effective nuclear charge for removal of the nth electron (Z_n^*): fluorine and oxygen. When these curves are extrapolated back (dotted lines) a value for the ionization energy of the negative ion (E_0) is obtained.

The reason a fluorine atom can capture an electron is suggested by our one-electron interpretation of the ionization energy. Figure 2-10 shows a plot of the effective nuclear charge felt by the successive valence electrons of fluorine as they are removed one by one. The plot is extrapolated back to "E_0." This extrapolation shows that the most easily removed electron from the negative ion feels an effective nuclear charge of about 1.0. This shows that even for the negative ion, the nuclear attraction exceeds the electron repulsion. Figure 2-10 also shows a Z^* plot for oxygen. This time, the extrapolated value of

H 7.4																	He
Li 14	Be											B 7	C 29	N	O 34	F 79.5	Ne
Na 23	Mg											Al 11	Si 32	P 18	S 48	Cl 83.3	Ar
K 10	Ca	Sc	Ti 9	V 22	Cr 23	Mn	Fe 13	Co 22	Ni 29	Cu 41.5	Zn	Ga	Ge	As	Se	Br 77.6	Kr
Rb >5	Sr	Y	Zr	Nb	Mo 23	Tc	Ru	Rh	Pd	Ag	Cd	In	Sn	Sb	Te	I 70.9	Xe
Cs >4	Ba	La	Hf	Ta	W 12	Re 4	Os	Ir	Pt	Au	Hg	Tl	Pb	Bi	Po	At	Rn
Fr	Ra	Ac	Th	Pa	U												

Figure 2-11 Electron affinities (kcal/mole). (Parenthetical values are calculated, others are experimental.)

Z^* for O^- is only 0.65 and the electron affinity is 34 kcal/mole. We see that electron affinity, like ionization energy, drops off as we move to the left in the periodic table.

Electron affinities are extremely difficult to measure; yet they are as important as ionization energies if we try to understand chemical bonding. Figure 2-11 shows, via the periodic table, the few that are known.

An atom's electron affinity is important to its bonding because a positive E_0 implies that the energy of the neutral atom is lowered as another electron approaches. If this extra electron happens to be attached to a second atom, the two atoms will be held together—bound—by the energy lowering that accompanies this electron sharing. This is a chemical bond.

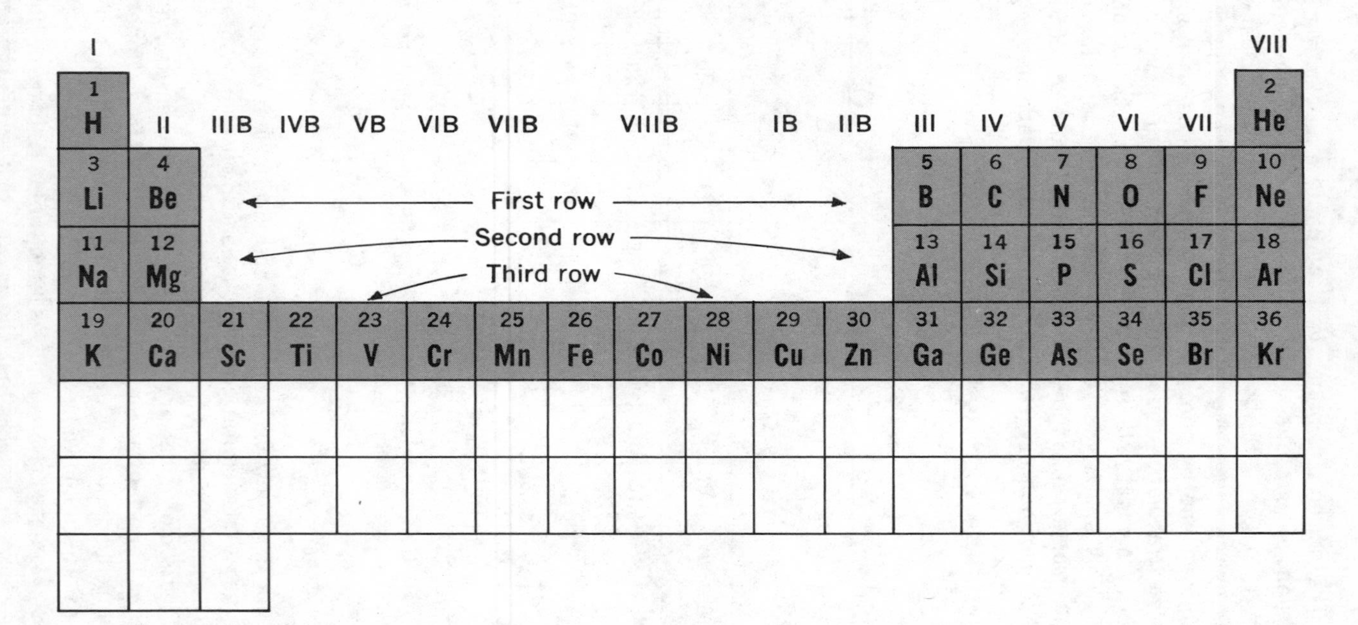

Figure 2-12 The first three rows of the periodic table.

(d) THE SECOND AND THIRD ROWS OF THE PERIODIC TABLE

Figure 2-12 shows the empirical grouping of the elements in the periodic table. Each vertical column relates elements whose chemistry is similar enough to define a familial relationship. We have been talking about the first row, the eight elements lithium to neon, whose chemistry is fixed by the 2s and 2p orbitals. Now let's see how the second- and third-row chemistries are fixed by ionization energies and orbital occupancies. For a starter, we note that the second row, like the first, encompasses eight elements, sodium through argon, whereas the third row includes eighteen elements, potassium through krypton.

Referring back to the hydrogen atom energy level diagram, Figure 1-15, we would conclude that the second row of the periodic table should have eighteen elements. The three sets of orbitals 3s, 3p and 3d have an electron capacity of eighteen. However, the eight-element second row shows that only the 3s and 3p orbitals act as valence orbitals. The reason for this behavior is evident in Figure 2-13. The electron shielding by the inner electrons (the 1s, 2s and 2p electrons) separates the 3s, 3p and 3d orbital energies. The d orbitals are least penetrating, hence they see a much smaller effective nuclear charge than either the 3s or 3p orbital electrons. The effect is so large that the 3d orbitals are raised in energy up to the 4s and 4p orbitals, leaving an energy gap between the 3s–3p orbitals and the 4s–4p–3d cluster.

The first ionization energies of the third-row elements confirm these energy relationships. Figure 2-14 extends Figure 2-4 to include the third row. In this figure, the orbital from which the first ionization occurs is indicated (see legend). The 4s orbitals fill first (elements potassium and calcium), as is to be expected from Figure 2-13. Next, the d orbitals fill and ionization energies remain reasonably constant all the way from scandium to zinc. The crosses indicate that the 3d orbitals remain higher in energy than the 4s orbitals from scandium to manganese, at which element they are just half filled. Thereafter, succeeding electrons enter 3d orbitals until they are completely filled (elements iron to zinc), but the 3d orbitals drop below the 4s orbitals.

Only after the 4s and 3d orbitals are completely filled at zinc, do valence electrons begin to occupy the 4p orbitals. Thereafter, from gallium to krypton, the ionization energies rise in the pattern evident in the first and second rows, and characteristic of the p orbitals. The ionization energy rises rapidly for the three elements gallium, germanium and arsenic. There is a jog at element selenium because of the first double occupancy of a

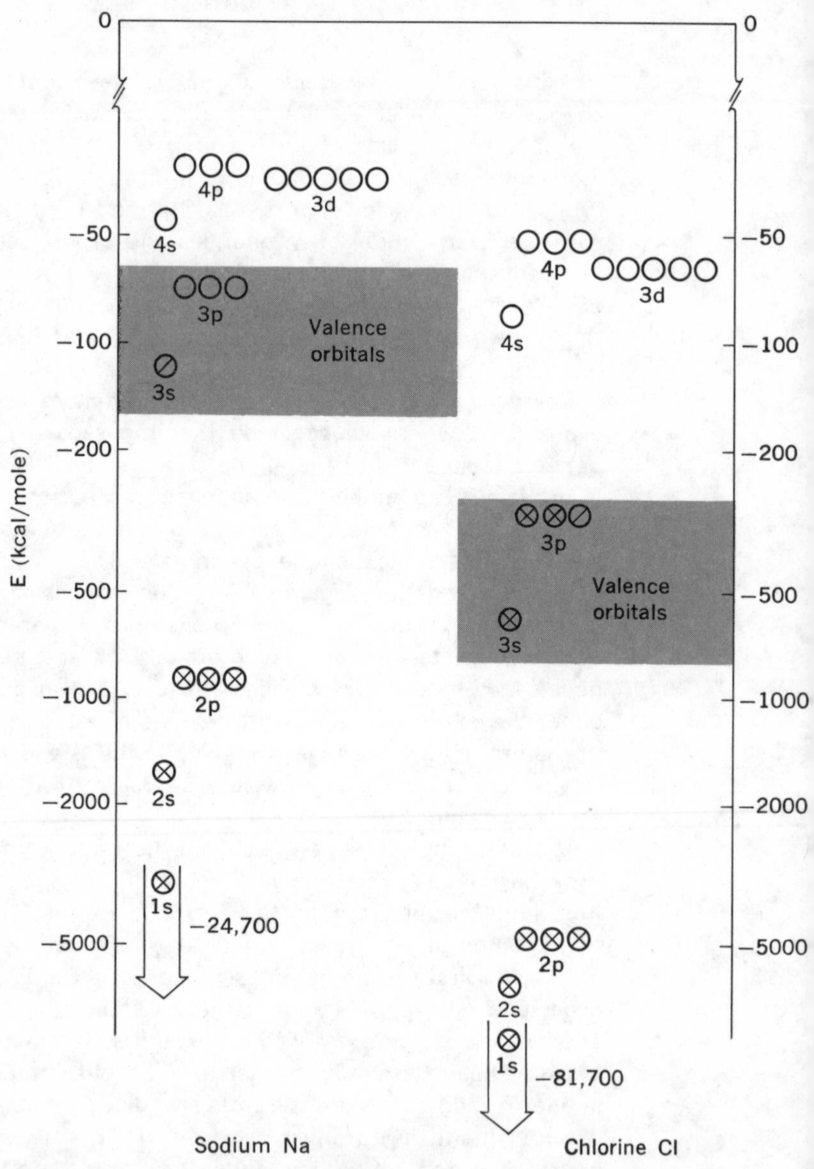

Figure 2-13 Energy levels and orbital occupancy for sodium and chlorine atoms. (Notice vertical scale compression as energy becomes more negative.)

4p orbital. Then the ionization energy rises rapidly again for bromine and krypton, completing the row.

Thus we see that the chemistry of the elements, as systematically represented in the periodic table, is determined by the energy clustering of orbitals. These orbitals can be directly related to those of the hydrogen atom provided electron repul-

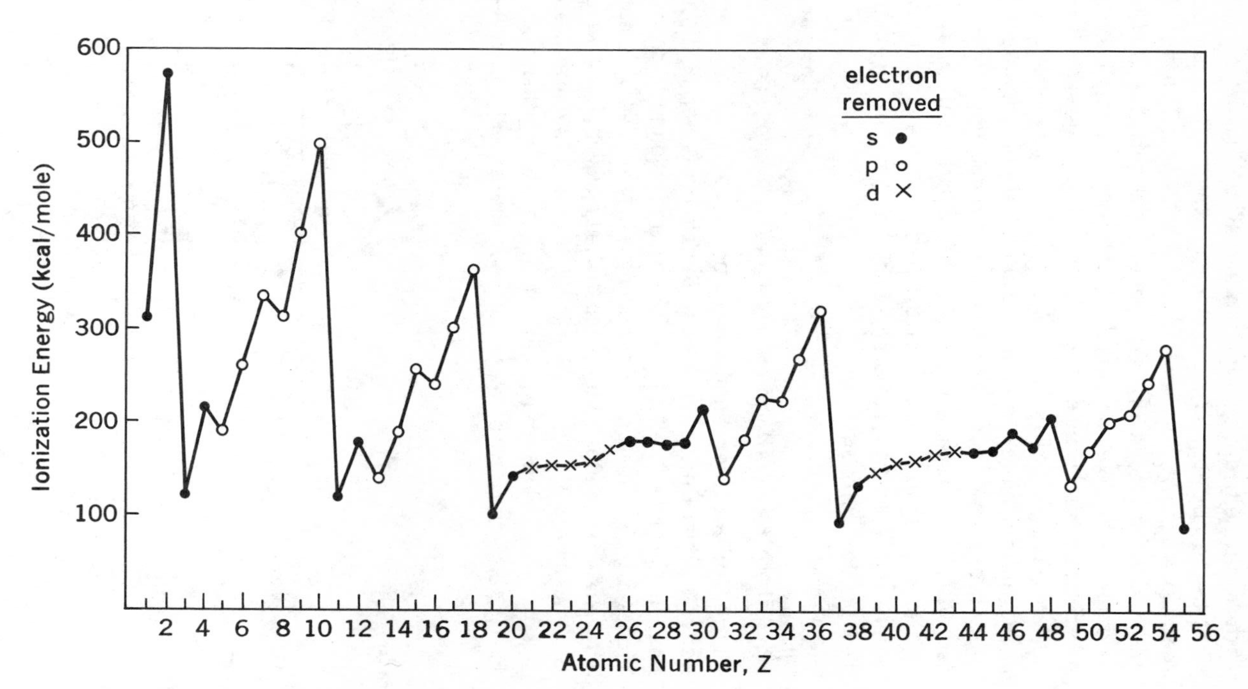

Figure 2-14 First ionization energies.

sions are considered. These electron effects are themselves understandable in terms of the spatial distribution and nodal patterns of the hydrogen atom orbitals. The s orbitals penetrate right to the nucleus, so the s electrons more effectively sense the nuclear charge, despite electron repulsions. The p orbitals and, more so, the d orbitals, have nodal surfaces that pass through the nucleus, so a p or d electron never gets very close to the nucleus. Hence, the electron repulsions shield the nuclear charge more effectively, and p orbital energies are above the s orbitals. In turn, the d orbital energies are above the p orbitals, so much so that the 3d orbitals cluster with the 4s and 4p orbitals, and the 4d orbitals cluster with 5s and 5p, and so on. The explanation for the periodic table is taking form before our eyes. Its cornerstone is the quantum mechanical view of the hydrogen atom.

(e) CHEMICAL FAMILIES

Figure 2-15 focusses attention on two vertical columns in the periodic table. On the left we see the elements beryllium, magnesium, calcium, strontium, barium and radium. These elements are said to form a chemical family, the *alkaline earths.* They are grouped together because they have similar chemistry. On the right, we have spotlighted the elements oxygen, sulfur selenium, tellurium and polonium. These are another chemical family, and they have their own characteristic chemistry—but

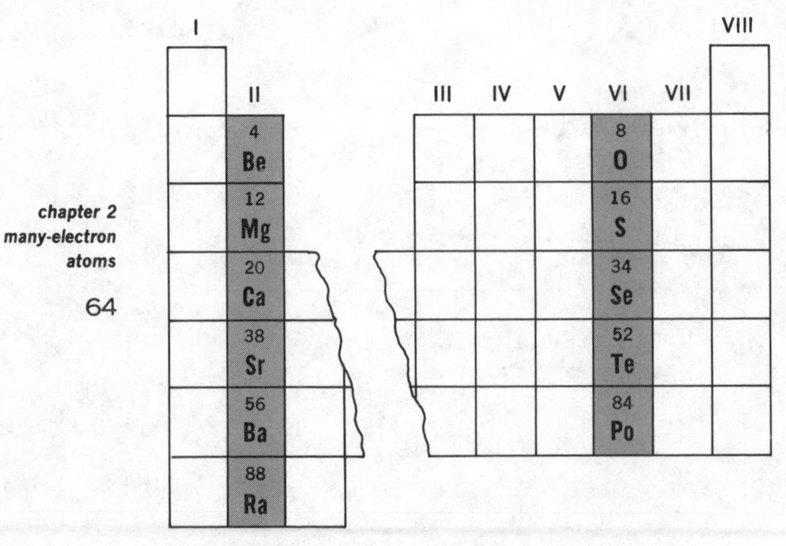

Figure 2-15 Two families in the periodic table.

Table 2-4 Orbital Occupancy and Ionization Energies of the Alkaline Earths

	Electronic Structure							E_1	E_2	E_3
								(kcal/mole)		
Be	$1s^2$	$2s^2$						215	420	3548
Mg	$1s^2$	$2s^22p^6$	$3s^2$					176	346	1847
Ca		$2s^22p^6$	$3s^23p^6$	$4s^2$				141	274	1181
Sr			$3s^23p^6$	$4s^23d^{10}4p^6$	$5s^2$			131	254	1005
Ba				$4s^23d^{10}4p^6$	$5s^24d^{10}5p^6$	$6s^2$		120	231	819
Ra					$5s^24d^{10}5p^6$	$6s^24f^{14}5d^{10}6p^6$	$7s^2$	122	234	—

Table 2-5 Orbital Occupancy and Ionization Energies of the Oxygen Family

	Electronic Structure						E_1	E_2	E_3
							(kcal/mole)		
O	$1s^2$	$2s^22p^4$					314	810	1266
S	$1s^2$	$2s^22p^6$	$3s^23p^4$				239	540	807
Se		$2s^22p^6$	$3s^23p^6$	$4s^23d^{10}4p^4$			225	496	738
Te			$3s^23p^6$	$4s^23d^{10}4p^6$	$5s^24d^{10}5p^4$		208	429	720
Po				$4s^23d^{10}4p^6$	$5s^24d^{10}5p^6$	$6s^24f^{14}5d^{10}6p^4$	194	447	630

nothing like that of the alkaline earths. Each vertical column in the periodic table identifies a group of elements that are alike. The valence orbital occupancies and ionization energies show why.

Table 2-4 lists the orbital occupancies (or, "electronic structures") and the first three ionization energies of the alkaline earths. Each element has only two electrons occupying the outermost orbitals—most of the valence orbitals are vacant. Each element has two electrons that are removable with moderate energy, and then the third ionization requires much more energy. Each element has two valence electrons. Even the ionization energies are similar. From beryllium, with nuclear charge +4, to radium, with nuclear charge +88, the first ionization energy changes only from 215 kcal to 122 kcal. No wonder these six elements are chemical look-alikes!

The oxygen family is quite different. Table 2-5 shows their orbital occupancies and the first three ionization energies. For this family, the valence orbitals are almost full; in each case, only two more electrons can be accommodated to reach the energy gap that defines the next row of the periodic table. The first ionization energies are again similar and each is about 50 percent higher than the alkaline earth in the same horizontal row. Again we have a group of elements with similar orbital occupancies, but this time having valence electrons that are rather difficult to remove. These elements are like each other and nothing like the alkaline earths.

Another view of these two families is provided by our one-electron, effective-charge model of these atoms. Table 2-6 lists the calculated values of Z^*, the effective nuclear charge, and $\bar{r}$, the equivalent average radius. In each family, both quantities increase as we move down the column. The average atomic size increases because the principal quantum number n increases. The effective nuclear charge increases because the true nuclear charge increases and because shielding is never complete. These factors compensate—the higher Z^* is felt at a large radius—so ionization energy remains reasonably constant within each family.

Table 2-6 Effective Nuclear Charge and Average Radius for the Beryllium and the Oxygen Families: One-Electron Approximation

	Z^*	$\bar{r}$(Å)		Z^*	$\bar{r}$(Å)
Be	1.66	1.92	O	2.00	1.32
Mg	2.25	3.18	S	2.62	2.53
Ca	2.68	4.74	Se	3.39	3.59
Sr	3.24	6.13	Te	4.07	4.74
Ba	3.71	7.69	Po	4.72	5.94
Ra	4.36	8.92			

Although the one-electron model exaggerates the size increase, atoms do follow that predicted trend.* We see that the atoms that make up a family have the same chemistry *despite* their size. The factors that cause the familial similarities are *equivalent valence orbital occupancies* and *similar ionization energies.* Each alkaline earth element has two, rather loosely bound valence electrons and lots of vacant valence orbitals. Each oxygen-like element has six, rather tightly bound valence electrons and no completely vacant valence orbitals. These are the factors that account for the chemical similarities within each family and for the great differences that exist between families. Within a family, there are secondary trends that can be connected to the other factor, atomic size.

Thus, with the hydrogen atom as a guide, we have an initial understanding of the periodic table and an open wedge on the principal subject of this book, chemical bonding.

* For example, the predicted increase in size between oxygen and tellurium is 4.74/1.32 = 3.6 whereas the actual ratio of single bond lengths is 2.86 Å/1.49 Å = 1.9.

three
quantum mechanics and chemical bonds

Quantum mechanics successfully explains the observable properties of single atoms and ions. Fortunately, it also applies to the stable aggregates of atoms, called molecules. Thus, quantum mechanics gives us a unifying theory by which we can understand the existence, stability and reactions of the many hundreds of thousands of compounds known to chemists.

We've already complained about the mathematical obstacles in the quantum mechanical treatment of many-electron atoms. Compared with atoms, even the simplest molecules are much more complicated. Exact calculations can be performed only for a handful of them, even with the aid of the largest computers. For these simple molecules, however, the results are in almost perfect agreement with experimental measurements of bond energies, bond lengths, vibrational frequencies, and the other observables that fix molecular properties. On this foundation, we confidently assert the general usefulness of quantum mechanics for explaining and predicting chemistry.

Even though the mathematics remains a serious block to the precise use of quantum mechanics, the theory permits us to refine and use more sensibly our existing schemes of chemical bonding. These schemes, though approximate and often empirical, have proved to be amazingly successful in leading chemists to new discoveries. Now they can be unified, interrelated, and made more quantitative with the benefit of a single guiding theoretical foundation.

3-1 What is a molecule?

Before proceeding, we'd best agree on the meaning on the word "molecule."

A molecule is an aggregate of atoms that possesses distinctive and distinguishing properties.

This definition is quite general—but deliberately so. There are some very important omissions:

—*It does not specify what properties shall be measured.* Any properties that suffice to identify and characterize a particular aggregate of atoms will also make it convenient and useful to recognize that aggregate as an entity—a molecule.

—*It does not limit the number of atoms in the aggregate.* There is no restriction that prevents a molecule from including 30,000 atoms, as long as the 30,000-atom aggregate has distinctive properties (as does the DNA molecule). It is a bit ambiguous when we consider a crystal of salt, a nugget of gold, or a strand of nylon. Any of these may be considered to be molecules if convenience dictates—our discussions of chemical bonding will not be disturbed. A single strand of nylon is usually called a molecule but a single crystal of salt is not. Yet they have the same ambiguity: each involves particular atoms in definite, simple ratios, and in special geometrical arrangements, but with widely varying total numbers of atoms. A strand of nylon can have a carbon—nitrogen skeletal chain 100 or 10,000 atoms long. A salt crystal can come through a salt shaker or weigh 500 grams. No important distinction is made if we insist either that the nylon and salt must be called molecules or that a semantic law be passed against such a usage.

—*It does not say that the molecule must exist under normal conditions*—at room temperature and one atmosphere pressure. We'll talk about garden-variety molecules like hydrogen H_2 and carbon dioxide CO_2, and in the same breath about molecules that have never been detected at room temperature, like gaseous LiF or the inert gas compound KrF_2. The diatomic LiF molecule is detectable only in ovens at temperatures around 1000°K, and the triatomic KrF_2 molecule decomposes spontaneously if it is as warm as melting ice.

—*It requires stability but not lack of reactivity.* Thus, the two molecules nitric oxide, NO, and methyl, CH_3, are both stable and each possesses many distinctive and identifying characteristics. In air, however, nitric oxide is extremely reactive, and an average NO molecule might join with oxygen to form nitrogen dioxide NO_2, another molecule, in a few thousandths of a second. Methyl, on the other hand, would react even in the absence of air and even more rapidly to form the more complex ethane molecule, C_2H_6.

Thus we distinguish "stable" and "unreactive." A molecule is considered to be stable if the molecular group of atoms does not spontaneously rearrange or fall to pieces; a molecule is said to be unreactive under a given set of conditions if it does not undergo rearrangements (chemical changes) involving itself and other molecules under these conditions. At room temperature, a molecular aggregate of eight hydrogen atoms flies apart into four stable molecules of H_2. In contrast, an aggregate of eight sulfur atoms remains together in a stable ring molecule S_8, with fixed molecular structure and distinctive properties. On the other hand, S_8 becomes unstable at an elevated temperature; the S_8 molecule will separate into four molecules of S_2 at a temperature of 1200°K.

With this definition of a molecule—a definition of convenience—we can proceed to discuss chemical bonding. We would like to understand why a particular molecule hangs together and why that atomic cluster displays its own peculiar properties. A complete theory of bonding must tell us which molecules can exist and which cannot. It must also tell us the properties each molecule will display, including the most important property, reactivity with other molecules. Such a theory would encompass chemistry: substances and their changes. We will begin our study of bonding with the simplest cases—diatomic molecules in the gas phase.

3-2 Why do chemical bonds form?

If gaseous fluorine is exposed to an intense burst of light, a large concentration of fluorine atoms can be produced. The fluorine atoms persist, however, for only a small fraction of a second. As collisions occur, fluorine atoms recombine to form stable molecules of F_2. A chemical bond is formed.

$$F(g) + F(g) = F_2(g) \tag{3-1}$$

(a) POSITIONAL OR MOTIONAL RANDOMNESS?

Thermodynamics tells us that chemical changes occur spontaneously only if the randomness, the entropy, of the universe increases. Entropy can rise either because of increased positional randomness or because of increased motional randomness. In a reaction like (3-1), the positional randomness of the system is always *decreased,* since two particles are being locked in the blissful state of chemical bondedness. Certainly it is more random to let each atom have its own carefree and independent existence. So chemical bonds do *not* form because of positional randomness.

The alternative, increased motional randomness, requires that the reaction be exothermic. Energy is then released and distributed among the myriad, chaotic motions we identify as heat, or thermal energy. A reaction is exothermic if the energy of the products is below the energy of the reactants. Therefore, *chemical bonds must form because the energy of the system decreases,* this energy appearing as heat in the surroundings.

(b) KINETIC OR POTENTIAL?

With that confident conclusion, we can proceed to a more detailed question. What *causes* the energy to be lowered? Is it connected with kinetic energy or potential energy or can we be that specific? It turns out that we can.

Quantum mechanics allows us to calculate the average kinetic energy and the average potential energy of the electrons and nuclei in a molecule. We will designate them as follows:

$$\bar{T} = \text{average kinetic energy}$$

$$\bar{V} = \text{average potential energy}$$

These two averages must, of course, sum to the average of the total energy $\bar{E}$.

$$\bar{E} = \bar{T} + \bar{V} \qquad (3\text{-}2)$$

Is it possible that one of these terms, $\bar{T}$ or $\bar{V}$, is more important than the other in explaining the formation of chemical bonds? If so, we could focus our consideration on that energy term.

To investigate the relative importance of electronic, kinetic, and potential energies, we can use a quantum mechanical theorem known as the *virial theorem*. The virial theorem says that there is a remarkably simple relationship between the average potential energy $\bar{V}$ and the average kinetic energy $\bar{T}$. When all the forces acting are simple electrostatic attractions and repulsions, this relation is

$$\bar{T} = -\tfrac{1}{2}\bar{V} \qquad (3\text{-}3)$$

Since the forces in atoms and molecules are dominated by the electric charges, equation (3-3) is applicable in chemistry.* It says that the average kinetic energy within the molecule is opposite in sign and only half as large as the average potential energy.

Let us apply this relation to the formation of a chemical bond between our two fluorine atoms. The initial situation, which we will call state 1, involves two widely separated fluorine atoms, each with an average kinetic energy $\bar{T}_F$ and potential energy $\bar{V}_F$. Since there are two of them, the state 1 kinetic and potential energies will be $\bar{T}_1 = 2\bar{T}_F$ and $\bar{V}_1 = 2\bar{V}_F$. The final situation in reaction (3-1), state 2, is the stable fluorine molecule F_2. This

* There are magnetic forces in atoms and molecules due to nuclear and electron spin but their energy effects are extremely small. The same is true, but more so, for the gravitational forces between the particles in the atom.

molecule determines the kinetic and potential energies after reaction, so $\bar{T}_2 = \bar{T}_{F_2}$ and $\bar{V}_2 = \bar{V}_{F_2}$.

For both the initial and final states, we can apply the virial theorem:

$$\bar{T}_1 = -\tfrac{1}{2}\bar{V}_1 \tag{3-4a}$$

and

$$\bar{T}_2 = -\tfrac{1}{2}\bar{V}_2 \tag{3-4b}$$

Now we can investigate the energy changes that occur as the system goes from state 1 to state 2, that is, as the fluorine atoms form a chemical bond in F_2.

$$\Delta\bar{T} = \bar{T}_2 - \bar{T}_1 \tag{3-5a}$$

$$\Delta\bar{V} = \bar{V}_2 - \bar{V}_1 \tag{3-5b}$$

Substituting (3-4*a*) and (3-4*b*) into (3-5*a*), we obtain

$$\Delta\bar{T} = \bar{T}_2 - \bar{T}_1 = (-\tfrac{1}{2}\bar{V}_2) - (-\tfrac{1}{2}\bar{V}_1) = -\tfrac{1}{2}(\bar{V}_2 - \bar{V}_1) = -\tfrac{1}{2}\Delta\bar{V}$$

$$\boxed{\Delta\bar{T} = -\tfrac{1}{2}\Delta\bar{V}} \tag{3-6}$$

This virial theorem result can be inserted into the total-energy expression to relate $\Delta\bar{E}$ to either $\Delta\bar{T}$ or to $\Delta\bar{V}$:

$$\Delta\bar{E} = \Delta\bar{T} + \Delta\bar{V}$$

$$\Delta\bar{E} = (-\tfrac{1}{2}\Delta\bar{V}) + \Delta\bar{V} = +\tfrac{1}{2}\Delta\bar{V}$$

or

$$\Delta\bar{E} = \Delta\bar{T} + (-2\Delta\bar{T}) = -\Delta\bar{T}$$

or summarizing,

$$\boxed{\Delta\bar{E} = +\tfrac{1}{2}\Delta\bar{V} = -\Delta\bar{T}} \tag{3-7}$$

The change in total energy must always carry the same sign as the change in potential energy. Therefore, $\Delta\bar{E}$ can decrease (a bond can form) only if $\Delta\bar{V}$ decreases. The kinetic energy will change at the same time, but it will always change in the opposite direction and by half as much. We can get to the heart of chemical bonding by investigating which changes can lower the potential energy as two atoms come together to form a bond. *Bonds can form only if the potential energy of the electrons and nuclei decreases as the atoms come together.*

3-3 The simplest molecule, H_2^+

There is no simpler molecule than two protons bonded together by one electron. This molecule (or, molecule-ion) H_2^+ exists in a high-voltage glow discharge through hydrogen gas. Since it is an ion, it is extremely reactive, but its spectral properties reveal its bond energy, the energy needed to pull the atoms apart, and its equilibrium bond length. The molecule H_2^+ is stable.

(a) POTENTIAL ENERGY OF H_2^+

Let's apply the virial theorem to this molecule to see why the potential energy drops and the bond forms. All the potential energy terms are electrostatic: each term depends on the charges on two of the particles, q_1 and q_2, and the distance between them, r_{12}.

$$V_{12} = \frac{q_1 q_2}{r_{12}} \tag{3-8}$$

Figure 3-1 shows graphically the reaction between a neutral hydrogen atom made up of a proton A and an electron, and a second proton B, to form H_2^+. All the charges are the same in magnitude, a positive or a negative q_e. Table 3-1 tabulates the averaged contributions to potential energy before and after bond formation. It is clear how $\bar{V}$ goes down as the bond forms:

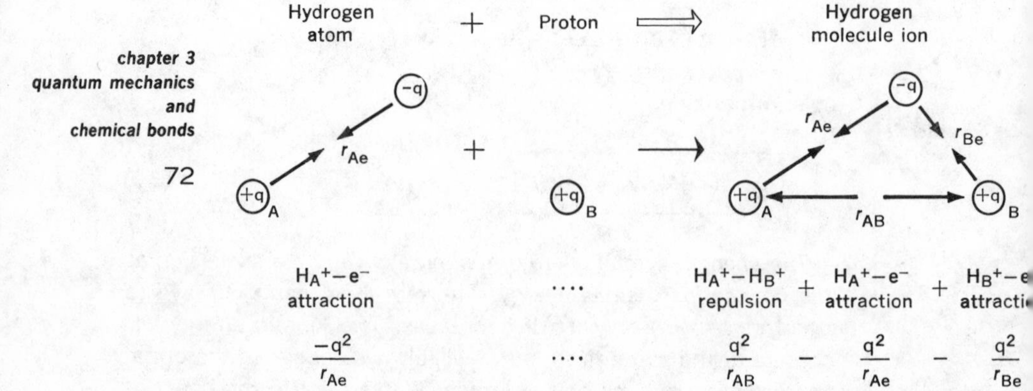

Figure 3-1 The formation of the simplest molecule—H_2^+.

Table 3-1 *Potential Energy Change* in the Formation of* H_2^+

	Before reaction	**After reaction**	
Terms that lower V	$V_{Ae} = -\frac{q^2}{r_{Ae}}$	$V_{Ae} = -\frac{q^2}{r_{Ae}}$	
		$V_{Be} = -\frac{q^2}{r_{Be}}$	
Terms that raise V	none	$V_{AB} = +\frac{q^2}{R_{AB}}$	
	Before reaction	**After reaction**	**Change**
$\bar{V}$	$\bar{V}_{Ae} = -627$ kcal	$\bar{V}_{Ae} = -534.5$	
		$\bar{V}_{Be} = -534.5$	
		$\bar{V}_{AB} = +313$	$\Delta\bar{V} = \bar{V}_2 - \bar{V}_1 = -129$
	$\bar{V}_1 = -627$	$\bar{V}_2 = -756$	
$\bar{T}$	$\bar{T}_1 = +313.5$	$\bar{T}_2 = +378$	$\Delta\bar{T} = \bar{T}_2 - \bar{T}_1 = +64.5$
	$\bar{E}_1 = -313.5$	$\bar{E}_2 = -378$	$\Delta\bar{E} = \bar{E}_2 - \bar{E}_1 = -64.5$ kcal/mole

* Note: Calculations refer to the experimental value of R_{AB} in H_2^+, 1.06 Å. The numerical magnitudes are averaged over all electron positions as dictated by the wave function, which gives the electron probability distribution.

there are two new contributions to $\bar{V}$, one which raises the energy (the nuclear–nuclear repulsion) and one which lowers the energy (the new electron–nucleus attraction). The only possible cause for the decrease in energy is the second term; *the electron is now near two nuclei at the same time.*

(b) THE FORCES IN H_2^+

It is also interesting to consider the forces that exist in the hydrogen molecule ion. Of course the electron position cannot be specified with certainty in a molecule any more than it can in an atom. All that quantum mechanics gives us is a probability picture. At any instant the electron might be in a position such as that shown in Figure 3-2(*b*); at another instant it will be in a new position, perhaps the one shown in Figure 3-2(*c*). Each of these configurations contributes to the average forces felt by the nuclei.

Figure 3-2(*a*) shows, first, the nuclear–nuclear force f_{AB}. It is directed along the AB axis and it tends to force the nuclei apart. This force must be offset somehow by the electron–nuclear forces if there is to be a stable molecule in which the average net forces are zero.

Consider the configuration (*b*). The electron is attracted to both nuclei with forces f_{Ae} and f_{Be} that become larger as the

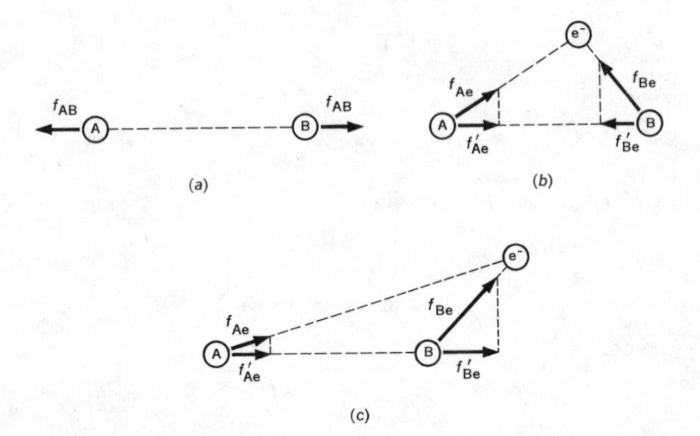

Figure 3-2 The forces in H_2^+. (a) The ever present nuclear–nuclear repulsion works against binding the nuclei together. (b) A possible electron position that contributes to binding the nuclei together. (c) A possible electron position that works against binding.

distance gets smaller. Each of these two forces has a component directed along the AB axis, f'_{Ae} and f'_{Be}. The force f'_{Ae} pulls A toward B and the force f'_{Be} pulls B toward A. This is the type of thing we need to counteract the nuclear–nuclear repulsion; configuration (*b*) contributes to the binding in the molecule.

Things aren't as good in configuration (c). When the electron is out on the periphery of the molecule, it exerts attractive forces that pull both nuclei to the right. Of course it pulls nucleus B more strongly than A because it is much closer to B. The forces along the AB axis, f'_{Be} and f'_{Ae}, give a net force that tends to pull B away from A. This hurts as far as neutralizing the nuclear–nuclear repulsion is concerned. This configuration works *against* binding.

Obviously these ideas can be made quantitative. A straightforward calculation indicates, for every possible electron position, whether that position causes the electron to pull the nuclei together or away from each other. Figure 3-3 shows the result for H_2^+. Anywhere in the shaded region, the forces bind the nuclei together—this is called the *binding* region. Outside the shaded region, the opposite is true: the net force pulls the nuclei apart. Since this works against binding, this region is called the *antibinding* region.

Examining Figure 3-3, we see that consideration of forces in a molecule leads to the same conclusion arrived at by the energy

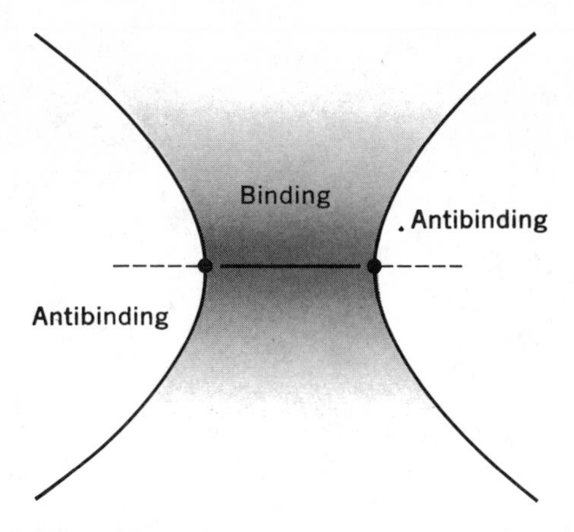

Figure 3-3 Binding and antibinding regic.ns in any homonuclear diatomic molecule.

arguments. If the nuclei are to be held together despite the repulsive forces between them, electrons must preferentially occupy the binding region, the volume between the nuclei. The closer the electrons come to the line between the nuclei, the better. This is, of course, exactly the region in which the electron is simultaneously near both nuclei. *Bonds form because electrons are simultaneously near two or more nuclei.*

(c) THE CORRELATION DIAGRAM FOR H_2^+

It is extremely informative to examine the evolution of each of the energy terms shown in Figure 3-1 as the two protons approach each other. In fact, we can go further and press the two nuclei right up against each other. This process requires a lot of muscle because of the proton–proton repulsion, but it is very educational, hence worth the effort. When the two protons are very close together, the electron sees a helium nuclear charge, so the energy levels must approach those of a helium ion, He^+. That is convenient. Ignoring the rising proton–proton energy, the energy levels must smoothly connect the levels of two separated hydrogen atoms to the levels of a He^+ atom. Both extremes are one-electron atoms, so their energy levels are exactly known, as shown in Figure 3-4.

On the right we have the energy levels of the separated atoms. We'll consider only the lowest energy levels, $1s_A$ and $1s_B$. These two orbitals correspond to the two possible electron occupancies when the atoms are separated by an infinite distance. The electron can either be near proton A, in orbital $1s_A$ with energy -313.6 kcal/mole, or it can be near proton B, in orbital $1s_B$, again with energy -313.6 kcal/mole.

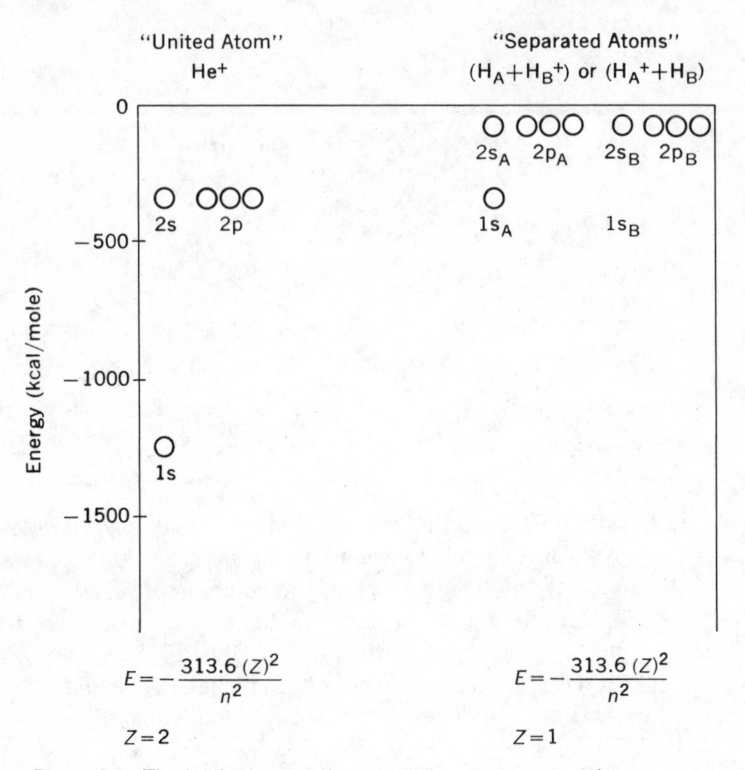

Figure 3-4 The beginnings of the correlation diagram for H_2^+.

As these two protons approach each other, the single electron has no basis for preference between occupying orbital $1s_A$ or $1s_B$. If we looked at many such pairs of protons, in some pairs we'd find the electron in $1s_A$ but in an equal number of others, we find the electron in $1s_B$. In any of the pairs, however, as the protons near each other, we would find the electron moving in a fashion that takes equal account of the presence of both nuclei. Now, instead of occupying one of the atomic orbitals, $1s_A$ or $1s_B$, the electron occupies a *molecular orbital* whose parentage lies equally in $1s_A$ and $1s_B$. Then, as the protons are pressed very close together, this molecular orbital must evolve again into an atomic orbital belonging to the helium ion.

To decide how this parentage reflects into the resultant molecular orbital, we must "think waves." Waves are characterized by nodal patterns and these nodal patterns furnish a guide to the manner in which the separated atom orbitals, $1s_A$ and $1s_B$, connect, through molecular orbitals, to particular atomic orbitals of the "united atom" He^+. Consider, for example, the nodal properties of the lowest He^+ orbital, 1s. It has

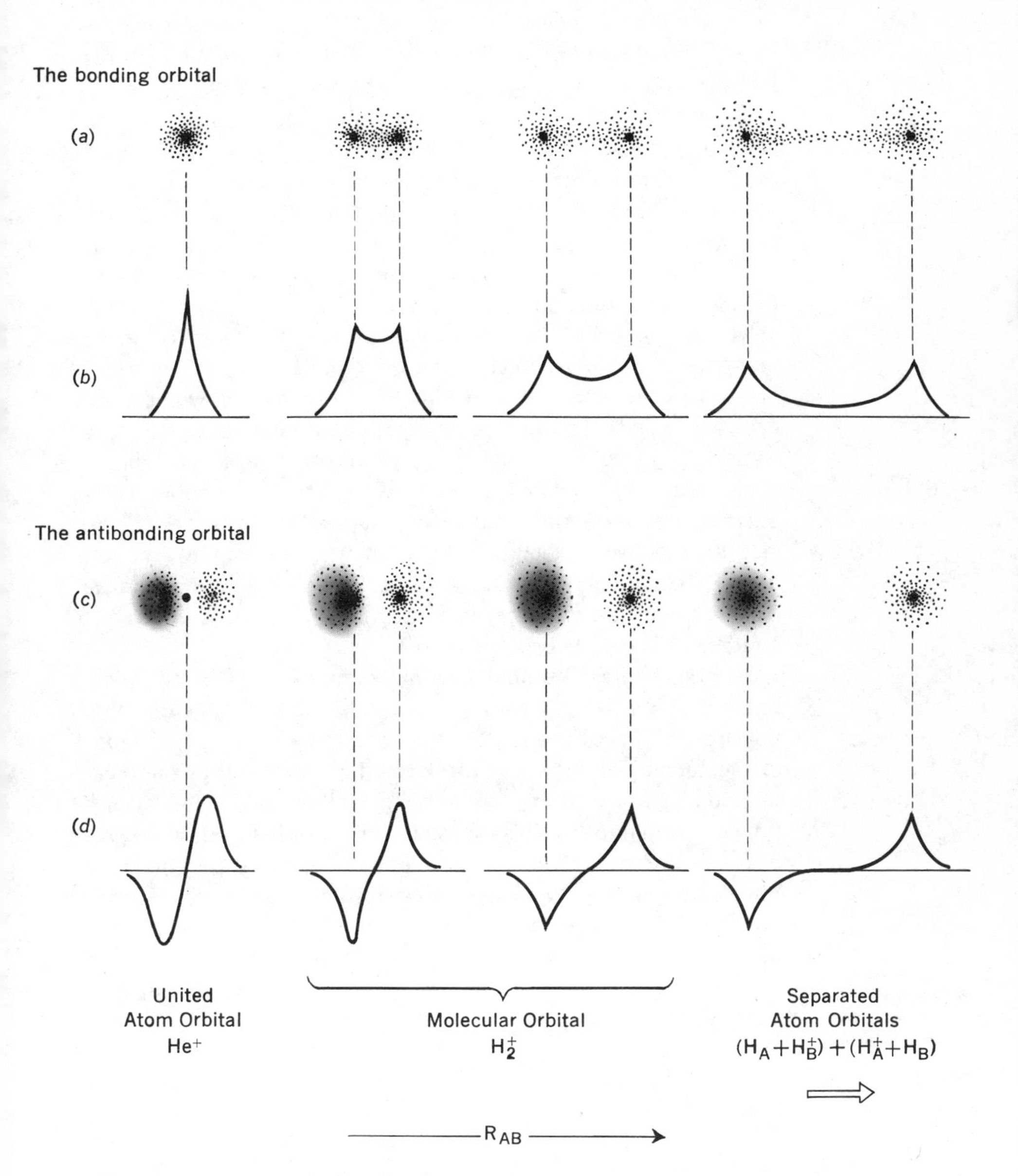

Figure 3-5 The evolution of orbitals as He^+ is separated into two protons plus an electron: (a) and (c) are probability patterns; (b) and (d) show magnitudes and phases. (Phase in the probability patterns is shown by shading.)

no nodal surfaces other than the inevitable one at infinity. As the protons separate, first just a little, then more and more, this nodal pattern is retained. At intermediate separations, we must think of a molecular orbital that has no nodal surfaces except the one at infinity. The upper part of Figure 3-5 pictures this evolution as it must occur. Figure 3-5(*a*) shows the probability representation and Figure 3-5(*b*) shows the magnitude of ψ along the line of the nuclei. The latter representation is the molecular orbital counterpart to the representations of atomic ψ's in Figure 1-13(*b*). We see that there is no nodal surface between the two protons at any separation. As the separation approaches infinity, the wave function begins to resemble two 1s functions with the same phase.

Now we note that the He^+ 1s orbital evolved through a molecular orbital into an in-phase combination of *two* separated atom orbitals. Reversing the direction of our thinking, two proton orbitals, $1s_A$ and $1s_B$, are parentally related to the single He^+ 1s orbital. It is intuitively reasonable to question how two orbitals could suddenly become only one. Even more reasonable, however, is that we should expect two separated atom orbitals to give rise to *two* molecular orbitals and connect to *two* united atom orbitals. This proves to be a general quantum mechanical principle—as atoms approach each other, orbitals do not suddenly disappear or appear.

So our in-phase combination of $1s_A$ and $1s_B$ (that connects to He^+ 1s) is only one of two molecular orbitals with the $1s_A$, $1s_B$ parentage. This phase relationship gives us a clue to the nature of the second. If one molecular orbital is like the in-phase combination of $1s_A$ and $1s_B$, perhaps the other is like the out-of-phase combination. This possibility is represented in Figure 3-5(c) and 3-5(*d*). The opposite phase relationship implies a nodal surface halfway between the protons. If this nodal pattern persists as the protons approach each other, it implies that the united atom will have a nodal plane through the nucleus. Of course, there is always a phase change at a nodal surface. Hence the united atom distribution is just like a $2p_x$ orbital (see Fig. 1-13) of He^+.

These are the two molecular orbital offspring of $1s_A$ and $1s_B$; they have the nodal properties of and they connect to the 1s and $2p_x$ orbitals of the united atom He^+. These nodal properties immediately tell us how the energy will change as the molecular orbital forms. The in-phase combination concentrates the electron probability between the nuclei (see Fig. 3-5(*b*)) in the binding region. This will lower the energy, tending to form a bond between the protons. This is a *bonding molecular* orbital. In contrast, when the orbitals interact out-of-phase, they possess a nodal surface between the nuclei and the probability that the electron will be found there is zero. This moves the electrons out of the binding region to the periphery of the molecule and into the antibinding region. The energy is raised, working against forming a bond between the protons. This is an *antibonding orbital.*

These two molecular orbitals, though having different nodal properties (one has no new nodal surface, the other has one between the nuclei) are both directed along the line AB. Such orbitals are designated σ (sigma) orbitals, the Greek letter for s. The antibonding orbital is identified by an asterisk as σ^*. The nodal properties tell us how to connect energy levels at the extremes shown in Figure 3-4. The in-phase orbital σ has no nodal surfaces (except at infinity). The He^+ ion has only one such orbital, its 1s orbital. Hence σ must connect to the He^+ 1s orbital. The out-of-phase orbital has a nodal surface in the

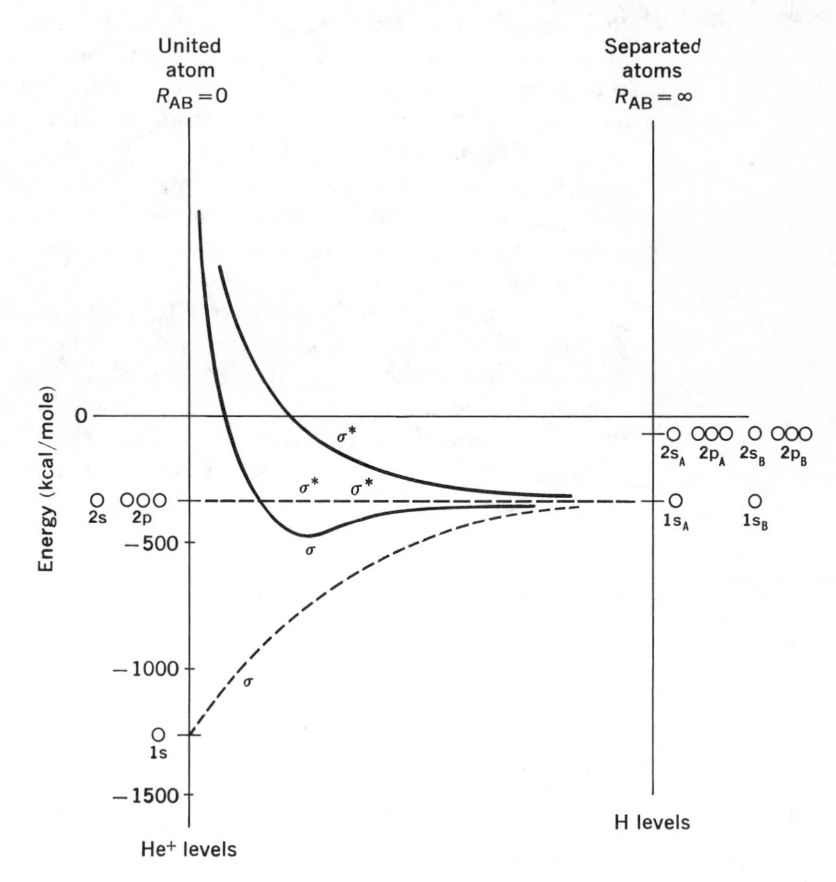

Figure 3-6 Correlation diagram for H_2^+: (---) without nuclear–nuclear repulsion; (——) with nuclear–nuclear repulsion.

yz plane, exactly like the He^+ 2p orbital. It must connect to that tie-point. The result is shown in Figure 3-6, first omitting the nuclear–nuclear repulsion (dashed curves) and then adding it on (solid curves).

Plainly, σ has the energy properties needed for chemical bonding. The energy drops as the nuclei near each other because the electron is near two nuclei simultaneously. At too close range, however, the energy rises because of nuclear–nuclear repulsion. There is a value of r_{AB} at which the energy is a minimum. This is the equilibrium bond length.

In contrast, σ^* has no tendency for the energy to drop as the nuclei approach and, instead, the nuclear–nuclear repulsion causes it to rise steadily. At any value of r_{AB}, the energy would be lowered if the nuclei were to separate again.

(d) THE ENERGY LEVEL DIAGRAM FOR H_2^+

Now we can draw a pigeonhole diagram for H_2^+ like those for atoms. For zero energy, we can take the separated atoms. On such a scale, the σ orbital lowers the energy and σ^* raises it, as shown in Figure 3-7. If the electron occupies σ, a bond should be able to form. Experiment shows that H_2^+ is stable; 64.5 kcal/mole is required to break the bond and pull the nuclei apart to form a proton and a neutral hydrogen atom. The experimental bond length is found to be 1.06 Å. Both these numbers can be calculated exactly by the methods of quantum mechanics.

There are, of course, higher energy pigeonholes associated with H_2^+; states with their parentage in the higher orbitals of the component hydrogen atoms (2s, 2p, 3s, etc.). The most important pigeonholes, however, are the lowest in energy and their occupancy will determine the important properties of the molecule formed: the bond energy, the bond length, and, as we shall see, the chemistry.

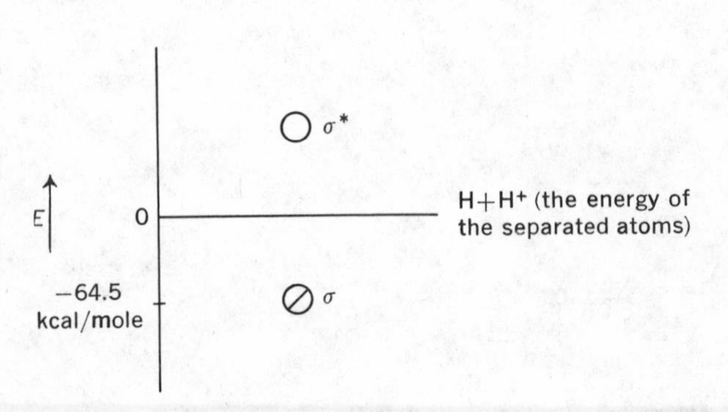

Figure 3-7 The lowest orbitals of H_2^+.

3-4 Molecules and the Pauli Principle

The molecular parallel to the quantum mechanics of atoms has been so close, it is natural to expect it to continue. For example, consider the **Pauli Principle**—*only two electrons can occupy a given atomic orbital.* We can expect this principle to apply to molecules as well. Let's see what it implies.

(a) THE HYDROGEN MOLECULE

Figure 3-7 shows the pigeonhole diagram for H_2^+ and, for the orbital occupancy shown (one electron in the bonding σ orbital), the energy is lowered by 64.5 kcal. If the Pauli Principle applies, this orbital should be able to accommodate a second electron. The molecule would then be converted to a neutral hydrogen molecule H_2, and the orbital diagram suggests that the bond between the two protons should be strengthened. If one electron in the σ orbital lowers the energy by 64.5 kcal, two electrons might lower the energy by double that amount, $2 \cdot 64.5 = 129$ kcal. Of course we recognize this again as our "turned-off" electron repulsion approximation. Actually, the energy won't be lowered as much as 129 kcal because of a new term in the energy box score due to the electron–electron repulsion. Experimentally, it is found that the bond in hydrogen, H_2, has an energy of 108 kcal/mole, not quite double the 64.5 kcal but close to it. The difference is due to the new potential energy term—a term that works against bonding. Nevertheless, to a first approximation, the second electron in a bonding molecular orbital strengthens the bond by a factor of nearly two.

(b) MORE ELECTRONS AND THE PAULI PRINCIPLE

We see that the H_2^+ orbital diagram, with the aid of the Pauli Principle, explains the bonding in H_2. Perhaps a more severe test would be to consider the potential bonding in the helium analogues to H_2^+ and H_2, that is, to the possible molecules $(He)_2^+$ and $(He)_2$.

A naive application of the Pauli Principle (two electrons, at most, per orbital) to the molecular orbital pigeonhole diagram is pictured in Figure 3-8. For $(He)_2^+$, the third valence electron must occupy the σ^* antibonding orbital. Instead of strengthening the bond, this electron should weaken it, effectively negating the beneficial effect of one of the bonding electrons. Sure enough, experiment shows that the bond energy of the molecular ion He_2^+ is 72 kcal/mole, quite close to that of H_2^+. Continuing on to the hypothetical molecule $(He)_2$, the fourth electron also

	H_2^+	H_2	He_2^+	He_2
Antibonding orbitals: σ^*	○	○	⊘	⊗
Bonding orbitals: σ	⊘	⊗	⊗	⊗
Number of bonding electrons: N_b	1	2	2	2
Number of antibonding electrons: N_a	0	0	1	2
Bond order $\frac{1}{2}(N_b - N_a)$	$\frac{1}{2}$	1	$\frac{1}{2}$	0
Experimental bond energy (kcal/mole)	64.5	108	72	0.02

Figure 3-8 Electron occupancy of the lowest H_2^+ molecular orbitals for some simple molecules.

must occupy the σ^* antibonding orbital. Now, with two bonding electrons and two antibonding electrons, there should be no bond at all. Again experiment is consistent with this view. Two helium atoms attract each other very weakly—effectively there is no chemical bond. We see that the Pauli Principle is valid for molecules as well as for atoms. It helps us understand and predict chemical bonds.

(c) BONDING ELECTRONS AND BOND ORDER

The molecule H_2, with two electrons in a bonding orbital, proves to be a prototype of the bonding in many familiar, garden-variety substances. Molecules like H_2O (water), NH_3 (ammonia), and Cl_2 (chlorine) have bonds each associated with a pair of electrons. About 40 years ago, G. N. Lewis postulated that a normal chemical bond was caused by the sharing of two electrons between two atoms. Such a bond became known as a "single bond" and it was assigned a "bond order" of one. Lewis' brilliant hypothesis can now be evaluated in the light of quantum mechanics. A "single bond" formed by the sharing of two electrons corresponds to the full use of a bonding molecular orbital in accordance with the Pauli Principle.

Extending this idea, then, a bonding orbital containing only one electron should be called a "half-order bond." Now a new idea appears; one that amplifies the Lewis concept of the bond —an electron in an antibonding orbital contributes a minus-one-half bond order. Hence, the combination of two electrons in a bonding orbital and one electron in the associated anti-bonding orbital again constitutes a "half-order bond." Figure 3-8 includes appropriate indications of the bond order for each of the molecules considered.

(d) EXPERIMENTAL MEASURE OF BOND ORDER

We have already been using bond energy as a measure of bond order. The H_2^+ molecule, with a one-half bond order, releases 64.5 kcal/mole when the bond is formed. With two bonding electrons and a full single bond, H_2 releases 108 kcal/mole when the bond is formed. In each case, as the reactants approach each other, the energy drops, but more in the case of H_2 than H_2^+. This is shown pictorially in Figure 3-9. To pull each

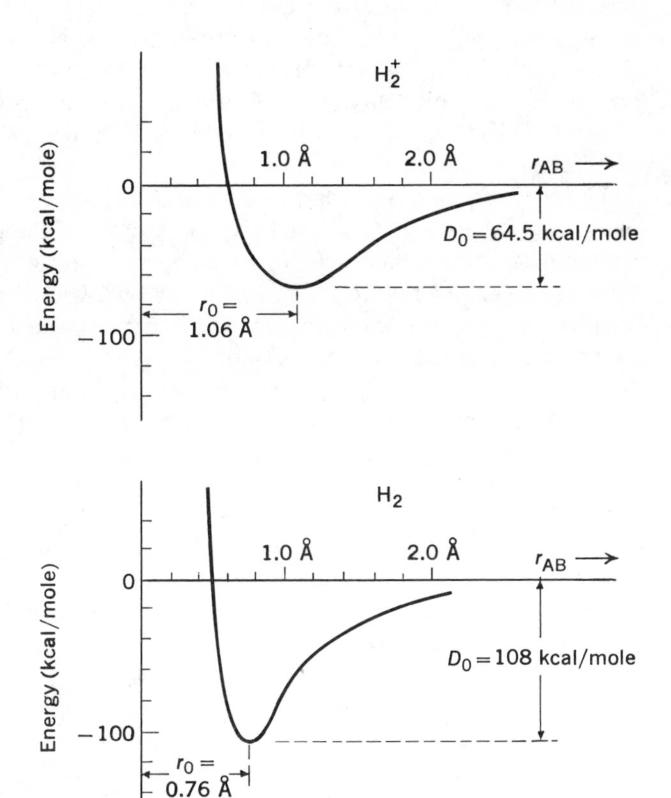

Figure 3-9 Energy versus internuclear distance for H_2^+ and H_2.

of these molecules apart again requires that we pay back the price to climb the energy hill—we must put in 64.5 kcal/mole to break the bond in H_2^+, and 108 kcal/mole to break the stronger bond in H_2. *These bond dissociation energies are a primary measure of bond order and bond strength.*

Figure 3-9 reveals another characteristic of chemical bonds. The stronger bond in H_2 pulls the two protons closer together than in H_2^+. Whereas in the half-order bond of H_2^+, the nuclei find their energy minimum (equilibrium bond length) at 1.06 Å, the single bond of H_2 moves the energy minimum to 0.76 Å. This is not too surprising, since two negative electron charges placed in the binding region between the protons provide more electrostatic "glue" to offset the nuclear–nuclear repulsion.

This relationship is a general and most useful one. For similar atoms, *as bond order increases, bond length decreases.* Bond length measures bond order, too.

Figure 3-9 has one more bit of gold for us to mine. There is one other obvious difference between the energy curves of H_2^+ and H_2; the curvature at the bottom of the energy valley. For H_2^+, the energy hills near the equilibrium distance rise gently. In contrast, the energy minimum for H_2 lies in a rather steep-walled valley. This curvature difference has an important experimental consequence which provides a third measure of bond order.

The significance of the energy curvature at $r = r_0$ can be seen with a simple analogue. It takes energy to change the proton–proton distance. If, for example, we try to stretch the H_2^+ bond by 0.1 Å, we must roll up its energy curve by 1.0 kcal/mole. If we stretch H_2 by the same amount, 0.1 Å, almost four times as much energy is required because of the steeper valley walls—this time 3.9 kcal/mole. Having stretched each of these two bonds, if we now let go of the molecule, the protons will spring back together. Of course, the H_2 will spring back faster because it is rolling down a steeper hill. Each molecule will reach the valley floor with some kinetic energy, and it will climb up the inside wall. Then the compressed bond will expand again, so that the atoms roll back and forth in the valley.

This is analogous to the behavior of two weights hooked together by a spring. There is an equilibrium distance at which the spring is neither stretched nor compressed. If the spring is stretched and then released, the weights vibrate back and forth at a frequency characterized by the masses of the weights and the strength of the spring. Molecules are just the same. The atomic masses determine the "masses of the weights" and the curvature of the bond energy plot determines the "strength of the spring." Figure 3-10 shows the connection chemists make

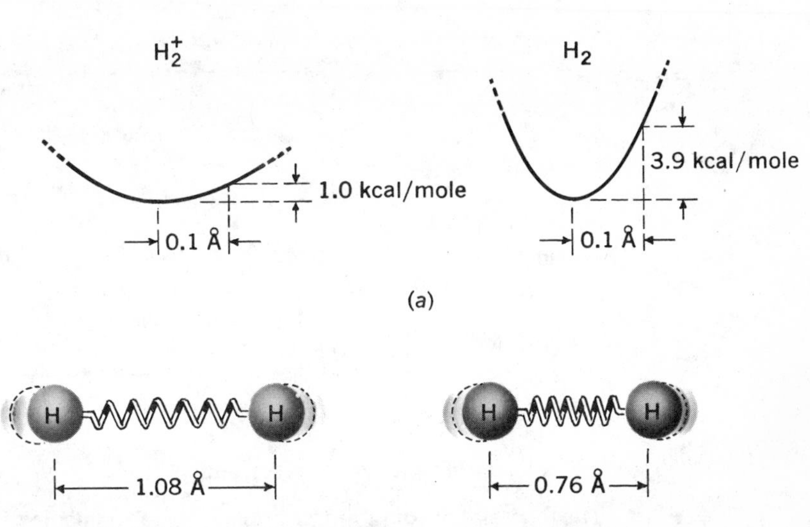

Figure 3-10 Molecular vibrations—a measure of bond order. (a) Greatly expanded view of the bottom of the energy curves of Figure 3-9 showing the energy required for a 0.1 Å change in bond length. (b) Ball-and-spring model shows a weak spring in H_2^+ and a stiff spring in H_2.

between the energy–distance relation and the ball-and-spring model.

It should be no surprise that the vibrational energy levels are quantized, just as are the energy levels in which electrons are excited to high-energy orbitals. The vibrational energies are quite low, however, compared with bond energies. The frequency at which H_2^+ absorbs corresponds to only 6.2 kcal/mole. Quanta with energies this low have frequencies (colors) well below the deepest red color discernable to the human eye. For this reason, the light absorbed by molecules as their vibrations are excited is called "infrared" (beyond-the-red) light. With a suitable "eye" (an infrared-sensitive detector) and appropriate optical equipment (a spectrometer), the infrared "colors" absorbed by a molecule can be measured. These colors tell us about the bonds in molecules. The information we desire is in the spring strength. With the known atomic masses, it is an easy matter to calculate the spring strength, or *force constant,* from the observed vibrational frequency. Force constants have the dimensions of force per unit length and can be expressed in dynes per centimeter. The dimensions are unimportant and we'll choose millidynes per Angstrom (mdyne/Å), because the magnitudes are then typically between 1 and 20.

Table 3-2 gives our three measures of bond order for H_2^+ and H_2. These will be useful as we proceed to the study of larger and more complicated molecules.

Table 3-2 Bond Order Measures in H_2^+ and H_2

	H_2^+	H_2	Ratio $\frac{H_2}{H_2^+}$
D_0, energy needed to break bond	64.5	108 kcal/mole	1.7 ($\cong 2$)
r_0, equilibrium bond length	1.06	0.76 Å	$\frac{1}{1.4}\left(\cong \frac{1}{\sqrt{2}}\right)$
k_0, force constant for vibration	1.4	5.1 mdyne/Å	3.6 ($\cong 2^2$)
bond order	$\frac{1}{2}$	1	2

3-5 Molecules with more electrons

The correlation diagram is applicable to molecules with many electrons, though it gets increasingly complex. The principles, however, guide us as we frame molecular orbitals and use the Pauli Principle to decide on their occupancy. This can be illustrated with some simple, first-row diatomic molecules, beginning with Li_2.

(a) DILITHIUM, Li_2

In the gas phase at moderate temperatures (say, 500–600°K), the vapor of lithium contains a substantial fraction of Li_2 molecules. Spectroscopic and thermodynamic studies show that Li_2 has the bond properties shown in Table 3-3. Let's see how the molecular orbitals and orbital occupancies in this molecule help us understand these properties.

Table 3-3 The Properties of Dilithium, Li_2

bond energy	$D_0 = 25$ kcal/mole
bond length	$r_0 = 2.68$ Å
force constant	$k_0 = 0.25$ mdyne/Å

Figure 3-11 shows the energy level diagrams of two lithium atoms, and how molecular orbitals form as the two atoms approach the equilibrium separation 2.68 Å. At the bottom we find that the 1s orbitals form a bonding molecular orbital 1σ, and an antibonding molecular orbital $1\sigma^*$. There is very little energy difference between 1σ and $1\sigma^*$ and the reason for this is apparent in Figure 3-12. This figure reproduces the size of the 1s orbitals as given in Table 2-2, $\bar{r} = 0.34$ Å. At the equilibrium distance, these two orbitals are so small they barely interact

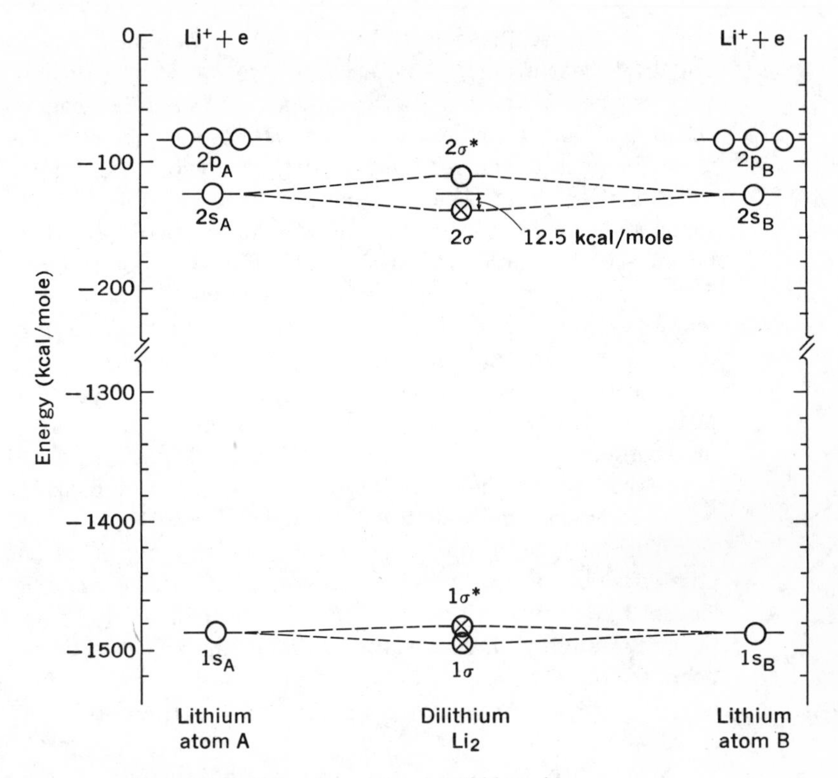

Figure 3-11 Molecular orbitals in dilithium, Li_2.

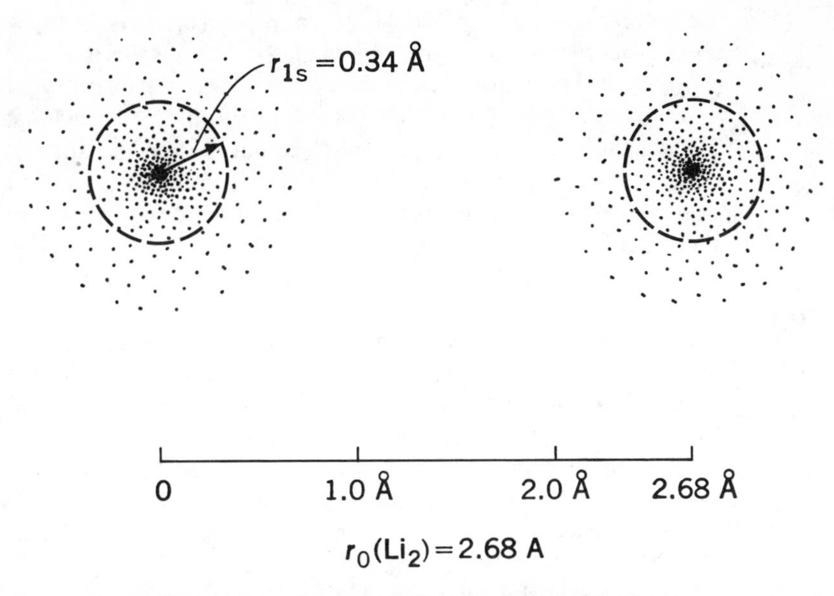

Figure 3-12 The size of the lithium 1s orbitals in Li_2. The atoms are placed at the internuclear distance of Li_2 and the average radius of the 1s orbitals is shown to scale by the dotted circles.

at all. Of course these two M.O.'s (M.O. = molecular orbital) are the lowest in energy so, according to the Pauli Principle, they will be occupied first. As shown, four electrons can be accommodated, but these electrons will not contribute to the bonding. Since 1σ and $1\sigma^*$ are both fully occupied, the antibonding orbital $1\sigma^*$ neutralizes any bonding tendency furnished by 1σ. Quite apart from this, however, the interaction is very slight because of the small orbital size relative to the bond length. This illustrates our first guiding rule in predicting bonding in many-electron atoms. *Orbitals inside the valence orbitals must be occupied* (two electrons each) *but they do not affect the bonding.*

Proceeding upward in Figure 3-11, we next encounter the M.O's formed predominantly from the in-phase and out-of-phase interactions of the 2s orbitals. The 2p orbitals are about 43 kcal higher and they, too, interact, but to a first approximation, we can consider only the 2s orbitals as we discuss 2σ and $2\sigma^*$. The M.O. 2σ accommodates the remaining two electrons possessed by Li_2 and it, then, must account for the bonding. Figure 3-11 shows this doubly occupied orbital to be 25 kcal below the energy of the separated atoms, as indicated by the bond energy (Table 3-3). Counting all the electrons, we find, net, one pair of electrons in a bonding orbital. The bond in Li_2 is a single bond.

The validity of this description has been confirmed by detailed computer calculations. Figure 3-13 shows probability-density contours calculated for the 1σ, $1\sigma^*$ and 2σ orbitals, and these three superimposed, giving the total electron distribution for Li_2. Each contour line shows a line along which electron probability density is constant. These plots show, for individual M.O.'s, the spatial extent of density contours differing successively by factors of four. The diagrams for 1σ and $1\sigma^*$ show that over 90 percent of the probability density is contained within the 0.053 $e^-/Å^3$ density contour, which has a radius of 0.7 Å. For the 2σ orbital, this same density contour occurs *between* the nuclei. The computer's plot shows that the 2σ orbital concentrates electrons between the nuclei and that, going outward, the density distribution drops off at a much lower rate. This orbital distributes electron probability both between and around the lithium nuclei, placed at the equilibrium distance 2.68 Å. Hence the 2σ orbital, occupied by two electrons (the valence electrons) determines the bonding.

With this conclusion, we can contrast the Li_2 single bond with the single bond in H_2. Contrasting the data in Tables 3-3 and 3-2, we see that the single bond in Li_2 is much weaker than the single bond in H_2—it has a much lower bond energy, a longer bond length, and a lower force constant. Why is this? The an-

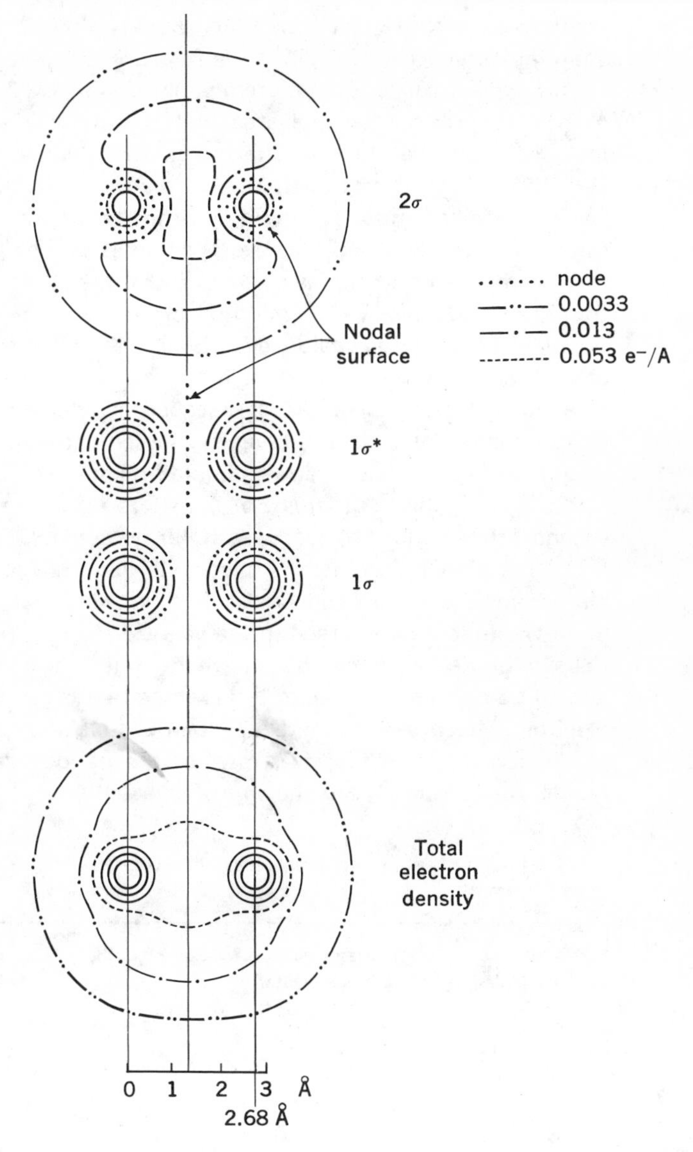

Figure 3-13 Probability density contours for Li_2. Identical contours are drawn in each diagram: successive contours differ by a factor of four. (Units are electrons per cubic angstrom.)

swer is found first in Figure 2-2 where the ionization energies are compared. We see that the H atom attracts its 1s valence electron quite strongly. It takes 313.6 kcal/mole to remove that 1s electron. Lithium, however, will let go of its 2s valence electron for only 124 kcal/mole. Now, bonding arises from the sharing of electrons as they occupy valence orbital space of two atoms at once. Clearly there is less to be gained when two lithium atoms share electrons than when two hydrogen atoms share electrons.

This explanation is given more meaning in Table 2-2. The one-electron approximation indicates that the lithium 2s electron feels an effective nuclear charge of 1.26, higher than the $Z = 1.00$ felt by the hydrogen 1s electron. However, this lithium valence electron must occupy a 2s orbital and, with principal quantum number $n = 2$, its average radius is 2.52 Å. This is to be compared with the average radius of the H atom 1s electron, 0.76 Å. Obviously the lithium electron will be more weakly bound at such a large radius, even though the effective nuclear charge is about the same as in the H atom. This is generally true: *as atomic size increases, chemical bonds tend to weaken.*

As a matter of fact, it is revealing to compare the Li_2 bond length to the one-electron size estimate for lithium. The average radius of the lithium atom is 2.52 Å. Where would we expect a second lithium atom to lodge as it forms a bond? The bond forms because lithium atom A places its valence electron near lithium nucleus B, and conversely. In other words, each lithium atom wants to be immersed in the valence electron probability density of its partner. This suggests that the bond length should be reasonably close to the average radius of the valence electron. This proves to be true, within 10 to 20 percent, for all single bonds in the first two rows of the periodic table. The results for Li_2 and H_2 are shown in Table 3-4.

Table 3-4 Bond Length and Average Valence Orbital Radius (One-Electron Approximation)

Molecule	Average radius of valence orbital, $\bar{r}$	Bond length, r_0	$\frac{r_0}{\bar{r}}$
H_2	0.79 Å	0.75 Å	0.95
Li_2	2.52 Å	2.68 Å	1.06

(b) LITHIUM HYDRIDE

So far, all the molecules considered are of the type called "homonuclear." Each contains two identical nuclei. Life becomes more interesting when we turn to the gaseous molecule, lithium hydride. It is the simplest example of a "heteronuclear" compound—one with two different kinds of atoms. This particular molecule is difficult to study—though stable, it is extremely reactive. Yet its properties are very well known. Because of the importance of this prototype, heteronuclear molecule, much experimental and theoretical effort has been focussed upon it. The properties listed in Table 3-5 are all well established.

Table 3-5 The Properties of Lithium Hydride, LiH

bond energy	$D_0 = 58$ kcal
bond length	$r_0 = 1.61$ Å
force constant	$k_0 = 0.96$ mdyne/Å

Consider first the orbital situation as shown in Figure 3-14. Obviously the lithium 1s orbital is in a class by itself. As is always the case, this 1s orbital plays no direct role in the bonding of lithium. These two electrons are tucked away and, except for their shielding of the nucleus, they are forgotten. The next orbitals, going upward in the diagram, are the hydrogen 1s and the lithium 2s pigeonholes. As usual, bonding and antibonding M.O.'s are formed. Orbital occupancy, given four electrons (three electrons from lithium and one electron from hydrogen),

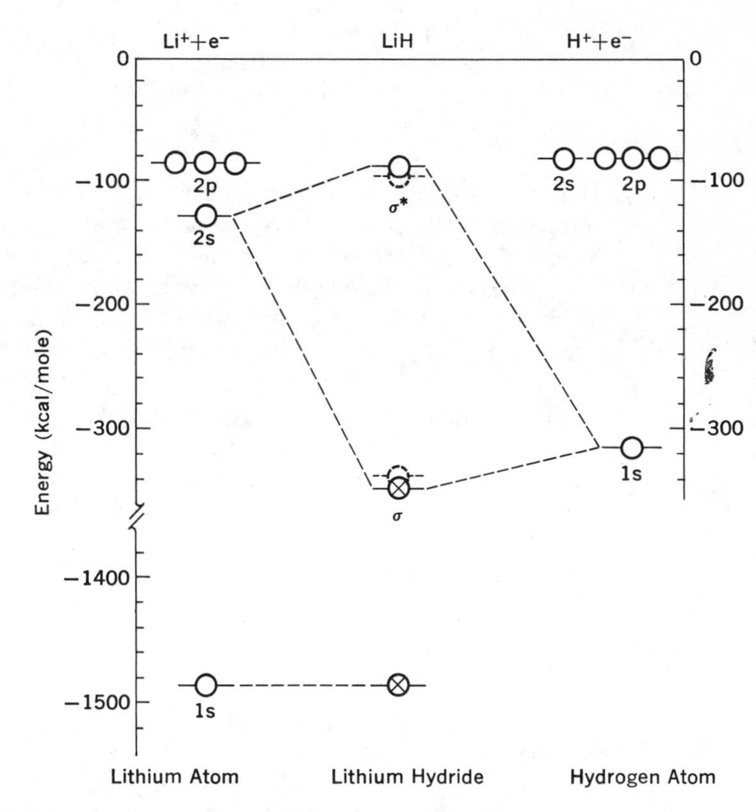

Figure 3-14 Molecular orbitals in lithium hydride, LiH.

places two electrons in the σ bonding orbital. A single bond is formed.

The experimental bond energy, 58 kcal/mole, is much stronger than that of Li_2, but not as strong as that of H_2. Well, that seems reasonable. The bond energy should be between that of Li_2 and that of H_2—the two bonding electrons, if perfectly shared, would look in one direction and think they are in Li_2, and in the other direction, the poor dumb things would think they are in H_2. As a first guess, we might estimate that the bond energy of LiH should be some average of the Li_2 and H_2 bond energies.* The average we want is called the geometric mean, defined as the square root of the product of the two bond energies. Let's call this average $\bar{D}(\mathrm{LiH})$

$$\begin{aligned}\bar{D}(\mathrm{LiH}) &= \sqrt{D_0(\mathrm{Li_2}) \cdot D_0(\mathrm{H_2})} \qquad (3\text{-}9)\\ &= \sqrt{(25)(103)}\\ &= 51 \text{ kcal}\end{aligned}$$

Actually $\bar{D}(\mathrm{LiH})$ underestimates the observed dissociation energy. The observed bond is stronger by 7 kcal.

$$D_0(\mathrm{LiH}) - \bar{D}(\mathrm{LiH}) = 58 - 51 \text{ kcal} = 7 \text{ kcal} \qquad (3\text{-}10)$$

Figure 3-14 shows the discrepancy (3-10) pictorially. The dashed pigeonhole shows where the bonding M.O. would be in the perfect sharing situation; the one that would give a bond energy $\bar{D}(\mathrm{LiH})$. It is actually down closer to where the hydrogen atom energy level wants it. What does this mean in terms of electron probability distribution and the shape of the σ M.O.?

Figure 3-15 shows the answer to our question in terms of plots of the amplitudes of the wave functions as a function of position along the Li–H line of centers. If the separated atom probability distributions (Fig. 3-15(*a*)) are superimposed at the observed internuclear distance (1.61 Å), the "perfect sharing"

* There are two ways to average two numbers X and Y:

$$\text{Arithmetic mean} = \frac{X+Y}{2}$$

$$\text{Geometric mean} = \sqrt{(XY)}$$

These two averages differ very little if X and Y have about the same magnitude. For example, if $X = 100$ and $Y = 104$, the two averages are, respectively, 102 and 101, only 1 percent different. They behave very differently, however, if X is large and Y is small. If $X = 100$ and $Y = 4$, the arithmetic mean is 52 and the geometric mean is only 20, a factor of $2\frac{1}{2}$ smaller. In fact, if Y approaches zero, the arithmetic mean approaches $X/2$, whereas the geometric mean approaches zero. Experience shows us that we want the latter.

molecular orbital is just their sum. This gives a first approximation to the bonding M.O., which always pulls the electrons toward the center of the bond, into the binding region (as shown by the shading in Fig. 3-15(*b*)). However, since the hydrogen atom attracts the bonding electrons more strongly than the

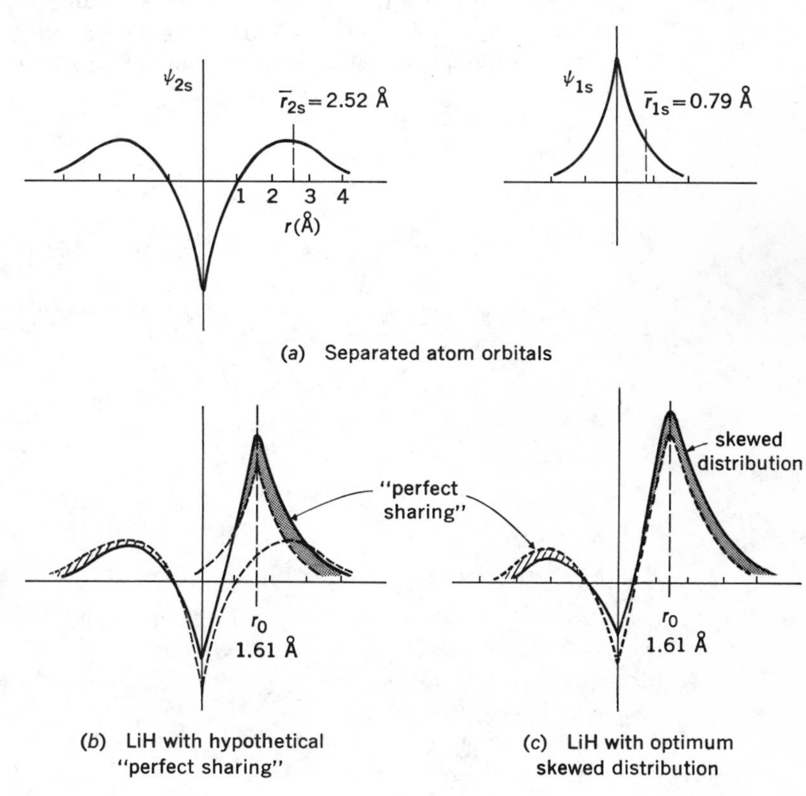

Figure 3-15 Skewing of the molecular wave function in LiH: (a) separated atom wave functions; (b) "perfect sharing" wave function (shading shows concentration of electron probability near the bond center); (c) skewed distribution (shading indicates skewing toward the hydrogen atom).

lithium, there may be an additional lowering in energy to be gained by skewing the probability distribution toward the hydrogen atom. Notice in Figure 3-15(c) that this does not require the valence electron to be removed from lithium. Its average radius was 2.5 Å before hydrogen arrived on the scene. Concentrating the electron probability near the H atom at the equilibrium distance tends to concentrate the bonding electrons even closer to the lithium than in the separated atoms. The extra 7 kcals of bonding (3-10) exist because this skewed distribution is energetically more favorable than the perfect sharing distribution. It tends to concentrate the electrons near the atom that attracts them more strongly.

(c) THE CHARGE DISTRIBUTION IN LITHIUM HYDRIDE

The lithium hydride molecule is sufficiently simple that modern computers can adequately cope with its quantum mechanical complexity. Consequently, we know quite a lot about the probability distributions partially represented in Figure 3-15. Again,

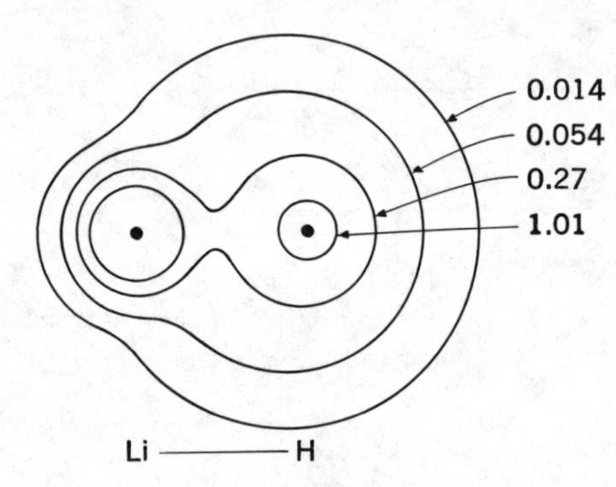

(*a*) Total electron distribution

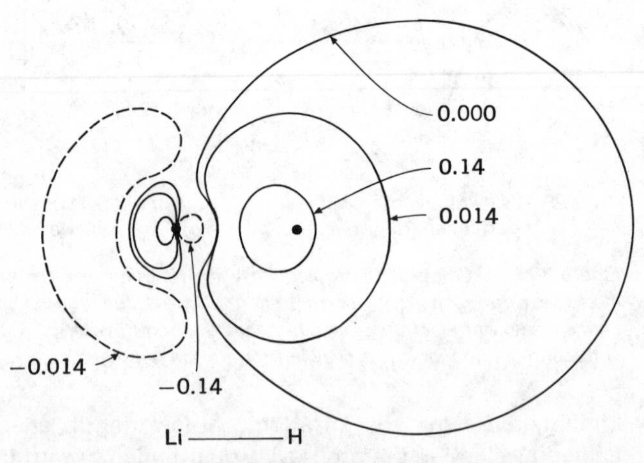

(*b*) Difference electron distribution (total distribution minus "perfect sharing" distribution)

Figure 3-16 The electron distribution in LiH (electrons per cubic angstrom). (a) Total distribution: the two 1s electrons cause the density to rise rapidly in the vicinity of the lithium nucleus. (b) Difference electron distribution: the dotted contours indicate less electron density in the molecule than predicted by the perfect sharing model. Electron density has moved from the regions indicated by these contours into regions marked by positive contours. Thus the distribution is skewed towards the hydrogen nucleus.

to give a more complete picture, a contour diagram is needed in which each contour line shows a line along which probability density is constant. Just as in reading a map, we can then see the regions in which electron probability is high and where it is low. Figure 3-16(*a*) shows such a contour map for LiH, taken from recent research literature. Even more informative, though, is Figure 3-16(*b*) in which something approximating the "perfect sharing" probability has been subtracted. This contour map shows where the LiH molecule has higher electron probability (solid contours) and lower probability (broken contours) than that dictated by superimposing the H atom and Li atom distributions. Figure 3-16(*b*) clearly shows that the electrons have

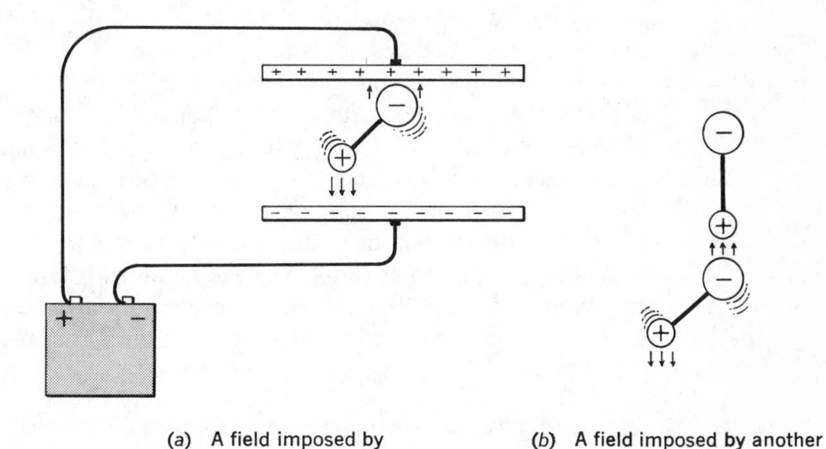

(*a*) A field imposed by an experimenter

(*b*) A field imposed by another molecular dipole

Figure 3-17 A molecular dipole responds to an electric field. (Molecules are represented by a highly simplified ball and stick model.)

moved towards the H atom and away from the more symmetrical "perfect sharing" distribution.

One would think that this migration of electrons toward the H atom would have electrical consequences. The positive charge, all located at the nuclei, hasn't moved, but the negative charge has. This skewing of electron distribution causes the center of negative charge to no longer be at the same spot as the center of positive charge. Such a charge distribution is called an *electric dipole*. Figure 3-17 shows how an electric field exerts a torque on this charge distribution. Whether the field is imposed by condenser plates with an external battery, or by another molecule, a molecular dipole tends to orient into the position most favorable from an energy point of view. This orienting response to an applied field lets us measure these electric dipoles. The corresponding response to a nearby molecule implies that these dipoles affect interactions between molecules. We already know that the skewed electron distributions also affect the bond energy. So molecular dipoles are extremely important in chemistry.

The magnitude of the torque felt by a molecular dipole depends on two factors: how much charge has been displaced and how far it has been moved. The product of these two, (amount of charge) times (distance moved), is called the dipole moment. These quantities are so important that scientists have developed a variety of ways to measure them. That of LiH is $5.9 \cdot 10^{-18}$ esu cm or, in more conventional units, 5.9D (1D = one "Debye" = 10^{-18} esu cm).

Now let's return to our immediate interest; the bond in lithium hydride. We should recall that the charge movement that caused the dipole moment also caused the bond to be stronger than first predicted. Hence the charge movement changes the character of the bond. Chemists say that a bond with a dipole moment has "ionic character." The amount of ionic character is connected to the deviation from the symmetrical sharing of the bonding electrons. We have two measures of it—the magnitude of the dipole moment and the energy excess $D_0 - \bar{D}$, given by (3-10). The estimating of extent of ionic character has been the subject of volumes and volumes of controversy—a measure of the importance of a concept.

(d) A CONTRAST OF THREE MOLECULES WITH IONIC CHARACTER

With the three kinds of atoms, hydrogen, lithium and fluorine, we can form three heteropolar, diatomic molecules, LiH, HF and LiF. The two lithium compounds are extremely reactive and can be obtained in the gas phase only at high temperature and under carefully controlled conditions. The third, hydrogen fluoride, is a common laboratory chemical. It is reactive compared with other compounds on the laboratory shelf, but gaseous HF is easily obtained. All three of these molecules have been well studied, again both experimentally and theoretically. We might compare them.

As a starter, consider the ionization energies for the first electron in each atom: Li, 124 kcal/mole; H, 313.6 kcal/mole; F, 402 kcal/mole. Hydrogen holds its valence electron more tightly than lithium, but fluorine holds its valence electron more tightly still. If hydrogen can pull electrons toward itself in LiH, then fluorine should pull the electrons its way, and away from hydrogen, in HF. Even larger charge movement, or ionic bond character, is to be expected in LiF.

Table 3-6 Evidence of Skewed-Charge Distributions (Ionic Bonds) in LiH, HF and LiF

	D_0 (kcal/mole)	$D_0 - \bar{D}$ (kcal/mole)	μ (Debye)	$\delta = \frac{\mu}{r_0 e}$
LiH	58	7	5.9	0.76
HF	134	73	1.9	0.43
LiF	137	107	6.3	0.84

Table 3-6 shows the relevant quantities: $D_0 - \bar{D}$ and the dipole moments which are symbolized μ (mu). As expected, both $D_0 - \bar{D}$ and μ for lithium fluoride are higher than the corresponding values for both HF and LiH. Figure 3-18 completes the description with contour maps for HF and LiF, showing both the total electron probability distributions and the difference maps. The contours show that the dipole moments in the two

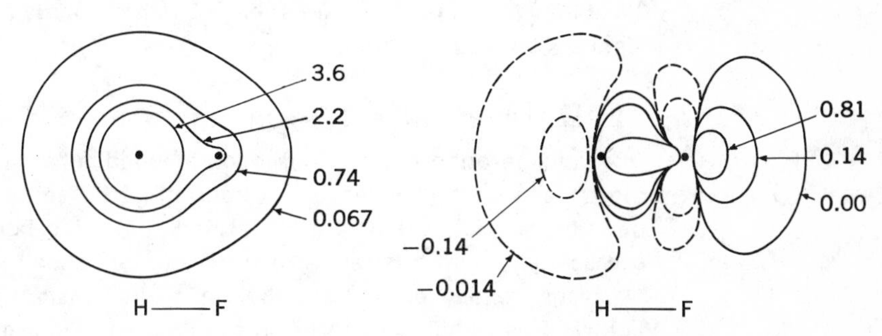

(*a*) HF total electron distribution

(*b*) HF difference: total—perfect sharing distribution

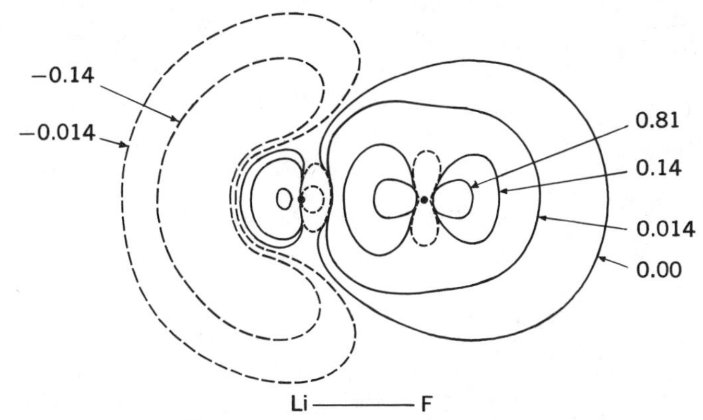

(c) LiF difference: total—perfect sharing distribution

Figure 3-18 Electron distributions in hydrogen fluoride and lithium fluoride (electrons per cubic angstrom).

fluorides are caused by charge movement towards fluorine and away from its lower-ionization-energy partner, H or Li.

As mentioned in the last section, the dipole moment measures the product of the charge displaced and the distance it is moved. An elementary way of expressing this product is to assume that the experimental dipole moment is caused by movement of a fraction of an electronic charge all the way from one atom to the other. This fractional charge is usually symbolized δ (delta), and it is calculated by dividing μ by the electron charge and the equilibrium bond length, r_0. The values of δ for our three molecules are given in the last column of Table 3-6. These can be taken as a measure of ionic character in these bonds. Symmetrical sharing would cause no charge displacement, so $\delta = 0$, and the ionic character would be zero. The most extreme situation would be that obtained with $\delta = 1$, in which both bonding electrons are centered on one of the atoms. We would call that a bond with 100 percent ionic character. Table 3-6 shows that on this scale HF has 43 percent ionic character, whereas LiF has 84 percent.

(e) ELECTRON SHARING IN AN IONIC BOND

Our consideration of the energetics of bond formation led to the conclusion that valence electrons must be simultaneously near two nuclei. Yet it is often stated that ionic bonds form *because* an electron is removed from an atom that releases its electrons readily and placed on an atom which will hold it tightly. This is not true—even in the skewed electron distributions, the bonding electrons must remain near both nuclei or the bond would not form. This can be seen for LiF, one of the most ionic bonds known, by considering its formation in steps.

$$\text{Li(g)} \rightarrow \text{Li}^+\text{(g)} + \text{e}^- \qquad \Delta H = +124 \text{ kcal} \qquad (3\text{-}11a)$$

$$\text{e}^- + \text{F(g)} \rightarrow \text{F}^-\text{(g)} \qquad \Delta H = -83 \text{ kcal} \qquad (3\text{-}11b)$$

step 1
formation of ions

$$\text{Li(g)} + \text{F(g)} \rightarrow \text{Li}^+\text{(g)} + \text{F}^-\text{(g)} \qquad \Delta H_1 = +41 \text{ kcal} \qquad (3\text{-}12)$$

step 2
bond formation
from ions

$$\text{Li}^+\text{(g)} + \text{F}^-\text{(g)} \rightarrow \text{LiF(g)} \qquad \Delta H_2 = ? \qquad (3\text{-}13)$$

overall process
step 1 + step 2

$$\text{Li(g)} + \text{F(g)} \rightarrow \text{LiF(g)} \qquad \Delta H = -137 \text{ kcal} \qquad (3\text{-}14)$$

Since the two-step process (3-12) plus (3-13) gives the overall reaction, the heat effects in (3-12) and (3-13) must sum to the bond energy given in (3-14).

$$\Delta H_1 + \Delta H_2 = \Delta H$$

$$+41 + \Delta H_2 = -137 \text{ kcal}$$

$$\Delta H_2 = -178 \text{ kcal} \qquad (3\text{-}15)$$

Assessing these quantities, we see that the formation of ions (3-12) by no means lowered the energy of the system. Just the opposite! Forty-one kilocalories of energy are *absorbed* to make a mole of gaseous Li^+ and F^- ions. Even though lithium has a relatively low ionization energy, it is well above the electron affinity of the fluorine atom. So in our two-step process, the energy is lowered in the second step as the F^- ion moves its electrons back toward Li^+. In fact, at the equilibrium bond length, the F^- electrons are 1.56 Å away from Li^+, on the average. This is *closer* to the lithium nucleus, on the average, than is the 2s valence electron in an isolated lithium atom. In Table 2-1, the average radius of lithium was calculated to be 2.52 Å. So the effect of skewing electron distribution in ionic bonds definitely does not *remove* valence electrons from the vicinity of either atom. The redistribution tends to center them near the atom that holds them more tightly, while keeping them simultaneously near both nuclei. *All bonds form because electrons are simultaneously near two or more nuclei.*

3-6 On to bigger game

We have considered six molecules: H_2^+, H_2, Li_2, LiH, HF and LiF. These simple examples display most of the ideas needed to understand all chemical bonding.

(i) When any bond forms, it is because the potential energy drops as the atoms come together.

(ii) Potential energy can drop as atoms come together because the electrons can then be near two or more nuclei simultaneously.

(iii) As in atoms, it is useful to describe the movement of electrons in a molecule in terms of orbitals. However, an electron moving simultaneously near two nuclei occupies an orbital that is molecular in character; that is, a *molecular orbital* (or, an M.O.).

(iv) The Pauli Principle applies to M.O. occupancy just as it does to atomic orbitals: at most, two electrons can occupy a given M.O. (and those electrons must have opposed spins).

(v) The nature of an M.O. is determined by its nodal surfaces, which can be deduced from the parent atomic orbitals of the separated atoms. These parent orbitals can be considered to contribute to the M.O. either in-phase with each other, or out-of-phase, as governed by their wave nature. The in-phase combination concentrates electrons in the binding region and hence lowers the energy. This is a bonding M.O. The out-of-phase combination has a nodal surface between the nuclei and concentrates electrons in the antibinding region. This is an antibonding M.O.

(vi) Bond order is defined to be the difference between the number of pairs of electrons in bonding M.O.'s minus the number of pairs in antibonding M.O.'s.

(vii) For a given pair of atoms, as the bond order goes up, both the bond energy and the stretching force constant go up, and the bond length goes down.

(viii) When two different kinds of atoms are bonded, the electron distribution will tend to be skewed toward the atom that holds the electrons more strongly, while still remaining close to both nuclei. Such a skewed electron distribution strengthens the bond and it gives rise to an electric molecular dipole. A molecule with an electric dipole is said to have a *dipole moment* and its bond is said to have ionic character.

So far, though, we've been careful to consider only simple atoms in simple molecules. Turning to atoms with more electrons and to molecules with more atoms, we find the chemical-bonding problem more difficult to handle. Chemists then use approximate methods based upon the principles we've just reviewed—those established for the simpler molecules. The next chapter shows how.

four simple molecular orbitals

The Schroedinger equation can be solved exactly for the electronic energies of the hydrogen molecule ion H_2^+. With high-speed digital computers, chemists have correctly calculated properties for other small molecules—molecules like H_2, Li_2 and LiH. However, even the last example, LiH, contains only two atoms and four electrons. Meanwhile, experimentalists are busily preparing and working with hundreds of thousands of molecules, most of which contain atoms by the dozen and electrons by the gross. These chemists evidently have some straightforward and quite effective ways of predicting and rationalizing molecular structures. Now, since quantum mechanics has proved itself, it permits us to evaluate and understand the earlier, empirical models of bonding. More important, it guides us as we address the as-yet unsolved problems of chemical bonding and as we improve the existing approximate theories.

The molecular orbital treatment is an approximate theory. To be sure, the M.O. concept is exact in a molecule containing only one electron, as in H_2^+, but in a many-electron molecule, it becomes a useful approximation, in precise analogy to the atomic orbital concept applied to many-electron atoms. The electrons are considered to occupy M.O.'s that are about the same as they would be if electron repulsions could be turned off. The Pauli Principle is applied as the orbitals are filled and, of course, we expect to see effects due to electron–electron repulsions when finer details are considered.

The diatomic molecules across the first row of the periodic table provide some more good examples. However, we'll return for a moment to the M.O.'s in H_2 to illustrate the approximate methods needed.

4-1 H_2^+: A source of approximate M.O.'s for H_2

Figure 4-1 is one way to look at the interaction of two 1s orbitals as a bond forms between two hydrogen atoms. The solid line in Figure 4-1(*a*) indicates how the wave function varies along the line connecting the two protons. The upward direction represents one phase and the down-

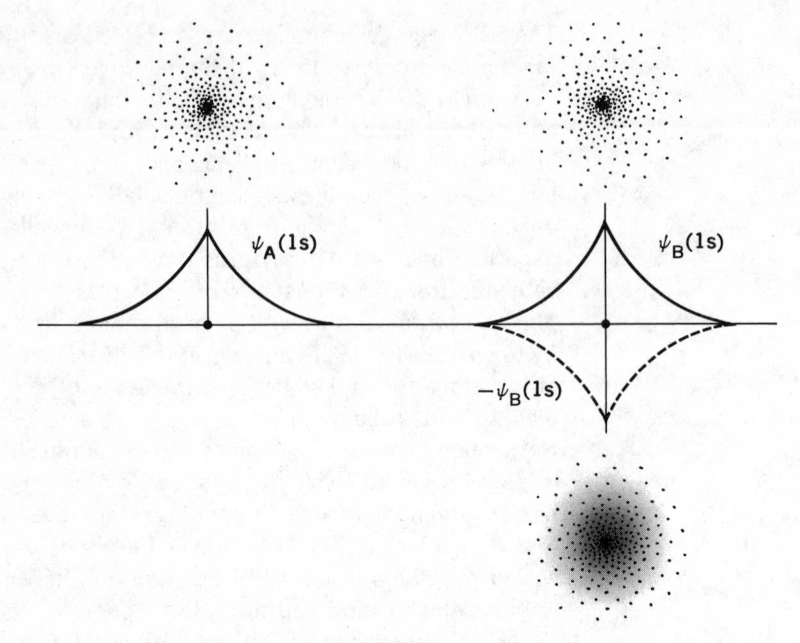

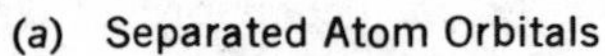

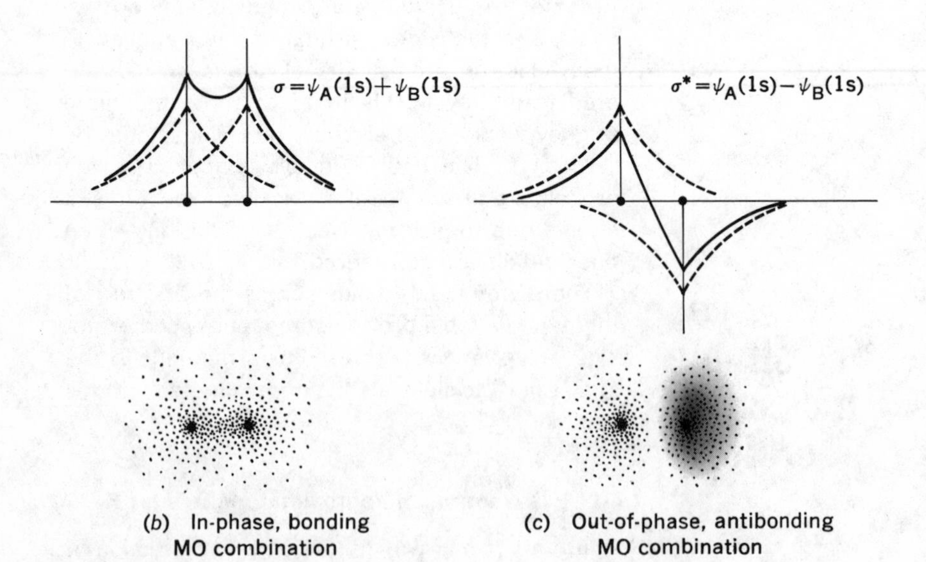

Figure 4-1 Approximate molecular orbitals in H_2. (Phase relationships are shown by shading in the probability plots.)

ward direction the opposite phase (think of the vibrating guitar string). Figures 4-1(*b*) and 4-1(*c*) show how the M.O.'s form as the two possible phase combinations occur. A possible first description of the σ bonding M.O. is $\psi_A(1s) + \psi_B(1s)$; simply the superposition of the two 1s orbitals at the equilibrium bond length, with the in-phase relationship. The σ^* antibonding M.O. can be roughly described as $\psi_A(1s) - \psi_B(1s)$, the out-of-phase superposition. The approximations to σ and σ^* are not sufficiently accurate to furnish quantitative estimates of energy, bond length, and stretching constants, but they certainly do give a correct qualitative picture. The most important qualitative feature they display is the *nodal pattern* of the molecular orbital.

We have already used these nodal patterns to develop the molecular orbitals of H_2^+ (in Section 3-3c). One of the principal values of the correlation diagram is to provide tie points (at either end of the diagram) at which the nodal pattern is known. Thus, the connection to the He^+ 1s orbital indicates that one of the H_2^+ molecular orbitals has no nodal surfaces except the one at infinity. The phase relationships then give us the clue to the nodal pattern of the next M.O. It connects to the $2p_x$ orbital, with a nodal plane perpendicular to the line along which the two H atoms join to form the united He atom. While the two atoms are still spatially separated, this central nodal plane identifies this M.O. as an antibonding orbital.

Having now decided on the nodal patterns of the two lowest-energy M.O.'s of H_2^+, we used them to discuss the two-electron molecule H_2. In conformity with the Pauli Principle, both of the electrons can occupy the lower orbital, the bonding one. Even though electron repulsions can be expected to affect the molecular properties, to a first approximation the bond in H_2 should be about twice as strong as that in H_2^+. It should require about twice the energy to break the H_2 bond (as it does), the atoms should be closer together (as they are), and the "spring constant" (the vibrational force constant) should be higher (as it is).

Evidently this simple scheme works well—at least qualitatively. As the two atoms H_A and H_B come together, their lowest orbitals $1s_A$ and $1s_B$ are considered to offer parentage to two molecular orbitals. What are these two M.O.'s like? That's easy; the scheme says they have the nodal patterns of the in-phase and out-of-phase combinations,

$$\sigma(1s) = 1s_A + 1s_B$$

$$\sigma^*(1s) = 1s_A - 1s_B$$

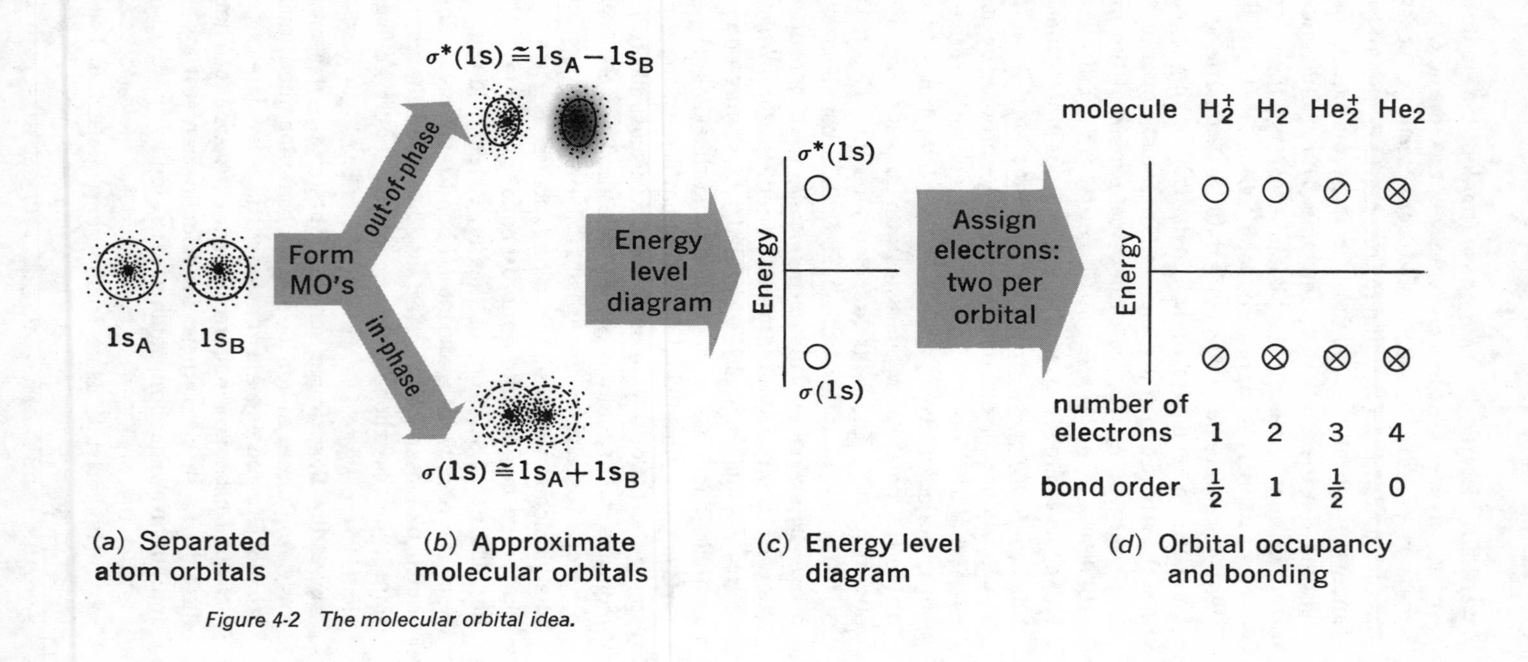

Figure 4-2 The molecular orbital idea.

The nodal patterns, in turn, identify $\sigma(1s)$ as a bonding orbital (it concentrates electron probability in the region between the nuclei) and $\sigma^*(1s)$ as an antibonding orbital (it moves electron probability away from the region between the nuclei). The energy level diagram is qualitatively defined. The number of electrons to be accommodated and the Pauli Principle take it from there. This approach is summarized in Figure 4-2.

4-2 Approximate M.O.'s for 2p orbitals

We might exercise these ideas by imagining the bonding questions two H atoms would ask themselves if they approached each other in energetically excited, 2p states. There are two situations to consider, 2p orbitals directed along the molecular axis ($2p_x$ orbitals) and 2p orbitals directed perpendicular to the molecular axis ($2p_y$ and $2p_z$ orbitals). We'll take the axial case first.

(a) AXIAL $2p_x$ M.O.'s

Figure 4-3(*a*) shows two separated H atoms, each excited to a $2p_x$ state, with the $2p_x$ orientation along the internuclear axis. These axially directed p orbitals give σ type molecular orbitals.

Figures 4-3(*b*) and 4-3(*c*) use the same ideas introduced in Figure 4-1. The "in-phase" combination concentrates electron probability between the nuclei to give a bonding orbital, $\sigma(2p_x) \cong \psi_A(2p_x) + \psi_B(2p_x)$. The out-of-phase combination has a nodal plane between the two nuclei. It is an antibonding orbital, $\sigma^*(2p_x) \cong \psi_A(2p_x) - \psi_B(2p_x)$.

(b) PERPENDICULAR $2p_y$ AND $2p_z$ M.O.'s

Consider now two H atoms, excited to the 2p state, approaching each other but with their 2p orbitals oriented perpendicular to the bond axis. Either the $2p_y$ or the $2p_z$ orientation may be pictured—they are exactly equivalent except for a 90° rotation around the bond axis. Figure 4-4 shows the M.O. formation in this case, as suggested by the in-phase and out-of-phase combinations of the parent orbitals.

There is, however, an obvious difference between these M.O.'s and those pictured thus far in Figures 4-1 and 4-3. In Figure 4-4, no electron probability is found along the bond axis. Instead, the M.O. retains the *xz* nodal plane possessed by the parent $2p_y$ orbitals. Here is another nodal characteristic with which to classify our M.O.'s. The sigma (σ) orbitals, concentrated along the bond axis, have no nodal surface passing through the

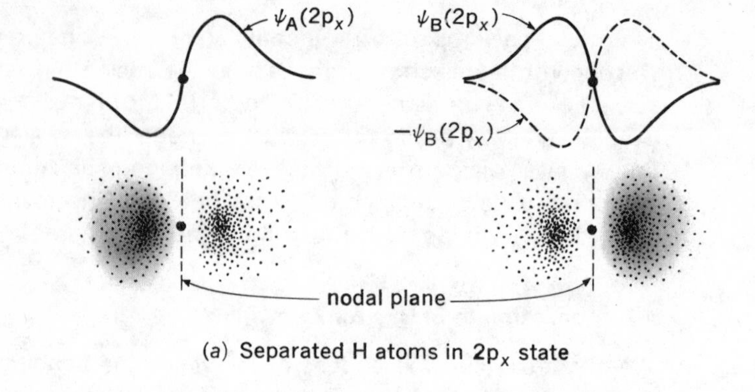

(a) Separated H atoms in $2p_x$ state

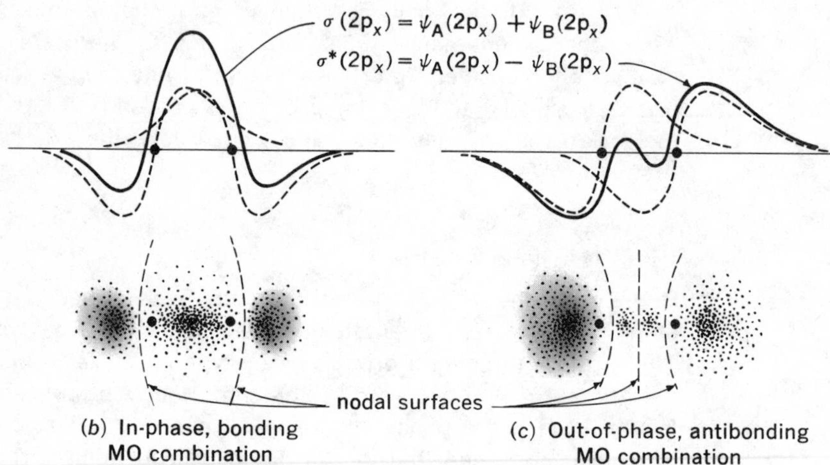

(b) In-phase, bonding MO combination

(c) Out-of-phase, antibonding MO combination

Figure 4-3 Sigma molecular orbitals from axial p orbitals. (The relative phases of the wave functions are shown by shading.)

bond axis, whether they are formed from s atomic orbitals or axially directed p orbitals. An M.O. formed from the perpendicular p orbitals does have such a surface. Such perpendicular M.O.'s, parented by p orbitals, are called π (pi) orbitals. Because of the difference in spatial distribution, sigma (σ) and pi (π) M.O.'s do not generally have the same energy. On the other hand, there are two bonding M.O.'s that are exactly equivalent; one formed from the $2p_y$ orbitals and one from the $2p_z$ orbitals. Since they are equivalent (except for a rotation through 90° around the bond axis), their energies are identical. The same is true for the two antibonding M.O.'s formed from the $2p_y$ and $2p_z$ orbitals.

Figure 4-5 summarizes the M.O.'s that can be considered to have their parentage primarily in the 2p orbitals of excited hydrogen atoms. All of the orbitals are placed on an energy scale appropriate to separated 2p orbitals. The bonding M.O.'s

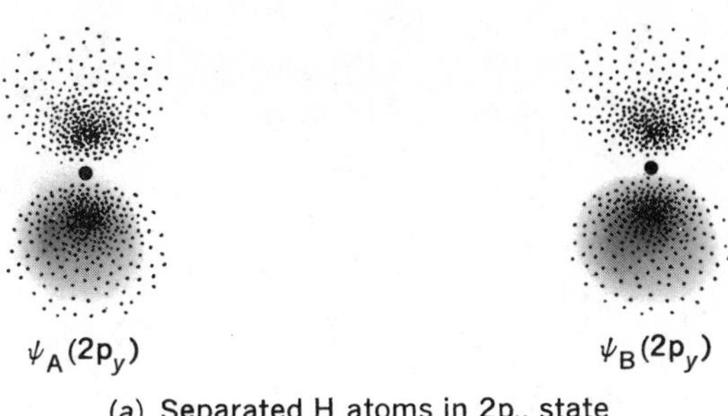

(a) Separated H atoms in $2p_y$ state

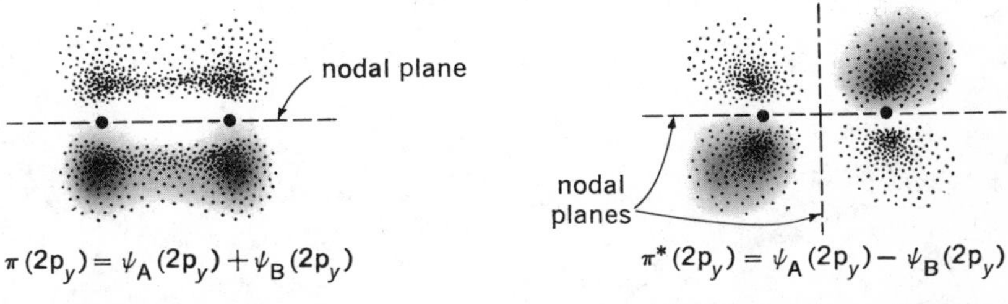

(b) In-phase, bonding MO combination

(c) Out-of-phase antibonding MO combination

Figure 4-4 Pi molecular orbitals from perpendicular p orbitals. (The relative phases of the wave functions are shown by shading.)

are lower and the antibonding orbitals higher in energy than the energies of the isolated 2p orbitals. The σ M.O.'s are shown below the π M.O.'s, although it is not necessarily that way. We can always expect the axial and perpendicular orbitals to have different energies, however, since their geographic distributions are so different.

4-3 M.O.'s for the oxygen molecule

section 4-3 for oxygen

All of this permits us to consider the M.O. energy level diagram for a molecule with quite a few electrons. Oxygen, O_2, is a good place to begin.

Figure 4-6 shows the energy levels of a single oxygen atom. There are large energy separations between the 1s and 2s orbitals, between the 2s and 2p orbitals, and again between the 2p and still higher orbitals. The implication is that the parentage of M.O.'s will be fairly simply related to these isolated atomic orbitals.

Figure 4-7 shows M.O. formation as two oxygen atoms approach. The M.O.'s all look familiar—the bottom part of the diagram looks just like that of dilithium, Figure 3-11. The upper

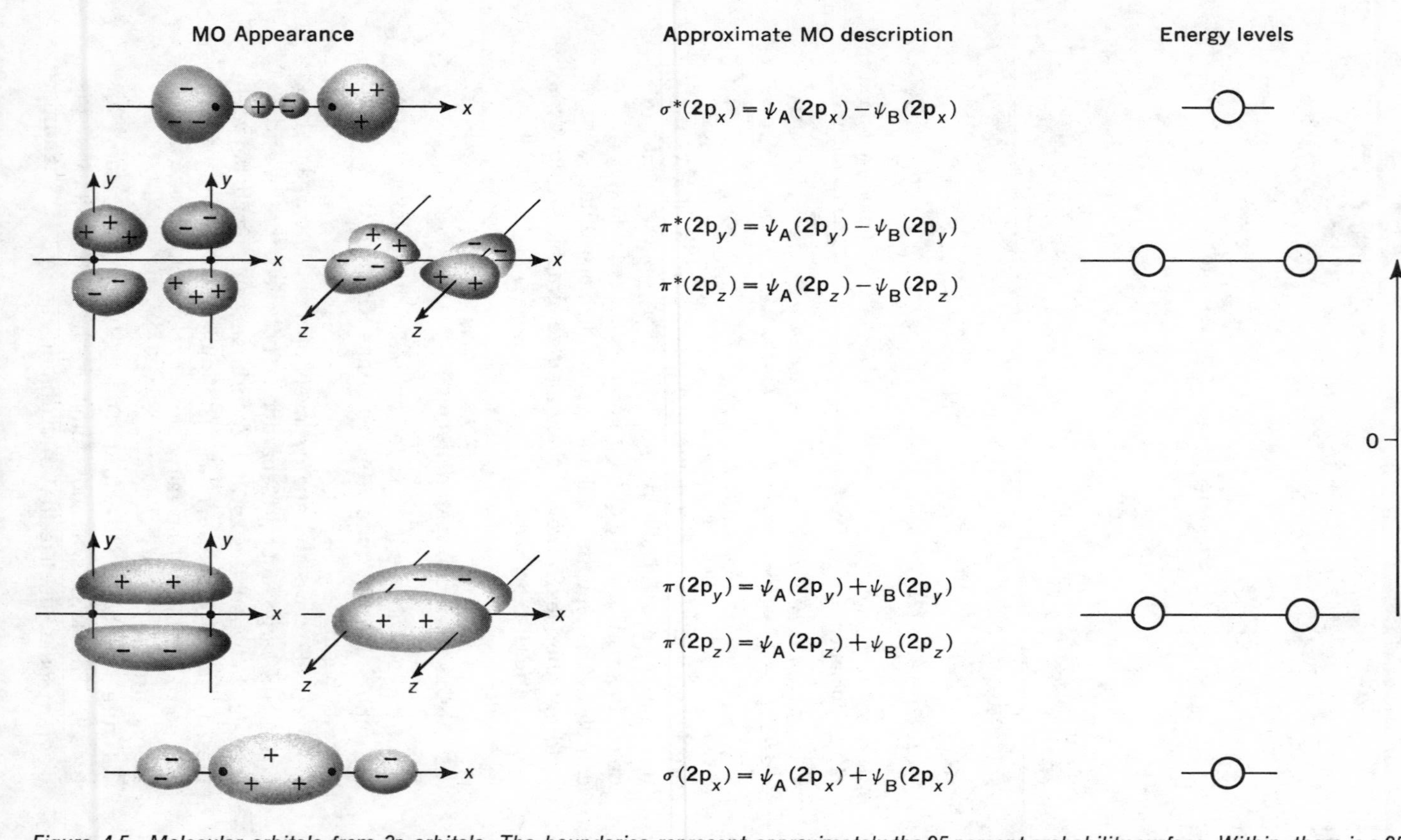

Figure 4-5 Molecular orbitals from 2p orbitals. The boundaries represent approximately the 95 percent probability surface. Within, there is a 95 percent chance of finding an electron. The relative phases are shown by plus and minus signs. These signs have nothing whatsoever to do with charge—they simply point out that the wave function which describes the orbitals has phase properties.

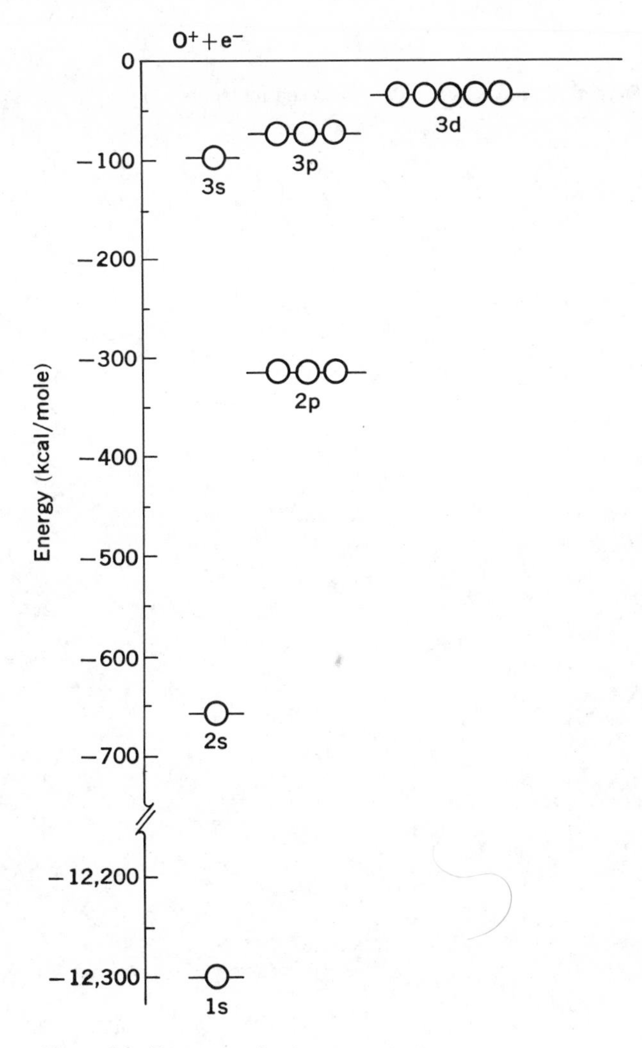

Figure 4-6 Energy levels of an oxygen atom.

section 4-3 for oxygen

part is the same as Figure 4-5. All we need do is decide orbital occupancy in accordance with the Pauli Principle.

Each oxygen atom gives us eight electrons. Four of these reside in the $\sigma(1s)$ and $\sigma^*(1s)$ orbitals which make no net contribution to the bonding. Four more electrons can be placed in the next orbitals, $\sigma(2s)$ and $\sigma^*(2s)$. Again, because both bonding and antibonding orbitals are filled, there is no net bonding. We now have $16 - 4 - 4 = 8$ electrons remaining. The energy level diagram shows three bonding M.O.'s—$\sigma(2p_x)$ and the pair $\pi(2p_y),\pi(2p_z)$. Six electrons into these bonding M.O.'s give three pairs of bonding electrons—sufficient for a bond order of three. However, we have disposed of only 14 of the 16 electrons. The

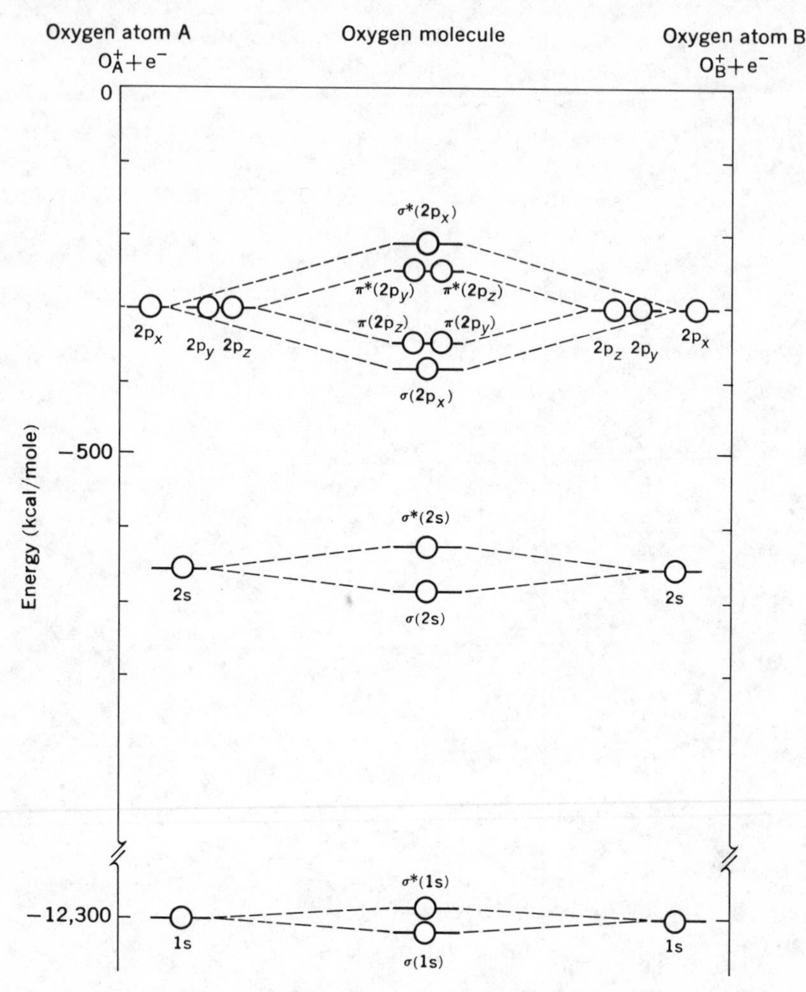

Figure 4-7 Molecular orbitals for O_2.

remaining pair of electrons must go into the antibonding orbitals, reducing the bond order, again, to two. Oxygen should have a double bond.

This last pair of electrons has two options. They could both enter one of the π^* orbitals, with opposite spin, of course. This orbital occupancy is shown in Figure 4-8(*a*). However, another possibility is shown in Figure 4-8(*b*). As far as the simplified, "turned-off" electron approximation is concerned, these two occupancies have the same energy.

It is not difficult to see how the electron repulsions will differentiate the two occupancies. Two electrons in the same orbital $\pi^*(2p_y)$, give large electron repulsions because the electrons occupy the same region of space. When one electron is moved to the second orbital $\pi^*(2p_z)$, it occupies a different region of space. Then the electrons are farther apart, so electron repulsions are reduced. The right-hand electron configuration has lower energy.

We conclude that there should be two states of O_2 reasonably close in energy. Both states have three pairs of bonding electrons and one pair of antibonding electrons in the uppermost M.O.'s, so both states have double bonds. The lower of the two states—lower because of reduced electron repulsions—can accommodate the last two electrons with parallel magnetic spins. This means the molecule is a tiny magnet and it should respond, somehow, to external magnetic fields. Because of this, certain isolated spectroscopic transitions appear as triplets in an imposed magnetic field. The higher of the two states, the one with the last two electrons in the same orbital $\pi^*(2p_y)$, cannot accept this last pair unless they have opposed spins. There will be no magnetic behavior—isolated spectroscopic transitions will remain as singlet lines even if a magnetic field is imposed.

Table 4-1 shows how well these expectations are realized. The M.O. picture is completely consistent with the known properties of the oxygen molecule in its two lowest energy states. In fact, this was one of the most important successes of the molecular orbital view, early in the development of our quantum mechanical view of bonding.

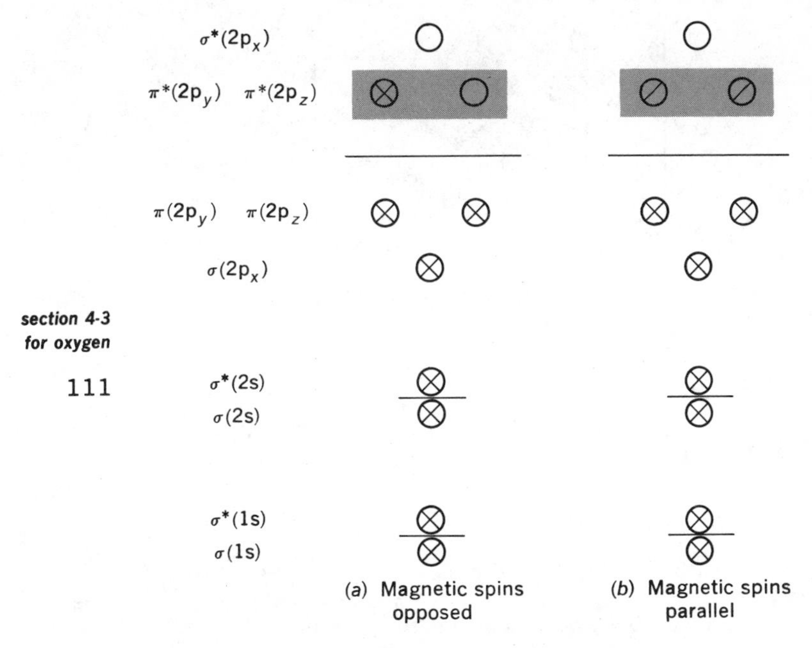

Figure 4-8 Possible orbital occupancies for O_2.

Table 4-1 *Two Lowest Energy States of Oxygen Molecule*

Upper orbital occupancy	$\sigma(2p_x)^2\pi(2p_y)^2\pi(2p_z)^2\pi^*(2p_y)\pi^*(2p_z)$	$\sigma(2p_x)^2\pi(2p_y)^2\pi(2p_z)^2\pi^*(2p_y)^2$
Name	"triplet"	"singlet"
Energy	(0)	+22.5 kcal/mole
Bond energy	118 kcal/mole	96 kcal/mole
Bond length	1.21 Å	1.22 Å
Force constant	11.4 mdyne/Å	10.7 mdyne/Å
Apparent bond order	2	2
Magnetic properties	magnetic	nonmagnetic

4-4 M.O.'s for other first-row, homonuclear, diatomic molecules

Figure 4-9 shows the energies of the 2s and 2p electrons for the atoms boron to neon. Fluorine and neon have a large energy gap between these orbitals: over 350 kcal—as does oxygen. These atoms should produce M.O. energy level diagrams qualitatively like that of O_2, like Figure 4-7. Boron, carbon, and nitrogen differ; their 2s and 2p states are only about 100 kcal apart. The 2s and 2p states no longer act independently and the energy level diagram is altered. The $\sigma(2p_x)$ M.O. is the one affected

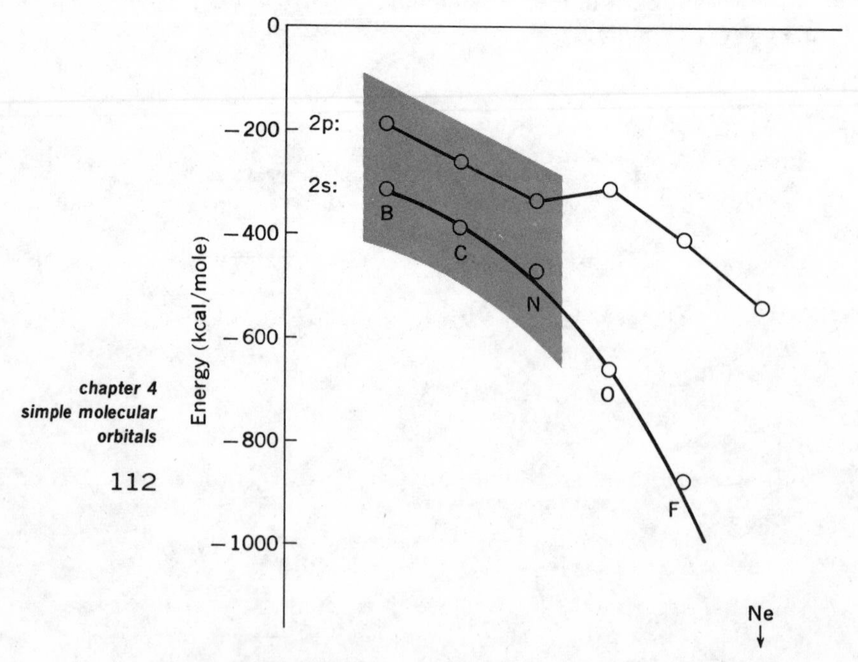

Figure 4-9 Atomic 2s and 2p energy levels for first-row atoms. The 2s and 2p orbitals of B, C and N (shaded) are close enough together to permit a significant interaction of their resultant M.O.'s.

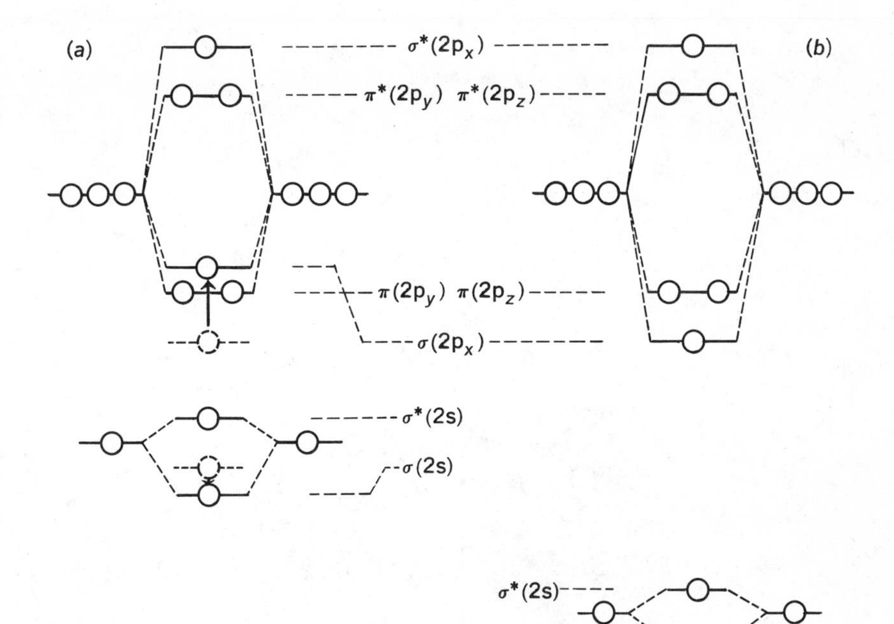

Figure 4-10 Molecular orbitals for first-row homonuclear diatomics: (a) B_2, C_2, N_2; (b) O_2, F_2.

most because it has a nodal pattern like that of the nearby $\sigma(2s)$ M.O. Experiment shows that the effect is sufficient to move the $\sigma(2p_x)$ M.O. upward in energy, above the $\pi(2p_y)$, $\pi(2p_z)$ pair.

Figure 4-10 shows the two kinds of energy level diagrams that result. The left-hand diagram, obtained when the 2s–2p energy separation is small, is found applicable to boron, carbon, and nitrogen. The right-hand diagram applies to oxygen, fluorine, and neon.

With this background, we can proceed with the game of counting electrons to determine orbital occupancy and bond order. Figure 4-11 does this and it includes the experimental data about each molecule so that we can test the utility of the M.O. scheme.

section 4-4 for other homonuclear atomic molecules

The bond orders are predicted to begin at B_2 with a single bond, to increase to a double bond at C_2 and to a triple bond at N_2. Then the bond order decreases progressively from O_2 and F_2 and, finally, to neon, where no bond at all is expected. All of the experimental criteria are in agreement, as shown in Figure 4-12. In a reasonably symmetric way, the bond energies and force constants peak at N_2 and the bond lengths are shortest at N_2. The double bond in O_2 is close to but weaker than that in C_2. The difference is, no doubt, due to the higher electron–

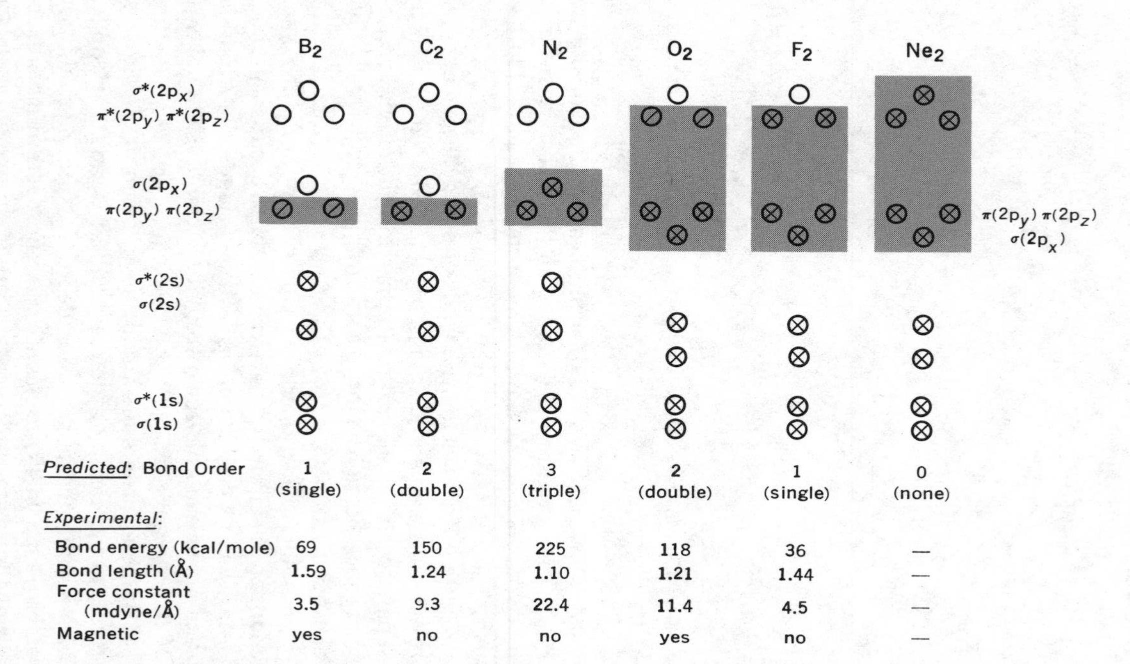

	B_2	C_2	N_2	O_2	F_2	Ne_2
Predicted: Bond Order	1 (single)	2 (double)	3 (triple)	2 (double)	1 (single)	0 (none)
Experimental:						
Bond energy (kcal/mole)	69	150	225	118	36	—
Bond length (Å)	1.59	1.24	1.10	1.21	1.44	—
Force constant (mdyne/Å)	3.5	9.3	22.4	11.4	4.5	—
Magnetic	yes	no	no	yes	no	—

Figure 4-11 Orbital occupancies and bond properties of first-row homonuclear diatomic molecules.

electron repulsions in O_2. The single bond in F_2 is close to but weaker than that in B_2, again because F_2 has many more electron repulsions affecting its bonds. No molecular species Ne_2 is predicted, nor is one observed.

Figure 4-11 displays one more opportunity for testing the M.O. description—the magnetic properties (see Fig. 4-13). Of the six

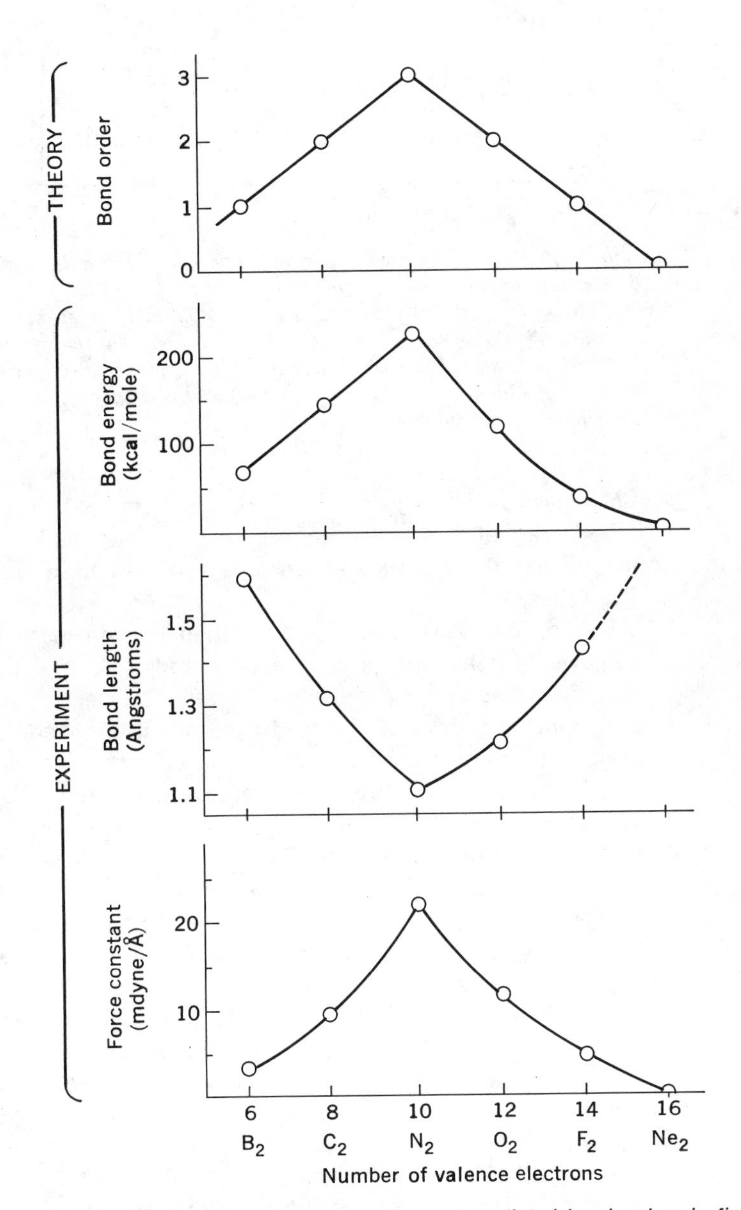

Figure 4-12 Trends in bond properties and predicted bond orders in first-row homonuclear diatomic molecules.

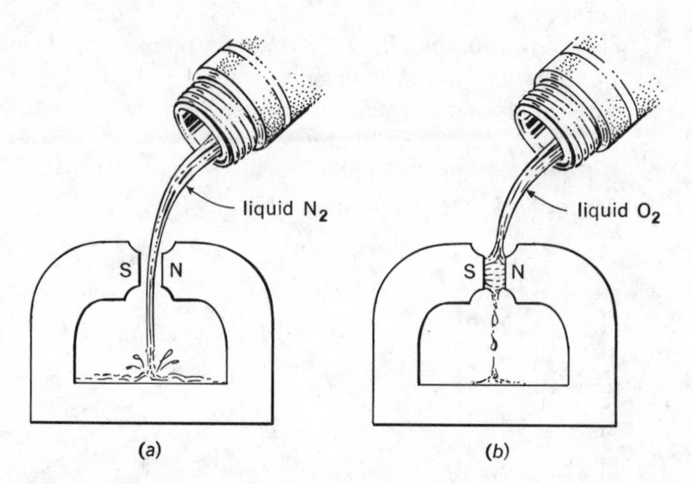

Figure 4-13 The oxygen molecule is magnetic, the nitrogen molecule is not. (a) Liquid nitrogen (bp = 77 degK) can be poured directly between the poles of a strong magnet. (b) Liquid oxygen (bp = 90 degK) is attracted by the magnet and fills up the gap between the two poles. This easily performed demonstration confirms the molecular orbital approach. Oxygen has two unpaired electrons (in antibonding orbitals) and is magnetic. Nitrogen has all its electrons paired and is not magnetic.

molecules, only two have orbital occupancies that permit unpaired magnetic spins; B_2 and O_2. Again this agrees with the observation that these are the only two that have magnetic properties.

This optimistic situation is strengthened even more by the properties of the two gaseous molecule ions N_2^+ and O_2^+. The M.O. diagrams suggest that the removal of an electron from N_2 should have quite a different effect from that caused by the removal of an electron from O_2. The uppermost electron in N_2 is a bonding electron. Its loss should weaken the bond. The uppermost electron in O_2 is an antibonding electron. Its loss should strengthen the bond! Figure 4-14 shows that this is exactly what is observed. The N_2^+ molecule has properties between C_2 and N_2—the nitrogen—nitrogen bond is weaker in N_2^+ than in N_2. In contrast, the O_2^+ molecule lies between N_2 and O_2—*the oxygen—oxygen bond is stronger in* O_2^+ *than in* O_2.

Also shown in Figure 4-14 are the orbital occupancies and stretching force constants for the gaseous ions N_2^- and O_2^-. (The bond energies are not yet known, and the bond length for O_2^- refers to the crystal KO_2.) The former, N_2^-, is the more interesting. Addition of an electron to N_2 requires occupancy of an antibonding orbital. Hence N_2^- has a bond order of $2\frac{1}{2}$, as do both N_2^+ and O_2^+. However N_2^- has the same orbital occupancy as O_2^+, so, from the molecular orbital point of view, N_2^- should be more

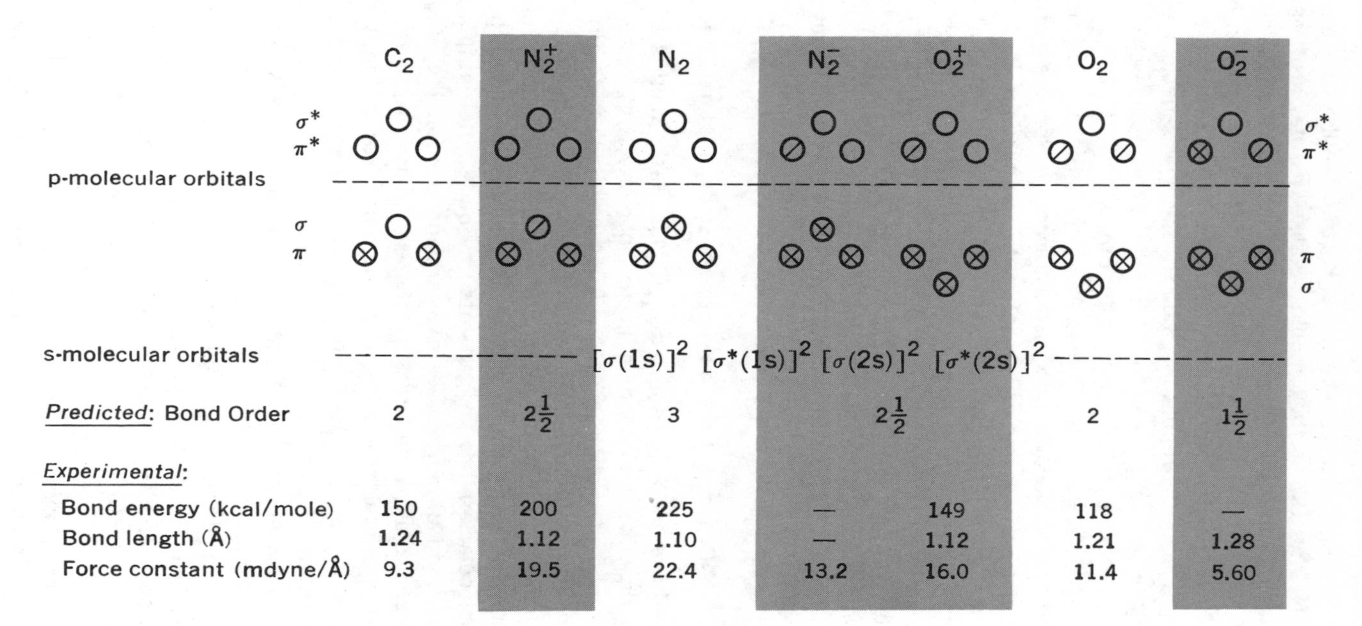

Figure 4-14 Orbital occupancy and bond properties of some homonuclear diatomic ions.

like O_2^+ than like N_2^+. The same electron repulsions are present in N_2^- and O_2^+, so they should be alike (as they are), and they should both have weaker $2\frac{1}{2}$ order bonds than the $2\frac{1}{2}$ order bond of N_2^+ (as they do). Even the small difference between N_2^- and O_2^+ is consistent and understandable. These two ions have the same electron repulsions, but N_2^- has lower nuclear charges with which to cope with them. Hence the bond weakening in N_2^- is a bit more severe than that in O_2^+.

We see that this discussion of the first-row homonuclear, diatomic molecules successfully accounts for their observed properties in terms of two concepts, the occupancy of a set of molecular orbitals considered in the light of electron repulsions. This success gives significant encouragement to the M.O. description. In particular, the fact that the *removal* of an electron from O_2 *strengthens* its bond (as in O_2^+), whereas the *addition* of an electron to N_2 and O_2 weakens their bonds (as in N_2^- and O_2^-) gives strong credibility to the concept of antibonding orbitals. Further exploration of this bonding model is warranted.

4-5 M.O.'s and first-row heteronuclear, diatomic molecules

We have seen that the p orbital M.O. description is sufficient to describe the bonding in the homonuclear, diatomics B_2 through the nonexistent Ne_2. The model took the "turned-off" electron approach—electron repulsions were ignored initially. Furthermore, except for noting the s–p energy difference, no attention was paid to the specific atoms involved. The same molecular orbitals are generated by quite different atoms, and the bond properties seem to depend only on the number of electrons available to occupy these orbitals. This behavior suggests that the atoms in a bond need not even be identical. Perhaps we can also predict the bond orders and bond properties of heteronuclear diatomic molecules.

Carbon monoxide, CO, and nitric oxide, NO, are two familiar heteronuclear diatomic molecules. A third example is CN, a well-known molecule, but less available because it is quite reactive. Figure 4-15 shows the orbital occupancies for these three molecules if all valence electrons are treated as community property. (Only the p orbital M.O.'s are shown—in addition, each molecule has eigh+ electrons held in the 1s and 2s M.O.'s.) Carbon monoxide is predicted to resemble nitrogen—it should have a triple bond. Both CN and NO should have $2\frac{1}{2}$ order bonds, respectively intermediate between C_2 and N_2 and N_2 and O_2. The experimental facts bear out these expectations quite well. This agreement can be assessed in Figure 4-16. The broken lines, transcribed from Figure 4-12, show the homonuclear

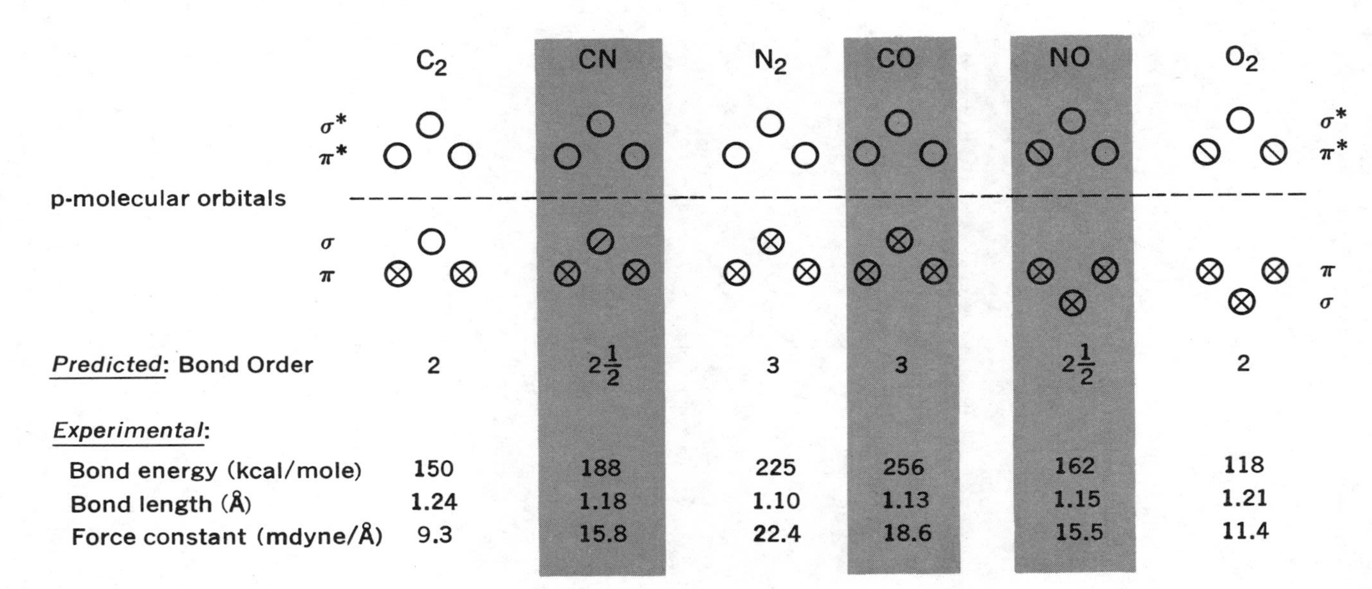

Figure 4-15 *Orbital occupancy and bond properties of some heteronuclear diatomic molecules.*

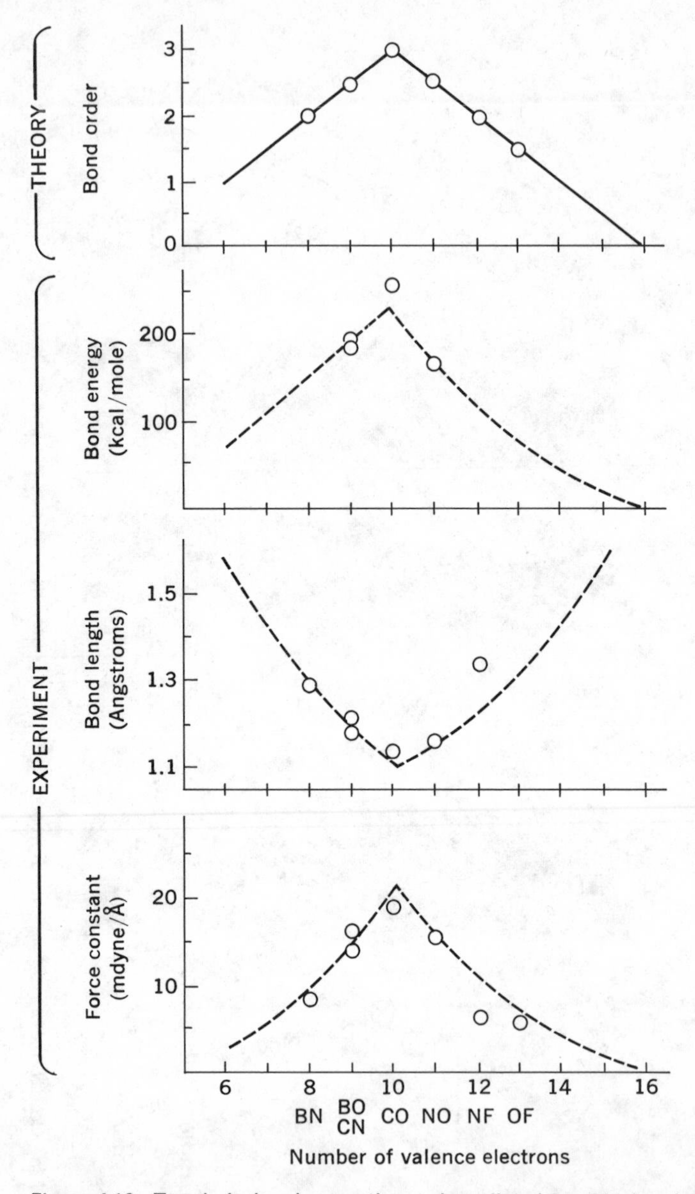

Figure 4-16 Trends in bond properties and predicted bond orders in first-row heteronuclear diatomic molecules.

molecular trends. The circles display the experimental facts for the heteronuclear diatomic molecules. Force constants and bond lengths for BN, NF and OF are also included; the bond energies, however, are not yet firmly established. There is noticeable scatter, but the bond properties are generally within about one half of a bond order of the expected values based upon the homonuclear trends. NO and CN do have $2\frac{1}{2}$ order bonds, while CO has a triple bond. The OF molecule, with a force constant of 5.41 mdyne/Å, is between F_2 and O_2, consistent with the $1\frac{1}{2}$ bond order prediction. Only NF seems out of line. The presently accepted values for its bond length (1.317 Å) and force constant (6.0 mdyne/Å) suggest that NF has a $1\frac{1}{2}$ order bond rather than a double bond like O_2.

The primary conclusion from Figures 4-12, 4-14, and 4-16 is that the M.O. concept provides a useful description of the bonding in the first-row diatomic molecules. Apparently electron repulsions and skewed charge distributions can be ignored in the first analysis. The method is sufficiently successful and easy to pursue that it permits us to apply quantum mechanical ideas of bonding to more complicated molecules, including polyatomic molecules far beyond the numerical capabilities of the largest computers. We'll proceed one more step in that direction by considering two simple triatomic molecules, CH_2 and CO_2.

4-6 Approximate M.O.'s for methylene, CH_2

Of all the molecules that have been definitely identified, methylene is one of the most reactive. It can be produced by photoexcitation and fragmentation of a number of more complicated molecules. After birth, it usually needs only two or three collisions with other molecular species to react and end its independent career. Nevertheless, we know its structure, its bond energies, its force constants, and much about its chemistry. It is a good example with which to pursue the consideration of bond and molecular orbitals.

The lowest energy state of methylene is found to be linear, with equal C—H bond lengths of 1.03 Å (see Fig. 4-17). The atoms lie in a line with the carbon at the center of the molecule. The two hydrogen atoms, H′ and H″, occupy equivalent positions; if the molecule were turned end for end, it would appear the same.

A correlation diagram approach is helpful. Consider the formation of CH_2 when a carbon atom is approached simultaneously by two H atoms, one at a distance R on the positive x axis and the other at the same distance R on the negative x axis. As R is decreased, always maintaining the final molecular symmetry, the orbitals begin to interact to form molecular orbitals. Just as in the diatomic cases, the orbitals on the molecular axis interact with each other but not with those perpendicular to the molecular axis. Each hydrogen atom contributes a 1s valence orbital on the x axis, 1s′ and 1s″, and the carbon atom contributes two such valence orbitals, 2s and $2p_x$. The carbon atom also

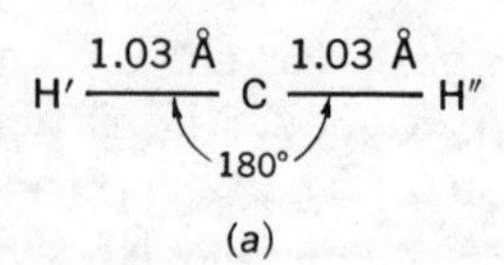

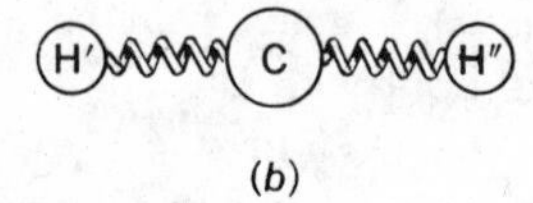

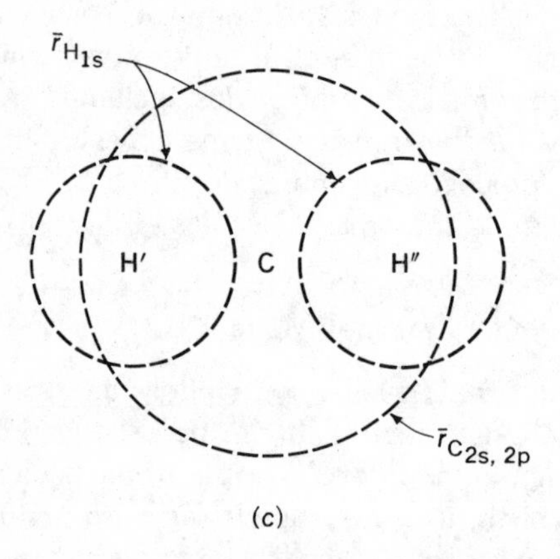

Figure 4-17 The structure of methylene, a reactive molecule. The structure of CH_2 is shown in (a) and, on the same scale, in the ball-and-spring model of (b). The average radii of the valence orbitals (see Table 2-1) are shown in (c) on the same scale as (a) and (b). The hydrogen nuclei are essentially "buried" in the carbon 2s and 2p orbitals.

has two perpendicular valence orbitals, $2p_y$ and $2p_z$, but they have different nodal properties (π nodal surfaces) so they won't be engaged in the axial (σ) M.O.'s.

As we frame M.O.'s from these four component orbitals, 1s′, 1s″, 2s and $2p_x$, it is helpful to consider the M.O.'s that would be formed from the hydrogen atom orbitals alone. If we pretend the carbon atom is not there, the M.O.'s become those of H_2^+ again, as pictured in Figure 4-2. For example, the $\sigma^*(1s)$ orbital has a nodal surface halfway between the atoms. In CH_2 that's exactly where the carbon atom is placed and, of course, the $2p_x$ orbital also has a nodal surface here. Having the same nodal surfaces, $2p_x$ and $\sigma^*(1s)$ can join to form M.O.'s. On the other hand, this nodal surface is quite incompatible with the nodal pattern of the centrally located carbon 2s orbital. The 2s orbital does not join in the $\sigma^*(1s)$ M.O.

The $2p_x$ and $\sigma^*(1s)$ orbitals join to form M.O.'s in the now-familiar manner. Figure 4-18 shows how $2p_x$ can join with $\sigma^*(1s)$ in an in-phase or an out-of-phase relationship. The in-phase relation concentrates electron probability between the atoms, in the binding region, to give a bonding M.O. The out-of-phase relation moves electron probability out of the binding region and an antibonding M.O. results. These two M.O.'s are both derived from the $2p_x$ and the antibonding $\sigma^*(1s)$ orbitals. The energy effects from the antibonding $\sigma^*(1s)$ parentage are relatively unimportant because in CH_2 the hydrogen atoms are too far apart to interact with each other strongly.

Returning to the H_2^+ M.O.'s, the $\sigma(1s)$ orbital has no nodal

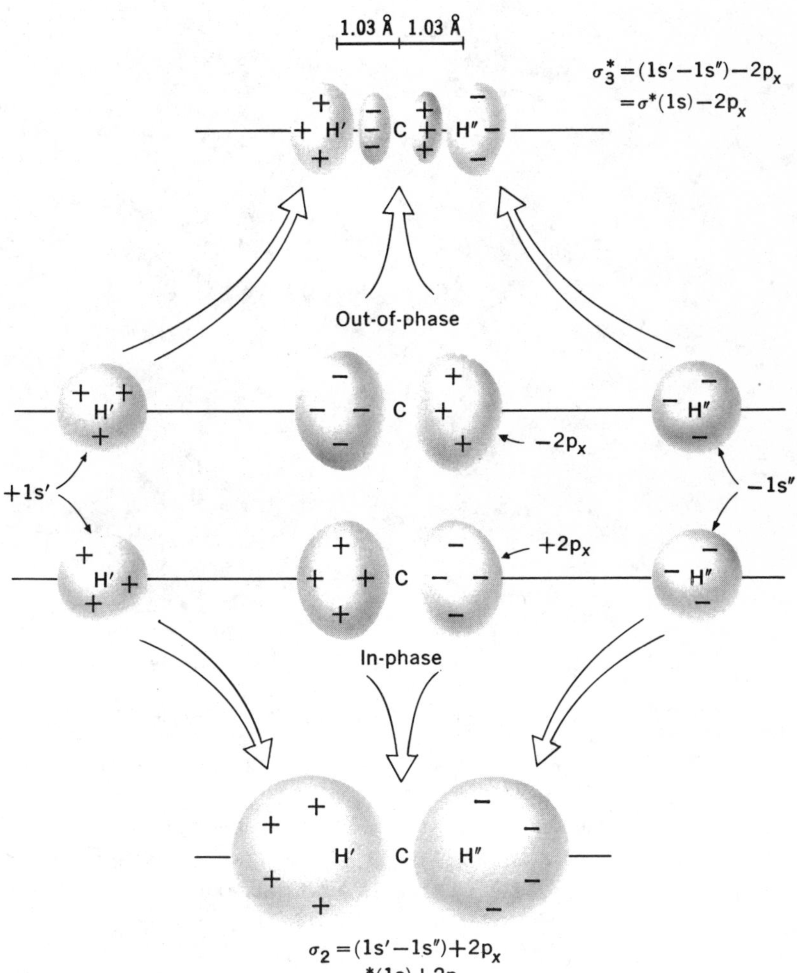

Figure 4-18 The CH_2 molecular orbitals from $\sigma^*(1s)$ and $2p_x$ (the atomic orbitals are shown to scale).

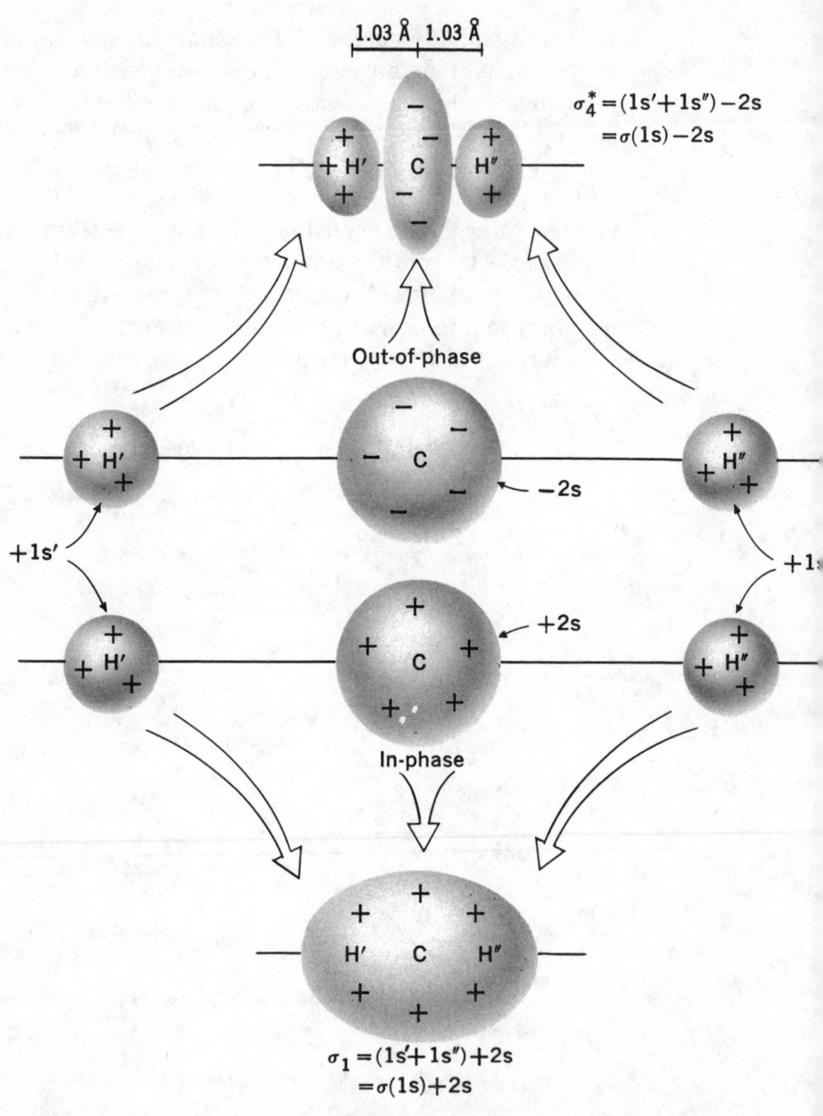

Figure 4-19 The CH_2 molecular orbitals from σ(1s) and 2s (the atomic orbitals are drawn to scale).

surface at the center of the molecule where the carbon atom lies. This nodal pattern is like that of the carbon atom 2s orbital. Hence the 2s and σ(1s) orbitals join to form M.O.'s. Figure 4-19 shows the resultant CH_2 M.O.'s—again, one bonding and one antibonding.

Thus the four atomic orbitals, 1s′, 1s″, 2s and $2p_x$, form four CH_2 M.O.'s, two bonding and two antibonding. The makeup of these four M.O.'s is readily deduced from the nodal patterns,

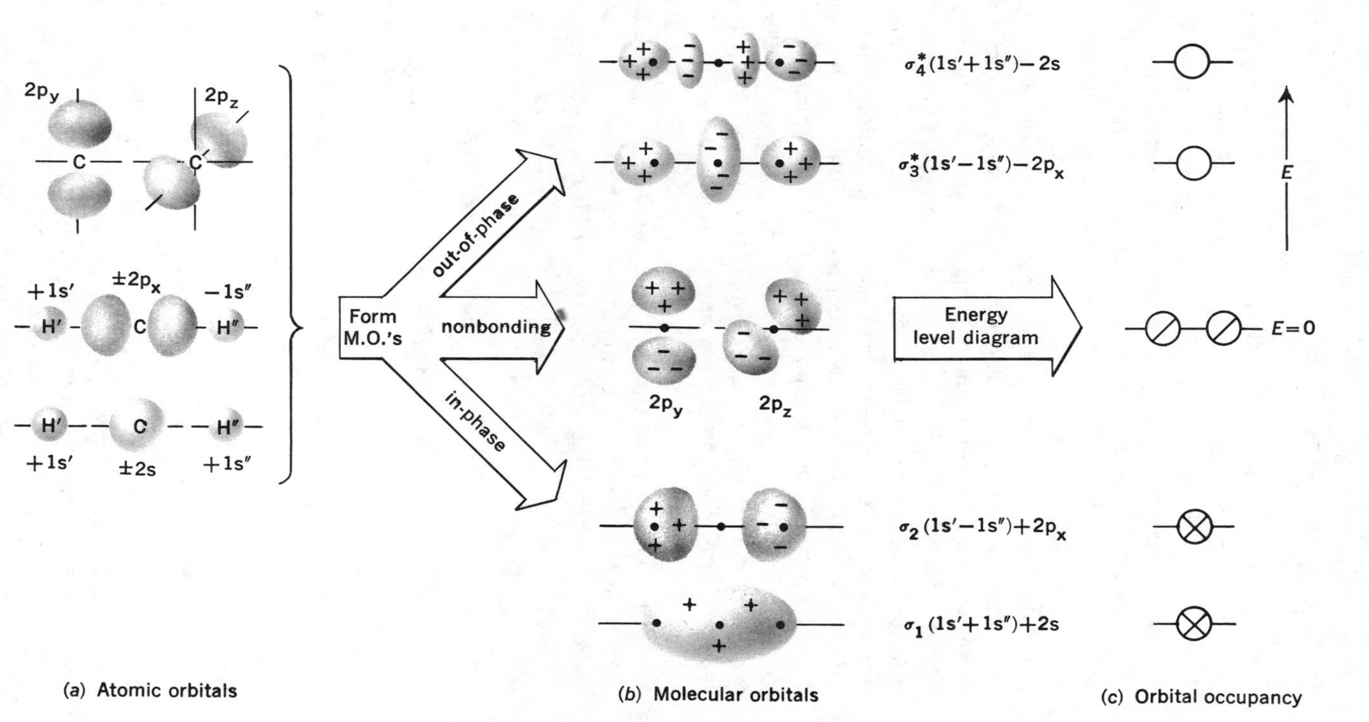

Figure 4-20 The molecular orbitals of linear methylene, CH_2.

taking account of the fact that the carbon atom is exactly halfway between the two hydrogen atoms in a linear arrangement.

Now we can construct the molecular orbital energy level diagram of CH_2, adding the $2p_y$ and $2p_z$ perpendicular orbitals of the carbon atom. They interact only weakly with the hydrogen atoms so they furnish essentially nonbonding orbitals (see Fig. 4-20). These nonbonding orbitals are shown at $E = 0$, the energy corresponding to the energy of atoms not interacting at all. They do not change the molecular energy as the atoms approach. The bonding orbitals σ_1 and σ_2 are at negative energies; they lower the molecular energy as the atoms approach. The antibonding orbitals σ_3^* and σ_4^* have the opposite effect; they are at positive energies.

It only remains to add the electrons; two per orbital. The carbon atom furnishes four valence electrons and each hydrogen furnishes one. That is enough to fill σ_1 and σ_2 and leave one electron each for $2p_y$ and $2p_z$. Note that two pairs of electrons, enough to form two single bonds, act to hold three atoms together. Each of these pairs occupies an M.O. that is shared by two C—H bonds, so each M.O. contributes half a bond order to each bond. The net result of adding the bond contributions of σ_1 and σ_2 is that each C—H bond has a bond order of $\frac{1}{2} + \frac{1}{2} = 1$, a single bond. This is in accord with the observed C—H bond length, 1.03 Å, appropriate to a normal, single bond.

The orbital diagram reveals one more expectation for linear CH_2. The last two electrons can have parallel spins, since they occupy different orbitals to minimize electron repulsion. With parallel spins the molecule should display a net magnetic behavior, and spectroscopic studies show that it does.

4-7 Approximate M.O.'s for carbon dioxide, CO_2

Carbon dioxide differs from methylene in the same way that O_2 differs from H_2: perpendicular 2p orbitals contribute to the bonding in CO_2, as they do in O_2, whereas they do not in CH_2 (because the hydrogen atom 2p orbitals are too high in energy). As for O_2, we must first consider the axial (sigma) M.O.'s and then the perpendicular (pi) M.O.'s.

(a) APPROXIMATE SIGMA M.O.'s FOR CO_2

Carbon dioxide has a linear, symmetric structure like that of CH_2. Hence, framing σ M.O.'s for CO_2 is little different from the process we just went through for CH_2. In fact, the only difference is that the end atoms, oxygen atoms, contribute to the σ M.O.'s with axial $2p_x$ orbitals instead of the 1s orbitals contributed by hydrogen. The oxygen atom 2s orbitals are so low in energy that they do not interact noticeably (see Fig. 4-9). That leaves the oxygen $2p_x$ orbitals with which the carbon atom axial orbitals, 2s and $2p_x$, can form M.O.'s. The results are entirely analogous to those of CH_2, as shown in Figure 4-21. There are two bonding sigma M.O.'s, σ_1 and σ_2, and two antibonding sigma M.O.'s, σ_3^* and σ_4^*.

(b) APPROXIMATE PI M.O.'s FOR CO_2

Since the terminal atoms possess 2p π atomic orbitals like those of the central atom, we must consider the pi M.O.'s as well. As in the case of O_2, it is only necessary to look at the $2p_y$ orbitals and the M.O.'s they form. Whatever the $2p_y$ orbitals do, the $2p_z$ orbitals will do as well, since they are equivalent.

Figure 4-22 shows how the $2p_y$ orbitals of carbon and the two oxygen atoms form M.O.'s as the molecule is formed, always in the linear, symmetric geometry. The carbon atom $2p_y$ orbital, directed along the *y* axis, forms a bonding M.O. and an antibonding M.O. with the combination of oxygen atom orbitals $(2p_y' + 2p_y'')$. There is, however, a combination of oxygen atom orbitals $(2p_y' - 2p_y'')$ with which it cannot interact. The combination $(2p_y' - 2p_y'')$ is a π combination with a nodal plane perpendicular to the molecular axis. The carbon atom has no π orbitals with such a nodal surface. Hence, the $(2p_y' - 2p_y'')$ combination is itself a molecular orbital but it only involves atoms that are far apart. In fact, the distance between the oxygen atoms is so great that there is no appreciable interaction or change in electron distribution. Without such interaction the energy is not changed, so this is neither a bonding nor an antibonding orbital. *A molecular orbital that involves no adjacent atoms is called* **a nonbonding** *M.O.* Its energy is the same as that of the separated atoms. We will identify its nonbonding character with a superscript zero, as in π^0.

(c) THE CO_2 ENERGY LEVEL DIAGRAM

We can now unite Figures 4-21 and 4-22 to give the CO_2 energy level diagram. Figure 4-23 shows the result. We need only plunk in the electrons, two per orbital, with opposite spins, as long as the electrons last. Each oxygen atom gives us six valence electrons, two of which are used by each atom to fill its low-energy 2s orbital. The remaining eight electrons (four per oxygen atom) are joined by four carbon atom valence electrons. So 12 electrons placed into the orbitals shown in Figure 4-23 fill all of the four bonding M.O.'s and the two nonbonding M.O.'s,

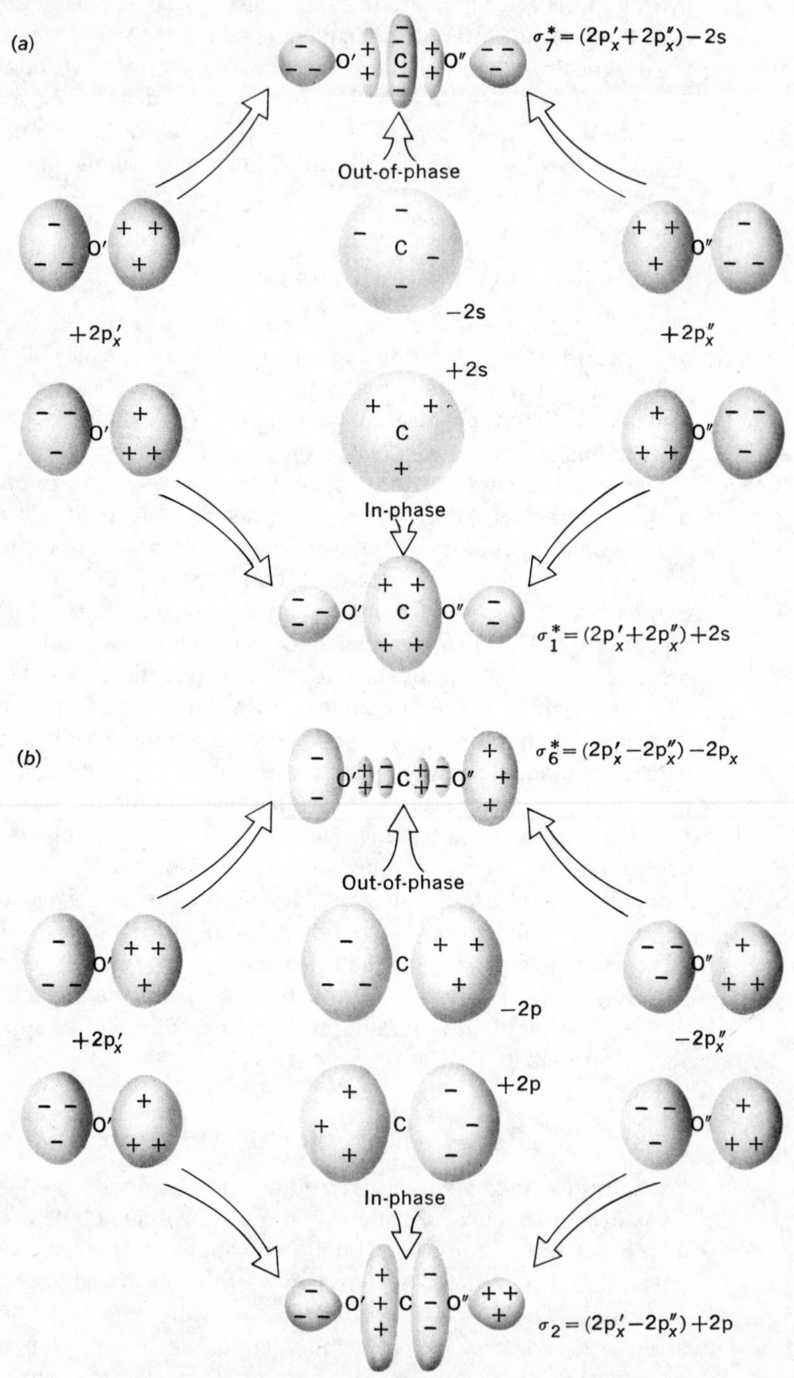

Figure 4-21 Axial (sigma) molecular orbitals for CO_2: (a) sigma molecular orbitals using the carbon 2s orbitals; (b) sigma molecular orbitals using the carbon 2p orbital.

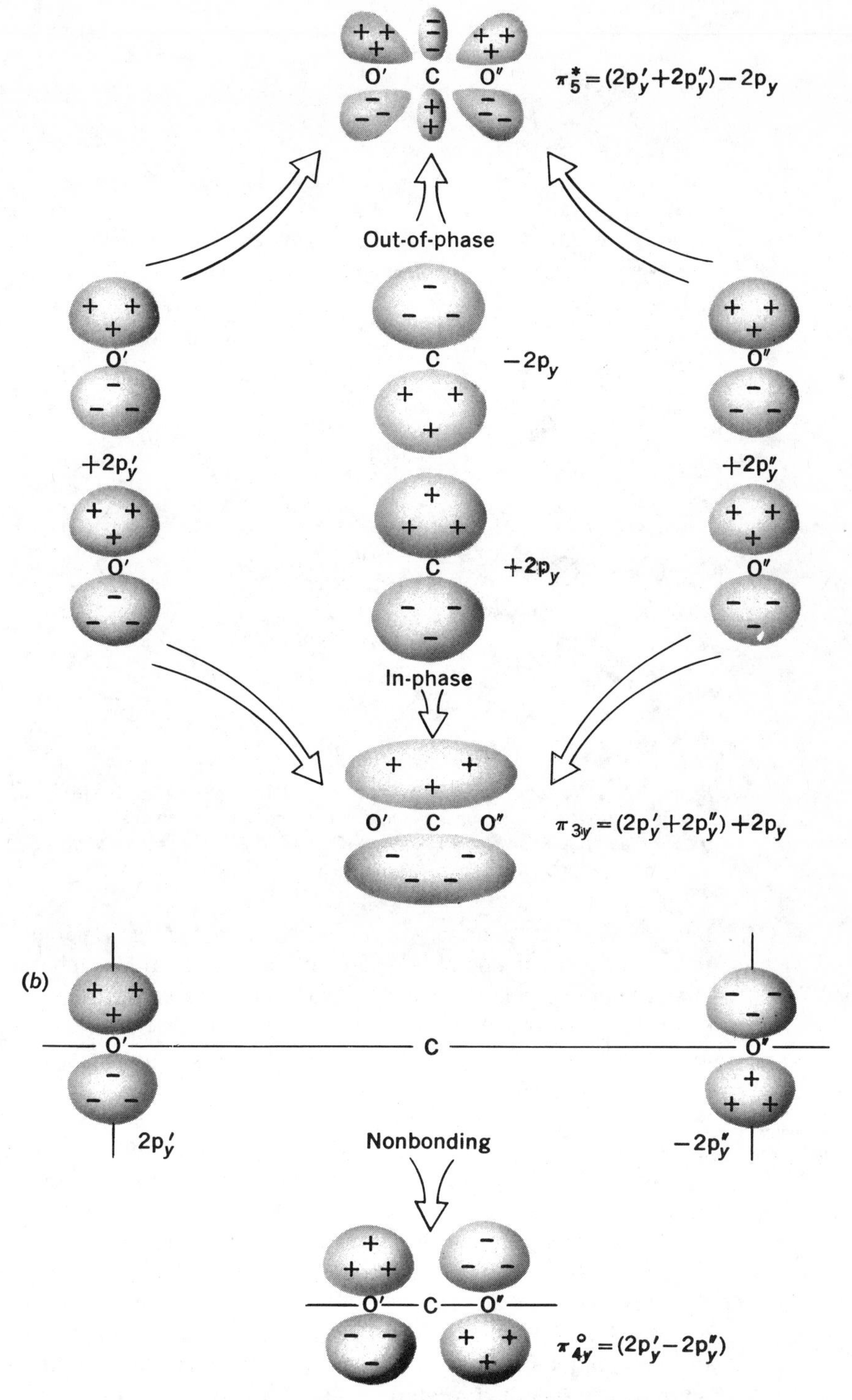

Figure 4-22 Pi (π) molecular orbitals for CO_2 formed from p_y atomic orbitals. (An identical set of orbitals can be drawn for p_z atomic orbitals.) (a) Bonding and antibonding orbitals. (b) Nonbonding orbitals.

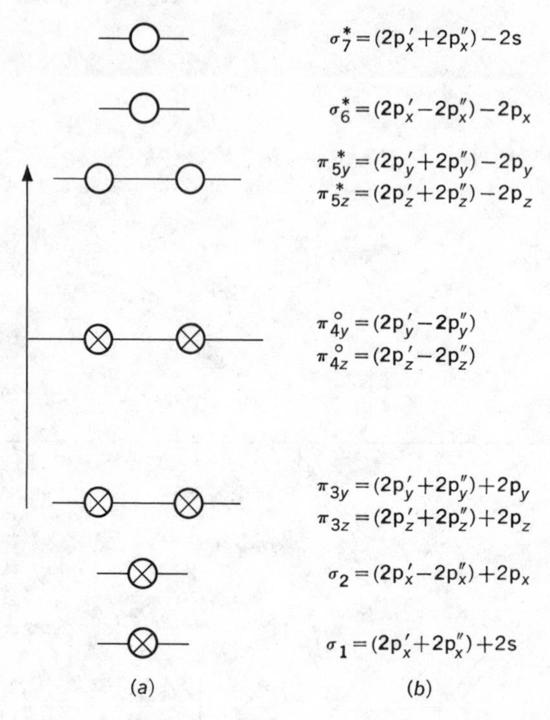

Figure 4-23 The energy level diagram and orbital occupancy of CO_2. (a) The twelve electrons occupy four bonding and two nonbonding orbitals, thus contributing a net bond order of two per C—O bond. (b) The approximate molecular orbital description of the energy levels.

but none are left over to go into antibonding M.O.'s. We find, then, four pairs of electrons in bonding M.O.'s, each M.O. sharing its bonding capacity between two bonds and contributing a one half bond order to each bond. We conclude that each bond in CO_2 will have a bond order of two and that the molecule should be nonmagnetic.

The experimental data for CO_2 are in rough agreement with this expectation. The average bond energy—one half the energy needed to pull the molecule apart into a carbon and two separated oxygen atoms—is 192 kcal/mole, the carbon–oxygen stretching force constant is 15.5 mdyne/Å, and the C—O bond length is 1.16 Å. These magnitudes compare rather well with the corresponding figures for CN (see Fig. 4-14) and suggest about a $2\frac{1}{2}$ order bond. As in the case of diatomic molecules, that is about the accuracy we can expect for this very simple approach, $\pm$ one-half a bond order.

And now we have nonbonding molecular orbitals added to our stable of ideas on bonding.

4-8 Wrap-up on M.O.'s

We have considered the molecular orbital approach first for H_2^+, then H_2, and we continued on to more ornate examples like CH_2 and CO_2. As we proceeded, more and more use was made of the word "approximate." As we continue to consider bonding, the molecules will become so complicated that all of our considerations will be approximate and they will become increasingly subservient to empirical facts. Yet, at this time in the evolution of the understanding of chemical bonding, quantum mechanical ideas finally unify and give coherence to most aspects of bonding. Chemists used to catalogue a dozen "bond types," empirically derived, and independently considered. Now we can see the origin of each type within a single framework that considers the parent valence orbitals, how they interact, and the orbital occupancy.

There is one important distinction to be made about the terminology of the molecular orbital approach. In its generic sense, the term molecular orbital indicates that, insofar as it is possible to discuss the properties of a molecule in terms of electrons in orbitals, each orbital is shaped by all of the nuclei in the molecule. Hence, in a molecule, the electrons certainly occupy *molecular orbitals*. In the same way that the electronic structure of an atom is soundly based on a consideration of its atomic orbitals, so also, the properties of a molecule can be expected to be evident and explainable through a consideration of its molecular orbitals. For both atoms and molecules, electron repulsion effects are sufficiently muted to be brought in only as a refining influence.

At another level of consideration we actually attempted to frame the molecular orbitals by taking parent atomic orbitals as our ingredients. Thus we wrote for H_2^+, and also for H_2, that there are two molecular orbitals, $\sigma(1s) = 1s_A + 1s_B$ and $\sigma^*(1s) = 1s_A - 1s_B$. However, the properties of H_2^+ and H_2 are different enough that these two molecules cannot have identical molecular orbitals. These specific representations are, then, only rough approximations. They give the principal atomic orbital parentage of each M.O. and they accurately show its nodal surfaces. But by no means can we assume that the simple superposition of two 1s orbitals at the equilibrium bond length of H_2^+ (or of H_2) accurately gives the electron distribution in the molecule.

With that caveat, let us turn to the prevailing representations of chemical bonds in complex molecules. We will see that they can now be given credence and meaning with the aid of our quantum mechanical point of view.

five approximate representations of chemical bonds

We now have a foundation to our understanding of chemical bonding. Chemical bonds form when the potential energy is lowered as atoms approach each other. Such energy lowering occurs when electrons can move near two nuclei simultaneously. There is a simple carry-over from the orbitals of atoms ("atomic" orbitals) to the orbitals of molecules ("molecular" orbitals). Even simple approximations to these molecular orbitals explain a variety of properties for the simple diatomic molecules: bond energy, bond length, vibrational frequency, and magnetism.

These considerations are all soundly based in quantum mechanics. They allow us to review the simpler empirical representations of chemical bonding that were developed by chemists to help predict and explain chemical formulas. There are several such schemes and they have proved to be extremely powerful aids. There is no arguing with success, and the preparation of a million and a half different chemical compounds shows that the chemist's empirical representations have provided an effective, guiding framework. Nevertheless, it is wise to reexamine these simple schemes in the light of quantum mechanics so that they can be used with judgement and insight.

5-1 General principles of bonding

It will be helpful to list again the principles that have been developed which, presumably, we will find implicit in the time-honored, empirical, bonding representations.

(i) Bonds form when one or more electrons can be simultaneously near two or more nuclei.

(ii) The bonds so formed are called chemical bonds when this electron sharing can involve "valence electrons" (the electrons most easily removed) shared in "valence orbitals" (the orbitals occupied, at least in part, by valence electrons).

(iii) Electrons that are simultaneously near two or more nuclei occupy orbitals that are molecular in scope, that is, molecular orbitals.

(iv) Electrons occupy molecular orbitals with the same occupancy rules as for atomic orbitals: two to an orbital, but with opposite spin!

(v) The strength of a bond between two atoms is related to the number of electrons shared by those atoms.

5-2 Inert gases, electron-pairs and chemical bonding

In the year 1916, G. N. Lewis formulated a theory of bonding that embodied a vast amount of empirical chemical information. Many chemists contributed, of course, to the development of the systematizing ideas that Lewis welded into a useful scheme. First the special stability of the inert gas atoms was noted. This seemed to be reflected in chemical formulas. Lewis saw that chemical formulas of many compounds could be understood with the simple rules that atoms strive to achieve inert gas electron distributions by sharing electrons, two at a time. It was as though making an atom look like an inert gas lowered its energy. Lewis developed this idea by writing chemical formulas with the aid of dots to represent the electrons around each atom. He did so without any distinction between s and p orbitals, possibly because they hadn't been invented yet.

(a) THE SIMPLEST EXAMPLES: H_2, F_2 AND HF

Consider the formation of the hydrogen molecule from two hydrogen atoms. This could be represented by electron dots as follows:*

$$\mathrm{H}\circ + \cdot\mathrm{H} \longrightarrow \mathrm{H}\overset{\circ}{\cdot}\mathrm{H} \qquad (5\text{-}1)$$

The formula shows that each H atom in H_2 is near two electrons, so each has the electron orbital occupancy of the inert gas helium. Hence a bond is formed. This is the total permitted occupancy of the 1s orbital, so no residual bonding capacity remains.

Fluorine is similar. Lewis wrote the reaction for its formation as follows:

$$\overset{\circ}{\circ}\underset{\circ\circ}{\overset{\circ\circ}{\mathrm{F}}}\circ + \cdot\underset{\cdot\cdot}{\overset{\cdot\cdot}{\mathrm{F}}}: \longrightarrow \overset{\circ}{\circ}\underset{\circ\circ}{\overset{\circ\circ}{\mathrm{F}}}\overset{\circ}{\cdot}\underset{\cdot\cdot}{\overset{\cdot\cdot}{\mathrm{F}}}: \qquad (5\text{-}2)$$

Notice that in the atom, the symbol F is shown surrounded by only seven electrons, though a neutral fluorine atom has nine. Two of these nine occupy the 1s orbital, and experience shows that they are quite ineffective in bonding. Hence these two are incorporated into the symbol for the nucleus (or, as it used to

* Electrons from different atoms will often be shown by different symbols (•, ∘, ×) to aid us in our bookkeeping. Of course, there is never any way of knowing one electron from another.

be called, the "kernel"). *In all electron dot formulas only valence orbital electrons are shown.*

As the two fluorine atoms come together to share two electrons, each nucleus "sees" eight electrons in its valence orbitals. Each atom now has the electron orbital occupancy of the inert gas neon, so a stable molecule results. No additional electrons can be accommodated, so no additional bonds will be formed.

The formation of hydrogen fluoride combines the "desire" of the H atom to achieve the helium electron configuration and of the F atom to achieve the neon electron configuration.

$$\mathrm{H}\circ + \cdot\underset{\cdot\cdot}{\overset{\cdot\cdot}{\mathrm{F}}}: \longrightarrow \mathrm{H}\overset{\cdot}{\circ}\underset{\cdot\cdot}{\overset{\cdot\cdot}{\mathrm{F}}}: \qquad (5\text{-}3)$$

We see that electron-dot formulas can be rationalized in terms of our guiding principles. The placement of electrons between the atoms signifies electrons that are simultaneously near two nuclei in an orbital that involves both atoms. Only the valence orbital electrons are shown: these occupy the orbitals in which significant energy lowering can result from electron sharing. The electron pairs tacitly express the Pauli Principle: two and only two electrons per orbital.

(b) COMPOUNDS OF OXYGEN: O_2, H_2O AND H_2O_2

By analogy to fluorine, we might picture the formation of O_2 as follows:

$$\overset{\circ}{\underset{\circ\circ}{\mathrm{\mathring{\mathrm{:}}O}}}\circ + \cdot\underset{\cdot}{\overset{\cdot\cdot}{\mathrm{O}}}: \longrightarrow \overset{\circ}{\underset{\circ\circ}{\mathrm{\mathring{\mathrm{:}}O}}}\overset{\circ}{\cdot}\underset{\cdot}{\overset{\cdot\cdot}{\mathrm{O}}}: \qquad (5\text{-}4)$$

The molecular formula on the right suggests the formation of a single bond (two electrons shared), but it also shows that neither oxygen atom has achieved the inert gas electron arrangement of neon. Hence the scheme indicates that another pair of electrons can be shared.

$$\overset{\circ}{\underset{\circ\circ}{\mathrm{\mathring{\mathrm{:}}O}}}\circ + \cdot\underset{\cdot\cdot}{\overset{\cdot}{\mathrm{O}}}: \longrightarrow \mathrm{:O}\!::\!\mathrm{O}\!\overset{\circ\circ}{\underset{\circ\circ}{}} \qquad (5\text{-}5)$$

Now the electron arrangement gives each oxygen atom eight valence orbital electrons (all four of the electrons between the atoms are shared) as in neon, and it correctly predicts that oxygen will have a double bond. The only failing of this simple scheme is that it predicts that the oxygen molecule will be diamagnetic. The paramagnetism which seems to require the single bond structure (5-4) has been quite an irritant to freshman chemistry lecturers for the last three decades until the molecular orbital approach became more widely used. In fact, the usual treatment was to avoid mentioning the paramagnetism of oxygen on the assumption that if it were ignored it would go away. The M.O. approach is needed to correct this failure of the Lewis scheme.

Turning to the reaction between hydrogen and oxygen, we find further developments. If a hydrogen atom shares an electron pair with oxygen, the hydrogen atom achieves the helium inert gas arrangement but oxygen does not achieve the neon arrangement.

$$\mathrm{H}\circ + \cdot\overset{\bullet}{\underset{\bullet\bullet}{\mathrm{O}}}: \longrightarrow \mathrm{H}:\overset{\bullet}{\underset{\bullet\bullet}{\mathrm{O}}}: \qquad (5\text{-}6)$$

The molecule OH is expected to be stable, but the oxygen atom has residual bonding capacity so it will be reactive. Unlike the oxygen molecule case, a single hydrogen atom cannot furnish a second electron to help out. Consequently OH can react with either another hydrogen atom or, indeed, with another OH molecule.

$$\mathrm{H}\circ + \mathrm{H}:\overset{\bullet}{\underset{\bullet\bullet}{\mathrm{O}}}: \longrightarrow \mathrm{H}:\overset{\mathrm{H}}{\overset{\circ\bullet}{\underset{\bullet\bullet}{\mathrm{O}}}}: \qquad (5\text{-}7)$$

$$\mathrm{H}:\overset{\bullet}{\underset{\bullet\bullet}{\mathrm{O}}}: + \mathrm{H}:\overset{\circ}{\underset{\circ\circ}{\mathrm{O}}}\,\overset{\circ}{\circ} \longrightarrow :\overset{\bullet\bullet}{\underset{\underset{\mathrm{H}}{\bullet\circ}}{\mathrm{O}}}\,\overset{\circ}{\circ}\,\overset{\overset{\mathrm{H}}{\bullet\circ}}{\underset{\circ\circ}{\mathrm{O}}}\,\overset{\circ}{\circ} \qquad (5\text{-}8)$$

Both reactions (5-7) and (5-8) give every atom an inert gas electron arrangement with no residual bonding capacity in any atom. Both molecules are expected to be stable, H_2O and H_2O_2. These are water and hydrogen peroxide, both familiar substances.

Thus, the Lewis scheme tells us the bonding capacity of an atom. As an atom forms bonds with other atoms, it still has residual bonding capacity as long as it has not achieved an inert gas population. A molecule that includes an atom having residual bonding capacity will be stable but extremely reactive.

(c) LINE REPRESENTATIONS

In each of the compounds O_2, H_2O and H_2O_2, we see that an oxygen atom needs to share two electron pairs with other atoms to reach the inert gas arrangement. To represent this bonding situation it isn't really necessary to show all the electrons. Chemists, lazy as they are, habitually show only the two bonding pairs, each pair indicated by a line drawn between the two atoms. One such line between two atoms means that they share one pair of electrons and hence they are connected by a "single bond." Two such lines between atoms means that two pairs of electrons are shared, so the bond has a bond order of two and is called a "double bond." Figure 5-1 compares this simpler

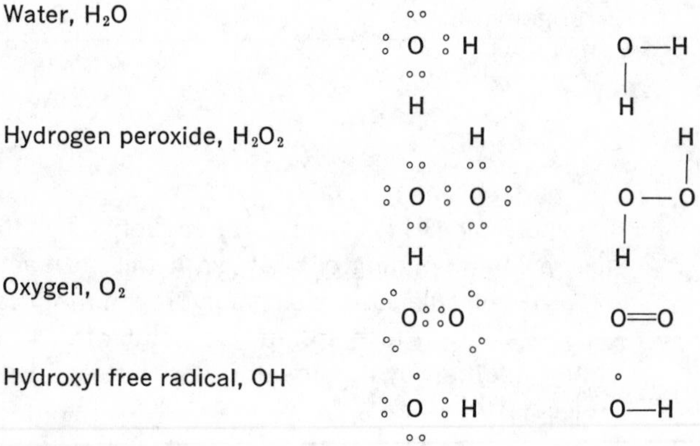

```
Water, H2O                          o o
                                  o O o H          O—H
                                  o o o            |
                                    H              H
Hydrogen peroxide, H2O2                H                H
                                    o o   o o           |
                                  o O o O o        O — O
                                    o o   o o      |
                                    H              H
Oxygen, O2
                                    O::O             O=O
Hydroxyl free radical, OH            o               o
                                  o O o H          O—H
                                    o o
```

Figure 5-1 Electron dots and line representations of bonds to oxygen.

representation with the electron-dot picture. In each of the first three relatively unreactive compounds there are two lines connected to each oxygen atom. In the last example, OH, one line is shown connecting the oxygen atom to the hydrogen atom and a dot displays the fact that oxygen has the capacity to share a second electron pair. This dot is the seventh, "unpaired" electron in the electron-dot representation. It indicates residual bonding capacity and hence a tendency toward extreme reactivity. Such a molecule, one with an unshared valence electron, is called a "free radical."

(d) CARBON COMPOUNDS WITH HYDROGEN

These ideas are readily exercised with the hydrogen compounds of carbon. If we consider first a single carbon atom, electron-dot formulas predict the *stability* of the molecules CH, CH_2, CH_3

and CH_4. The first three display residual bonding capacity for the carbon, so only the last, CH_4, should be unreactive.

$$\text{CH} \qquad \begin{array}{c} \circ \\ \circ\,\text{C} \,\colon\, \text{H} \\ \circ \end{array} \quad \text{or} \quad \begin{array}{c} \circ \\ \circ\,\text{C—H} \\ \circ \end{array} \qquad \text{stable, but reactive} \qquad (5\text{-}9)$$

$$\text{CH}_2 \qquad \begin{array}{c} \circ \\ \circ\,\text{C} \,\colon\, \text{H} \\ \circ\bullet \\ \text{H} \end{array} \quad \text{or} \quad \begin{array}{c} \circ \\ \circ\,\text{C—H} \\ | \\ \text{H} \end{array} \qquad \text{stable, but reactive} \qquad (5\text{-}10)$$

$$\text{CH}_3 \qquad \begin{array}{c} \circ \\ \text{H} \,\colon\, \text{C} \,\colon\, \text{H} \\ \circ\bullet \\ \text{H} \end{array} \quad \text{or} \quad \begin{array}{c} \text{H} \\ | \\ \circ\,\text{C—H} \\ | \\ \text{H} \end{array} \qquad \text{stable, but reactive} \qquad (5\text{-}11)$$

$$\text{CH}_4 \qquad \begin{array}{c} \text{H} \\ \bullet\circ \\ \text{H} \,\colon\, \text{C} \,\colon\, \text{H} \\ \circ\circ \\ \text{H} \end{array} \quad \text{or} \quad \begin{array}{c} \text{H} \\ | \\ \text{H—C—H} \\ | \\ \text{H} \end{array} \qquad \text{stable, unreactive} \qquad (5\text{-}12)$$

We recognize CH_4, of course, as methane, the household gas that is piped into our homes as a convenient fuel. The other three molecules, CH, CH_2 and CH_3, are all well known to chemists and they all have lifetimes of the order of microseconds at normal temperatures and pressures. Each of these species can, however, react with another identical species to form a molecule that does give an inert gas structure to each atom:

$$\text{CH} + \text{CH} \longrightarrow \text{C}_2\text{H}_2 \text{ (acetylene)}$$

$$\text{H} \,\colon \begin{array}{c} \circ \\ \text{C} \\ \circ \end{array} \circ \; + \; \bullet \begin{array}{c} \bullet \\ \text{C} \\ \bullet \end{array} \colon \text{H} \longrightarrow \text{H} \,\colon\, \text{C} \vdots\vdots\vdots \text{C} \,\colon\, \text{H} \qquad (5\text{-}13a)$$

or,

$$\circ \begin{array}{c} \circ \\ \text{C—H} \\ \circ \end{array} + \; \circ \begin{array}{c} \circ \\ \text{C—H} \\ \circ \end{array} \longrightarrow \text{H—C}\equiv\text{C—H} \qquad (5\text{-}13b)$$

$$\text{CH}_2 + \text{CH}_2 \longrightarrow \text{C}_2\text{H}_4 \text{ (ethylene)}$$

$$\text{H} \,\colon \begin{array}{c} \circ \\ \text{C} \\ \circ\bullet \\ \text{H} \end{array} \circ \; + \; \bullet \begin{array}{c} \text{H} \\ \bullet\circ \\ \text{C} \\ \bullet \end{array} \colon \text{H} \longrightarrow \begin{array}{ccc} \text{H} & & \text{H} \\ & \text{C} \colon\colon \text{C} & \\ \text{H} & & \text{H} \end{array} \qquad (5\text{-}14a)$$

or,

$$\begin{array}{c} \text{H} \\ \diagdown \\ \text{C} \\ \diagup \\ \text{H} \end{array} \begin{array}{c} \circ \\ \\ \circ \end{array} \; + \; \begin{array}{c} \circ \\ \\ \circ \end{array} \begin{array}{c} \text{H} \\ \diagup \\ \text{C} \\ \diagdown \\ \text{H} \end{array} \longrightarrow \begin{array}{ccc} \text{H} & & \text{H} \\ & \text{C=C} & \\ \text{H} & & \text{H} \end{array} \qquad (5\text{-}14b)$$

$$CH_3 + CH_3 \longrightarrow C_2H_6 \text{ (ethane)}$$

$$\begin{array}{ccccccc} & H & & H & & H & H \\ & \overset{\circ\bullet}{} & & \overset{\bullet\circ}{} & & \overset{\bullet\circ}{} & \overset{\bullet\circ}{} \\ H \,\vdots\, & C\,\circ & + & \bullet\, C \,\vdots\, H & \longrightarrow H \,\vdots\, & C \,\vdots\, & C \,\vdots\, H \\ & \underset{\circ\bullet}{} & & \underset{\bullet\circ}{} & & \underset{\circ\bullet}{} & \underset{\bullet\circ}{} \\ & H & & H & & H & H \end{array} \qquad (5\text{-}15a)$$

or,

$$\begin{array}{ccccccc} & H & & H & & H & H \\ & | & & | & & | & | \\ H- & C\,\circ & + & \circ\, C-H & \longrightarrow H- & C- & C-H \\ & | & & | & & | & | \\ & H & & H & & H & H \end{array} \qquad (5\text{-}15b)$$

All the carbon–hydrogen compounds predicted in this simple model, C_2H_2, C_2H_4 and C_2H_6, are familiar compounds commonly handled in tank-car amounts. Each has four bond-lines connected to each carbon atom.

Clearly the inert gas, electron population scheme is easy to apply and can predict lots of chemistry. Table 5-1 shows a variety of common molecular species, including some familiar ions, that have electron populations consistent with the Lewis electron-pair, inert gas structure model of bonding.

(e) TOO MANY STRUCTURES: "RESONANCE" SAVES THE DAY

Despite the many successes of the inert gas, electron-pair scheme for explaining chemical formulas, there are some difficulties. There are quite a few common substances whose existence is consistent with more than one acceptable inert gas arrangement. In every such case, any one of these structures implies that nonequivalent bonds could be formed, but they never are.

Sulfur dioxide, SO_2, is an example. Two electron arrangements can be drawn in which each atom has an inert gas arrangement nearby. Each structure implies one single bond and one double bond—one long bond and one short.

$$\begin{array}{cc} \overset{\circ\circ}{S} & \,\circ\!\!\!\!\!\circ\; \overset{\bullet\bullet}{\underset{\bullet\bullet}{O}} : \\ \underset{\bullet\bullet}{\overset{\circ\circ}{}} & \\ \cdot\!\cdot\, O \,\cdot\!\cdot & \end{array} \qquad \begin{array}{l} S-O \\ \| \\ O \end{array} \qquad (5\text{-}16a)$$

$$\begin{array}{c} \overset{\circ\circ}{S} \,\vdots\!:\! \overset{\bullet\bullet}{\underset{\bullet\bullet}{O}} \\ \overset{\circ\circ}{} \\ :\underset{\bullet\bullet}{O}: \end{array} \qquad \begin{array}{l} S=O \\ | \\ O \end{array} \qquad (5\text{-}16b)$$

Sulfur dioxide, however, is known to have two exactly equal sulfur—oxygen bond lengths in the structure (5-17)

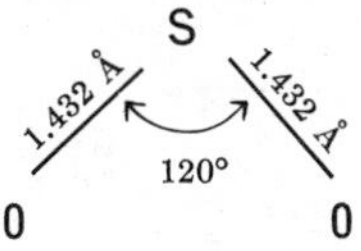

```
              S
   1.432 Å  /   \  1.432 Å
           / 120° \
          O        O
```
(5-17)

This discrepancy was not hard for chemists to resolve. The two structures (5-16*a*) and (5-16*b*) were already confusing because who was to say which was preferable? Energy-wise, the electrons ought to be just as comfortable in one as in the other. The solution is obvious: electrons are highly mobile, so they try to occupy *both* equivalent structures at the same time. Chemists introduce the term "resonance" to suggest that the electrons move back and forth between (5-16*a*) and (5-16*b*), as pictured in (5-18).

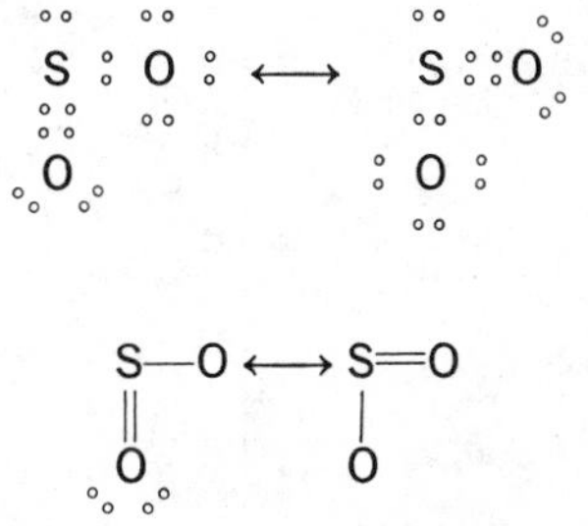

```
 oo   oo             oo    o
 S : O :    <--->    S ::O  o
 oo   oo             oo    o
 oo                 : O :
 O                    oo
o  o
```

```
S—O  <--->  S=O
‖           |
O           O
```
(5-18)

Now the two sulfur—oxygen bonds will be equal, averaged over the rapid electron movement back and forth between the two arrangements.* The actual molecule will not be like either of the "resonant structures," but rather, it will be like the superposition of the two. Instead of displaying one single and one double bond, the molecule will have two $1\frac{1}{2}$ order bonds.

Table 5-2 lists a number of examples that are in agreement with the known molecular structures *if* the resonance idea is used.

This rather awkward resonance idea is necessitated by a basic weakness in the electron-dot or simple electron-sharing view of bonding. In the light of our quantum mechanical principles, we can see that the individual structures (5-16*a*) and (5-16*b*) do not admit that the electrons move throughout the molecule, that is, in molecular orbitals. The superposition of

* This averaging over two (or more) hypothetical contributing structures is signified by the double-headed arrow in (5-18). That symbol should not be confused with the double arrow representation (⇄) used to designate an equilibrium situation between real and distinguishable chemical species.

Table 5-1 *Electron-Dot Formulas and Inert Gas Structures of Some Stable Compounds and Ions*

carbon monoxide CO	**carbon dioxide** CO_2	**sulfite ion** SO_3^{-2}	**sulfate ion** SO_4^{-2}
:C:::O:	:Ö::C::Ö:	$\left(:\ddot{O}:\ddot{S}(:\ddot{O}:):\ddot{O}: \right)^{-2}$	$\left(:\ddot{O}:\ddot{S}(:\ddot{O}:)(:\ddot{O}:):\ddot{O}: \right)^{-2}$
C≡O	O=C=O	$\left(O—S(—O)—O \right)^{-2}$	$\left(O—S(—O)(—O)—O \right)^{-2}$

ammonia NH_3	**nitrogen trifluoride** NF_3	**hydroxylamine** NH_2OH	**ammonium ion** NH_4^+
H:N̈:H with H above N	:F̈:N̈:F̈: with :F̈: above N	H:N̈:Ö: with H below N and H below O	$\left(H:\ddot{N}:H \right)^{+}$ with H above and H below N
H—N—H with H above N	F—N—F with F above N	H—N—O with H below N and H below O	$\left(H—N—H \right)^{+}$ with H above and H below N

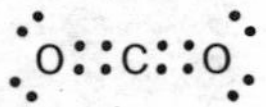

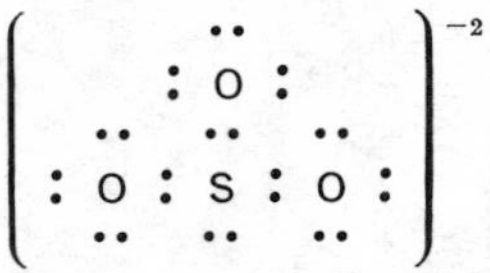

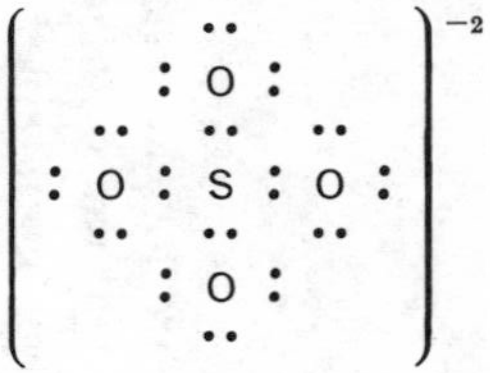

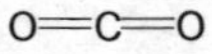

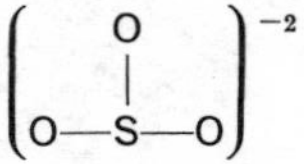

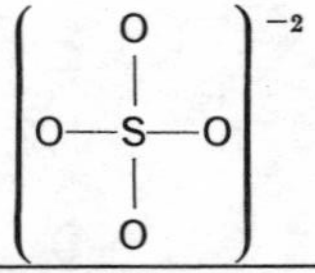

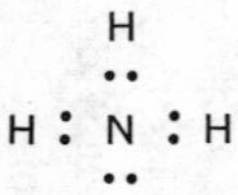

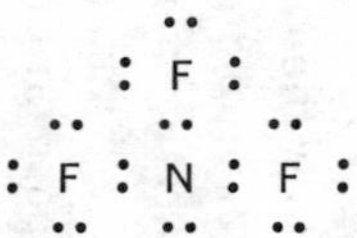

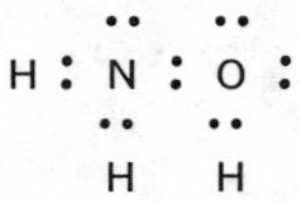

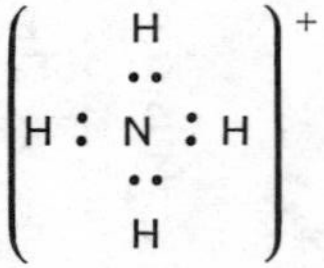

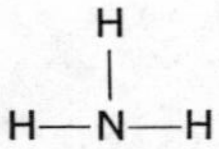

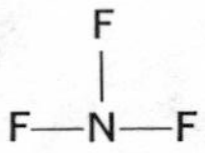

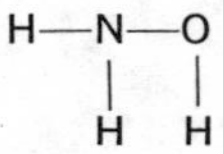

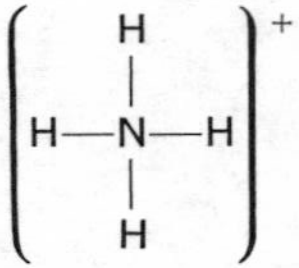

monofluoromethane	difluoromethane	trifluoromethane	tetrafluoromethane
CH_3F	CH_2F_2	CHF_3	CF_4

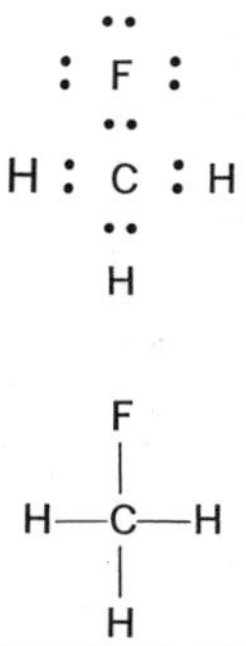

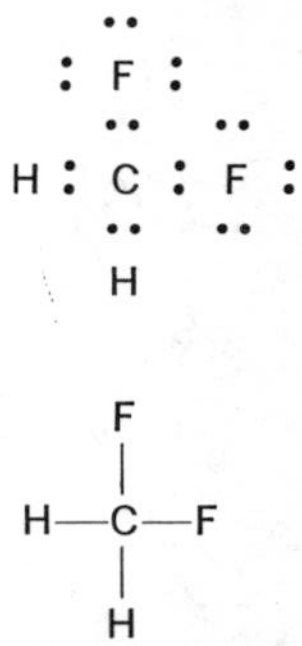

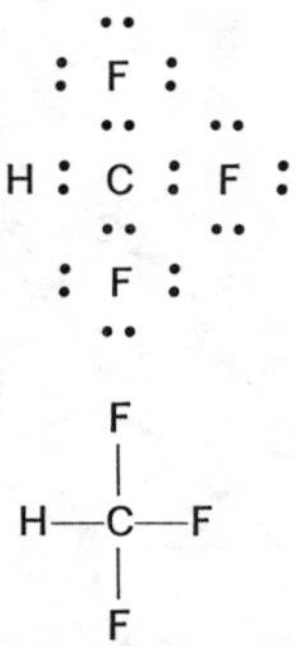

hypochlorous ion	chlorous ion	chlorate ion	perchlorate ion
ClO^-	ClO_2^-	ClO_3^-	ClO_4^-
$(Cl—O)^-$			

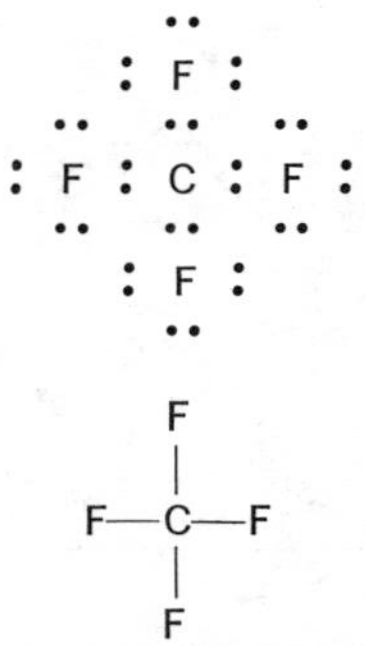

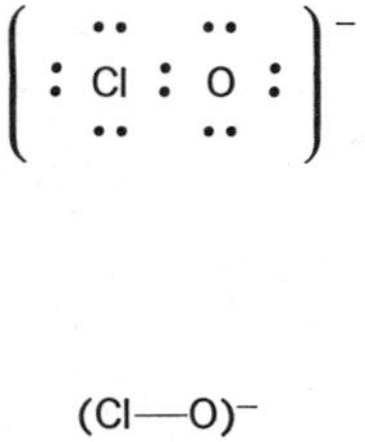

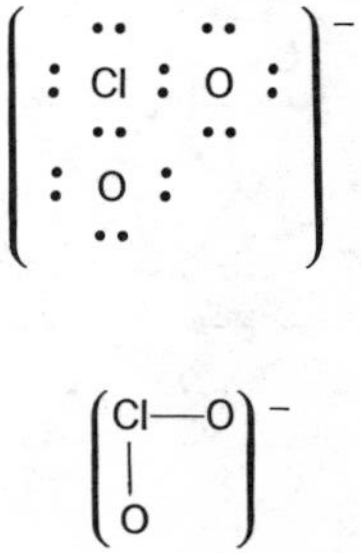

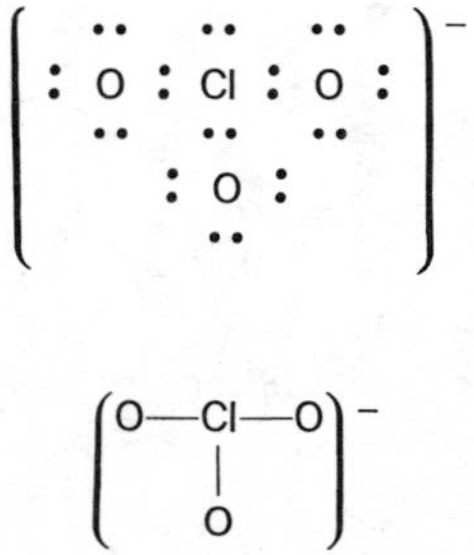

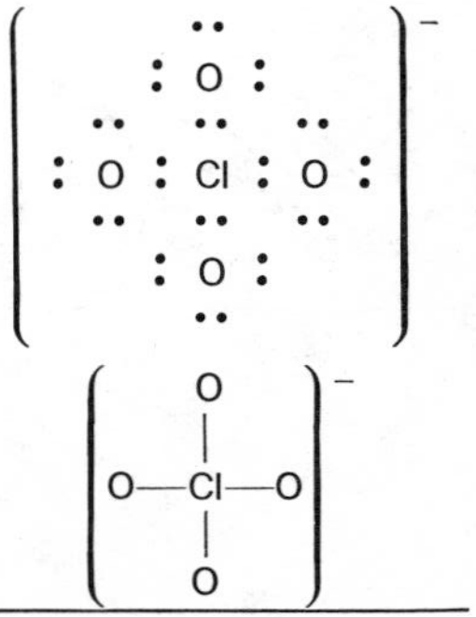

Table 5-2 Electron-Dot Formulas and Inert Gas Structures for Molecules That Involve Resonance

ozone
O_3

sulfur trioxide
SO_3

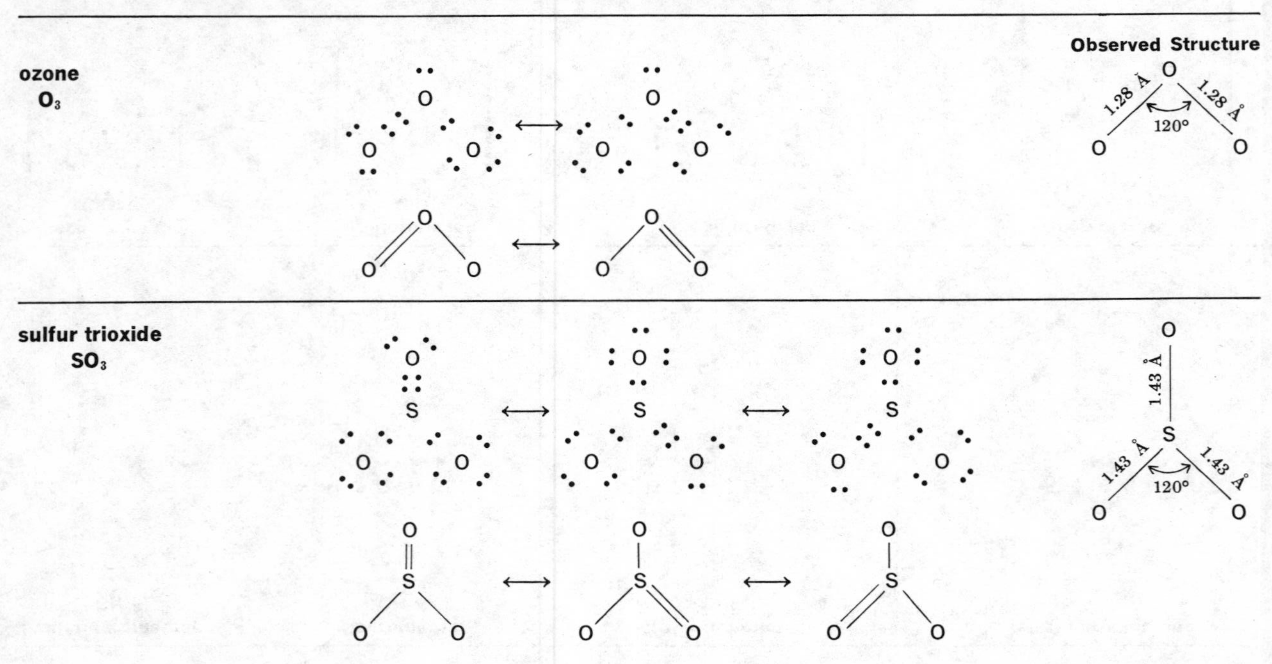

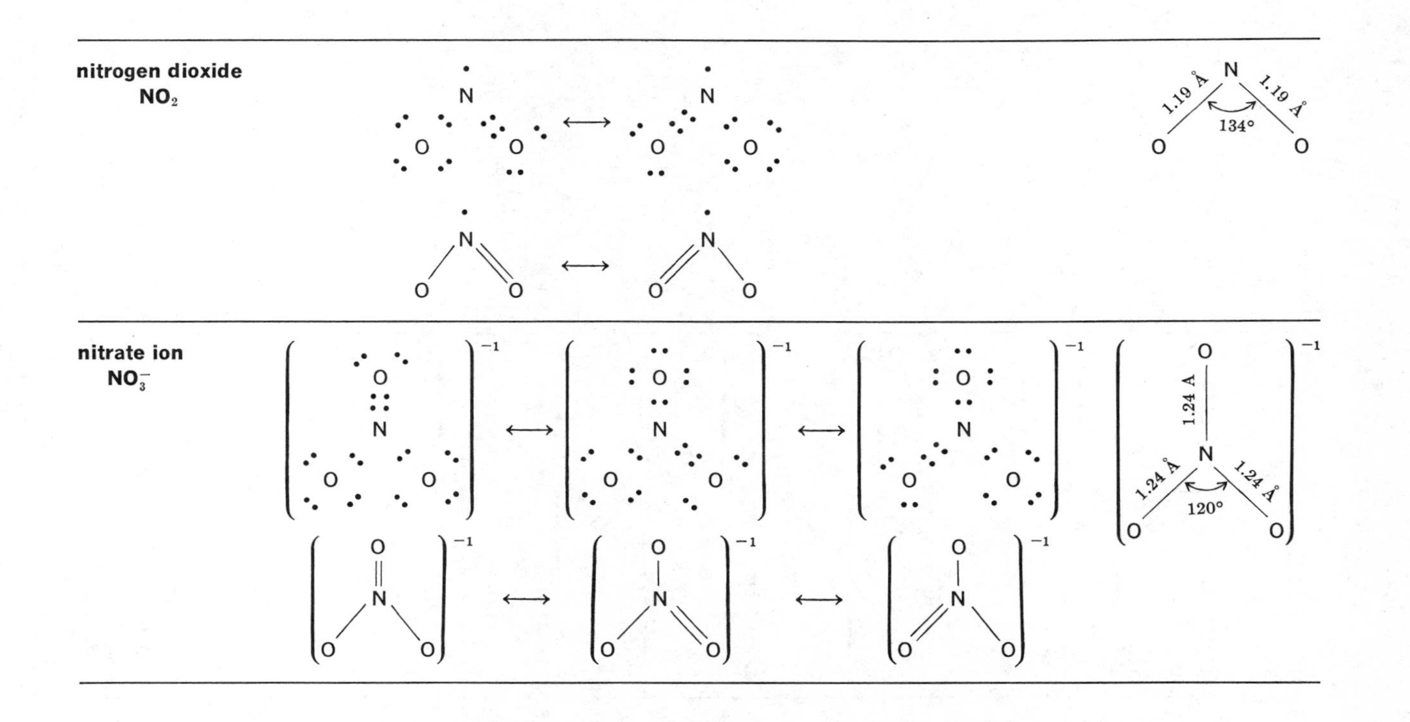

nitrogen dioxide
NO2
1.19 Å
134°
nitrate ion
NO3−
−1
1.24 A
1.24 Å
120°

Table 5-2 Electron-Dot Formulas and Inert Gas Structures for Molecules That Involve Resonance (Continued)

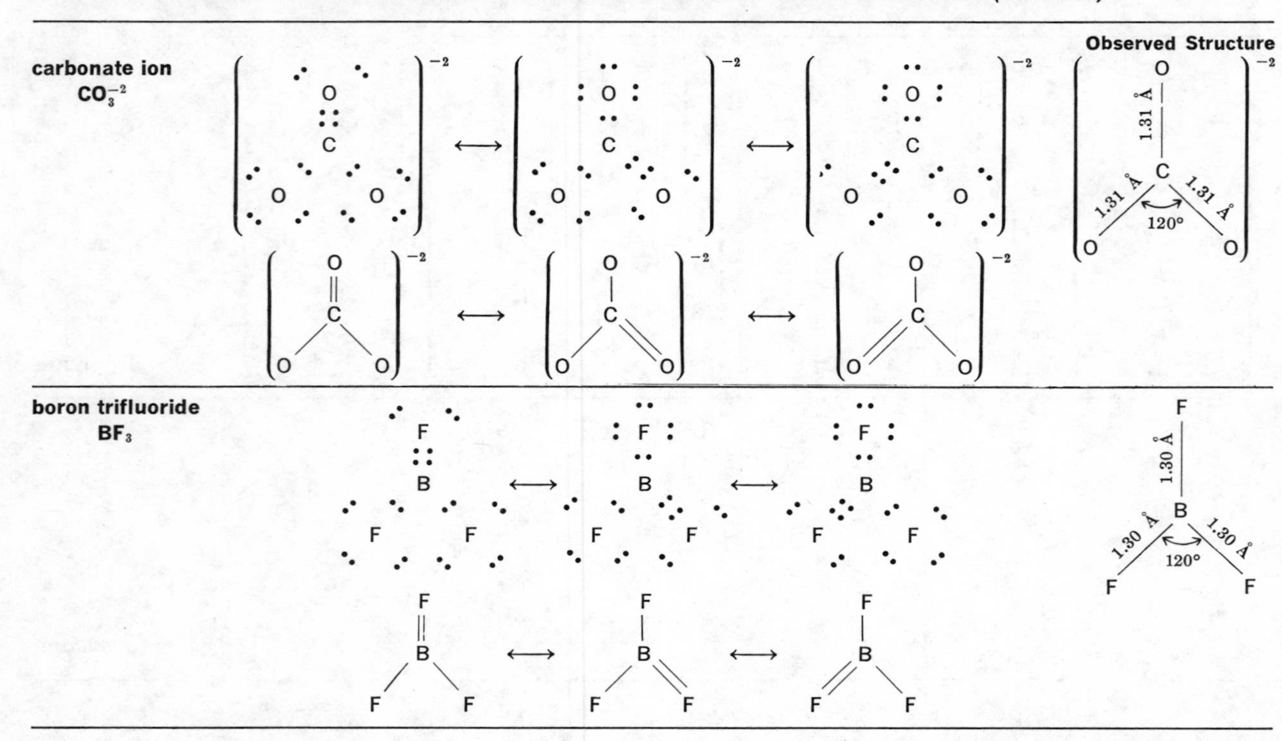

the two (or more) structures is necessary to correct for this deficiency.

In fact, a molecular orbital treatment of the bonding in any of the molecules shown in Table 5-2 (all of which are planar molecules) requires formulation of pi M.O.'s that extend over the whole molecule. Each of these π orbitals has a nodal surface in the plane of the molecule. Insofar as these π orbitals affect the bond strength, they affect all bonds in the same way, in accord with the experimentally observed bond equality. Quantum mechanics tells us that the two resonance structures shown in (5-18) (and those in Table 5-2) do not really exist. It was a fiction that the electrons move back and forth between the two electron arrangements. Yet we can see that the superposition of two (or more) fictional resonance structures does describe in a cumbersome way the nature of the real structure.

(f) ELECTRON STRUCTURES WITHOUT MOLECULES

Sometimes a perfectly reasonable electron structure can be drawn for a possible molecule but the molecule is not known. A simple example—until recently—was HOF, shown in (5-19)

hypofluorous acid HOF

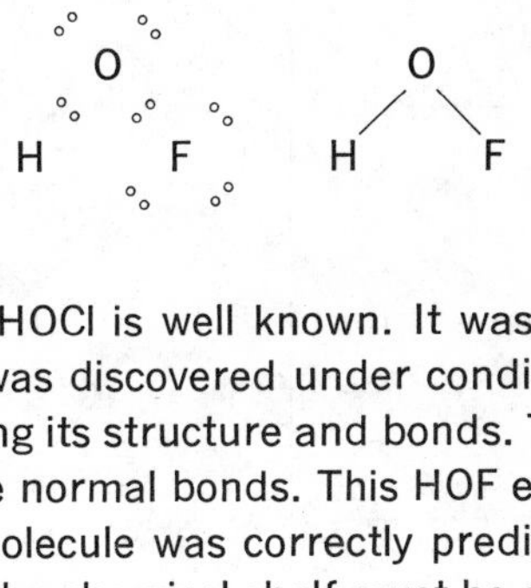

```
   °°  °°          O
    O          /      \
 °°   °°  °°      H        F
H      F
  °°   °°
```

(5-19)

The chlorine counterpart HOCl is well known. It was not until 1967, however, that HOF was discovered under conditions that permit inference concerning its structure and bonds. The molecule exists and it has quite normal bonds. This HOF example is reassuring because the molecule was correctly predicted, and our inability to stock it on the chemical shelf must be accounted for by the ease with which it converts into even more stable substances, as shown in (5-20)

$$2\,HOF(g) \rightarrow 2\,HF(g) + O_2(g) \qquad (5\text{-}20)$$

There are quite a few other examples, however, in which an electron structure predicts a molecular species that no one has been able to make. Table 5-3 lists a few. Some of these may be like HOF in that they exist as stable species but react readily to give more stable compounds. For example, both O_4 and H_2O_4 may exist at very low temperatures. The former, O_4, is at most 250 calories more stable than 2 O_2. With such small energy, it is

Table 5-3 Electron-Dot Formulas and Inert Gas Structures Predicting Some Molecules That We Don't Know (Yet?)

Molecular Formula	Electron-Dot Formula		Comments
FO_3^- (fluorate ion)	$[:\ddot{O}:\ddot{F}:O\ :\ddot{O}:]^-$	$[O{-}F(O){-}O]^-$	ClO_3^-, BrO_3^-, and IO_3^- all are well-known
H_2O_4	H O O O O H (electron-dot chain)	H—O—O—O—O—H	F_2O_4 known; synthesis of H_2O_4 claimed by Nekrasov
O_4	$:\ddot{O}:\ddot{O}:$ / $:\ddot{O}:\ddot{O}:$	O—O, O—O (square ring)	may exist, but if so, with energy of dissociation to 2 O_2 of only 250 calories
NO_3^{-3}	$[:\ddot{O}:\ddot{N}:\ddot{O}:\ :\ddot{O}:]^{-3}$	$[O{-}N(O){-}O]^{-3}$	nitrogen analogues to SO_3^{-2} and SO_4^{-2}; too much negative charge around a small atom? phosphorus counterpart, PO_4^{-3}, quite common
NO_4^{-3}	$[:\ddot{O}:\ :\ddot{O}:\ddot{N}:\ddot{O}:\ :\ddot{O}:]^{-3}$	$[O{-}N(O)(O){-}O]^{-3}$	

not clear that the O_4 aggregate will display the distinctive properties that justify its identification as a molecular species.

On the other hand, some of the molecules shown in Table 5-3 may not be stable at all. The fluorate ion FO_3^- may be energetically unstable because the central fluorine atom is too small to permit three oxygen atoms to crowd comfortably around it. The nitrogen analogues to sulfite and sulfate, NO_3^{-3} and NO_4^{-3}, probably don't exist because they require the accumulation of too much negative charge on a molecule with a small central atom.

In summary, when the inert gas, electron-dot representation predicts the existence of a molecule not found in the chemical catalogue, it may be waiting to be discovered under conditions that prevent its decomposition. On the other hand, it may be that the molecule really cannot exist, but then, chemists usually have at hand a rationale (as offered for FO_3^- and NO_3^{-3}). Nevertheless, these cases must be listed as liabilities of the Lewis bonding model.

(g) MOLECULES WITHOUT ELECTRON STRUCTURES

Far more serious are the instances of known molecular species that do not have any satisfactory inert gas, electron arrangement. For example, the triiodide ion I_3^- is a familiar species. Its existence accounts for the high solubility of I_2 in an aqueous iodide solution (despite the extremely small solubility of I_2 in pure water).

$$I_2(s) + I^-(aq) = I_3^-(aq) \qquad (5\text{-}21)$$

However, when we try to place electrons around the atoms, the inert gas arrangements do not encourage the ion to form:

$$:\overset{\circ\circ}{\underset{\circ\circ}{I}}:\overset{\circ\circ}{\underset{\circ\circ}{I}}: \quad :\overset{\circ\circ}{\underset{\circ\circ}{I}}:^{-} \longleftrightarrow {}^{-}:\overset{\circ\circ}{\underset{\circ\circ}{I}}: \quad :\overset{\circ\circ}{\underset{\circ\circ}{I}}:\overset{\circ\circ}{\underset{\circ\circ}{I}}: \qquad (5\text{-}22)$$

An unsatisfying escape from the absence of a suitable electron structure is to say that the electrons resonate between the two possible *unfavorable* structures, as suggested in (5-22).

Another case that is similarly disconcerting is the bifluoride ion HF_2^-. This species is also an easily identified constituent found in aqueous HF—F^- solutions:

$$HF(g) + F^-(aq) = HF_2^-(aq) \qquad (5\text{-}23)$$

Again, there is no satisfactory way to represent the joining of these atoms since the species coming together in (5-23) already have inert gas environments.

$$:\overset{\circ\circ}{\underset{\circ\circ}{F}}:H \quad :\overset{\circ\circ\,-}{\underset{\circ\circ}{F}}: \longleftrightarrow :\overset{-\,\circ\circ}{\underset{\circ\circ}{F}}: \quad H:\overset{\circ\circ}{\underset{\circ\circ}{F}}: \qquad (5\text{-}24)$$

Table 5-4 lists more molecules whose known molecular geometries are not readily explainable using inert gas electron arrangements. In every case, a molecular orbital approach helps us understand the molecular existence and properties.

(h) REPRISE ON THE INERT GAS, ELECTRON-PAIR REPRESENTATION

The Lewis scheme obviously has serious limitations. It provides no explanation for the paramagnetism of the oxygen molecule. Nitric oxide, NO, is another stockroom chemical that is not readily accommodated. The electron-dot picture displays residual bonding capacity and a double bond expectation, yet the substance is there on the chemical shelf and its nitrogen—oxygen bond is a $2\frac{1}{2}$ order bond. Then the molecules listed in Tables 5-2 and 5-3 further reduce our confidence in the electron-pair, inert gas description of bonds.

These shortcomings were not sufficient, however, to offset the usefulness of the Lewis electron-pair representation; it achieved universal adoption. They did, however, forecast the development of new and more generally applicable theories. One of these has recently been proposed by J. W. Linnett of Cambridge University.

5-3 Inert gases, electron quartets and chemical bonding

Linnett observed that incorporation of the empirical Pauli Principle into quantum mechanics places special restrictions on the movement of electrons with the same spin. In a given atom, all those electrons with the same spin correlate their motion so that, on the average, they tend to stay as far apart as possible. Electrons with opposite spin do not feel this spin-initiated effect. To be sure, any pair of electrons repel each other electrostatically, but those having the same spin are even further held apart.

Linnett attempted to introduce this rather subtle quantum mechanical result of the Pauli Principle into a simple bonding representation. He proposed that instead of considering the eight electrons around a neon atom as four pairs of valence orbital electrons, they should be considered as two quartets. Each quartet should be thought of as four electrons of the same spin placed approximately at the corners of a tetrahedron, the arrangement that places them as far apart as geometry allows.

Table 5-4 Known Molecules without Satisfactory Inert Gas Structures

Molecular Formula	Observed Structure	Molecular Formula	Observed Structure
chlorine trifluoride ClF_3	F–Cl (1.70 Å), Cl–F (1.70 Å), Cl–F (1.60 Å); 88°, 88° planar	iodine tetrachloride ion ICl_4^-	$[ICl_4]^-$: I–Cl 2.34 Å, 2.34 Å; 90°, 90° square, planar
xenon difluoride XeF_2	F–Xe (2.00 Å), Xe–F (2.00 Å); 180° linear	xenon tetrafluoride XeF_4	Xe–F 1.95 Å, 1.95 Å; 90°, 90° square, planar
diborane B_2H_6	B–H (terminal) 1.19 Å, 1.19 Å, 1.19 Å, 1.19 Å; B–H (bridge) 1.33 Å, 1.33 Å	aluminum chloride Al_2Cl_6	Al–Cl (terminal) 2.06 Å, 2.06 Å; Al–Cl (bridge) 2.21 Å, 2.21 Å, 2.21 Å, 2.21 Å
nitrosyl fluoride FNO	N–O 1.13 Å; F···N	difluorine dioxide O_2F_2	O=O 1.22 Å; F···O, O···F

The two tetrahedra in an isolated neon atom would have no set relationship to each other. Representing one spin by circles and the other by crosses, the neon atom would be the superposition of the following two pictures:

Ne and Ne (5-25)

Spin "up" Spin "down"

Thus two pictures are now needed to show the electron arrangement in the neon atom.

(a) SPIN QUARTETS AND SINGLE BONDS

Onward, Linnett ploughed, to diatomic molecules. When two atoms share electrons, they must still maintain around each atom the tetrahedral spacing of electrons of the same spin. This is readily pictured by drawing tetrahedra around each atom with shared corners, edges, or faces. Fluorine, for example, requires the following two pictures.

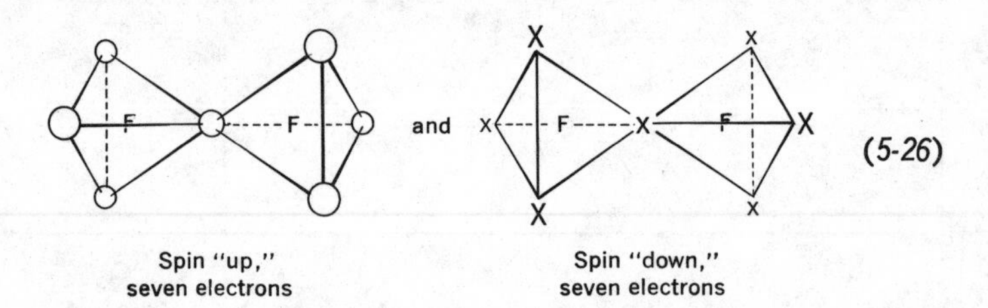

Spin "up," seven electrons — Spin "down," seven electrons

The superposition of the two electron arrangements in (5-26) accounts for all of the 14 valence-orbital electrons of F_2 and indicates one shared pair, hence a single bond in F_2.

Hydrogen, of course, is a special case. It has only one valence orbital and hence no tetrahedral tendencies. The molecules HF and H_2O would be simply as follows:

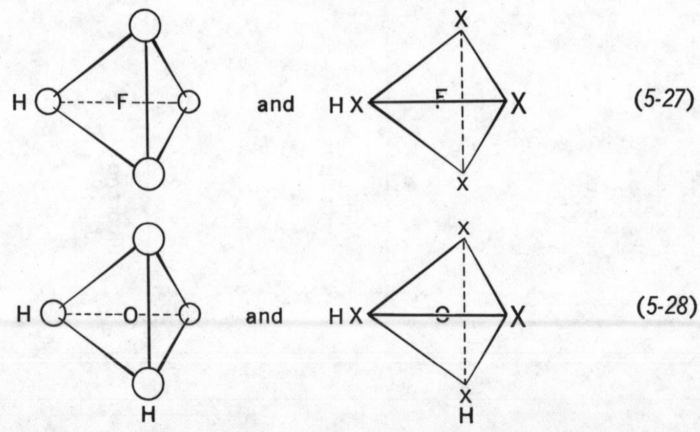

(b) SPIN QUARTETS AND THE MULTIPLE BONDS OF N_2, NO AND O_2

So far, the spin quartet has been nothing but a pain. To see its benefit, let's examine the nitrogen, nitric oxide, and oxygen molecules from this point of view. First consider N_2. This molecule has ten electrons, so each spin set must accommodate five electrons in intersecting tetrahedra. Two tetrahedra with a common face do the trick. The two spin pictures are as follows:

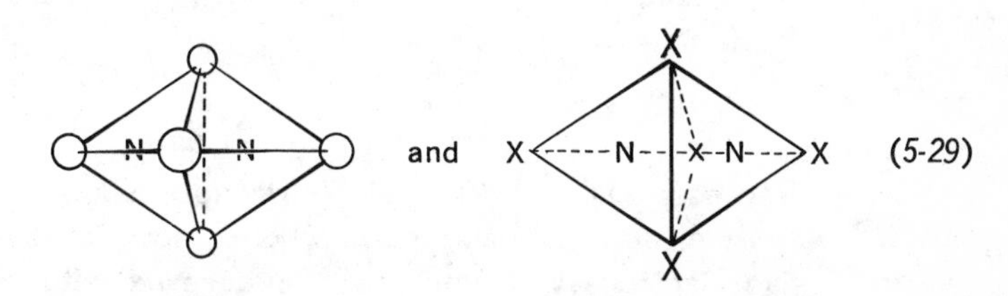

The superposition of these two representations places six electrons between the two atoms, so a triple bond is expected, as observed experimentally.

Nitric oxide, NO, has an odd number of electrons, so it cannot involve two identical representations. With 11 electrons, it requires a spin set of five to be coupled to a spin set of six. The appropriate diagrams are

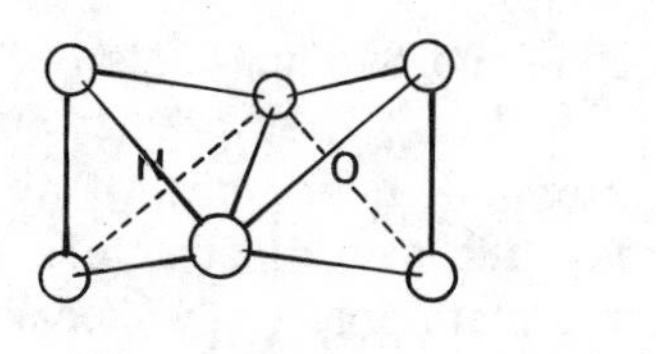

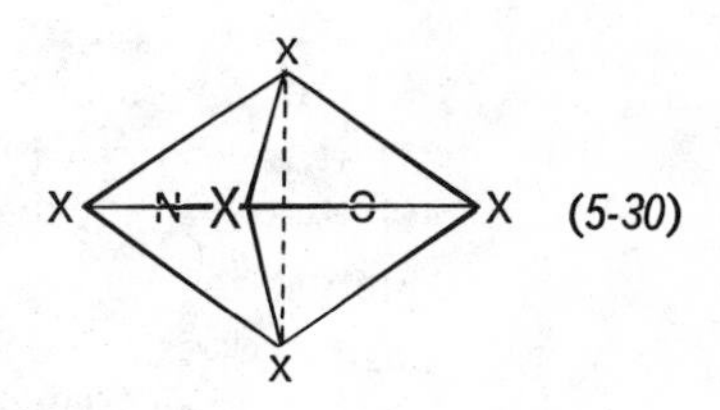

In the left-hand spin set, two electrons are placed between the nuclei, and in the right-hand set, three are so placed. With five electrons in the binding region, a $2\frac{1}{2}$ order bond is predicted, just as the doctor ordered.

Oxygen gives the opportunity for glory. With 12 valence electrons, the two spin sets can divide them equally, six electrons each.

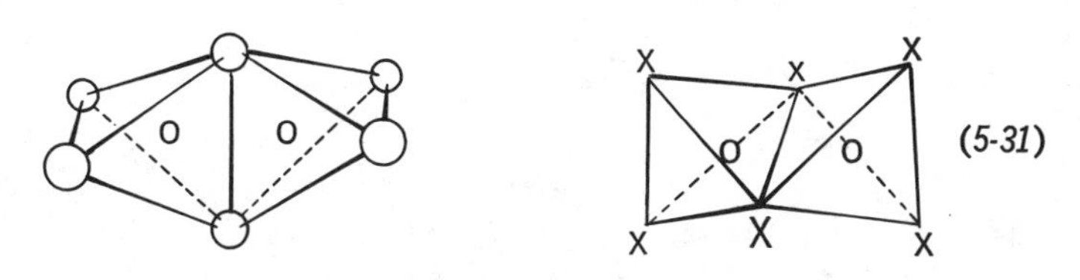

These two spin sets predict a double bond in O_2, but since there

are as many spins "up" as "down," the molecule would be diamagnetic. Not so fast, says Dr. Linnett. There is another way to handle twelve electrons in two spin sets. Perhaps one set has five electrons and the other has seven; one spin set would be like those in (5-29) and the other like those in (5-26)

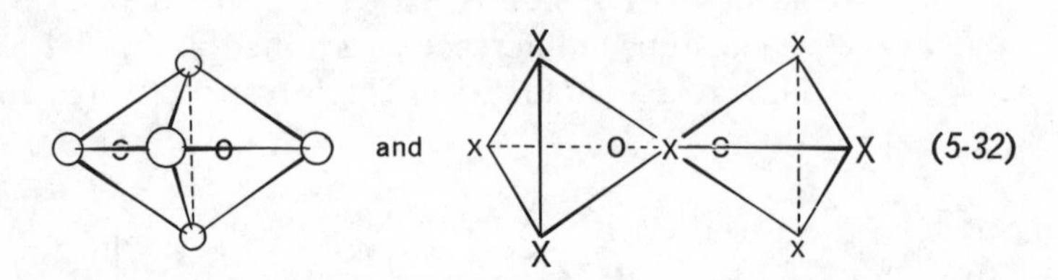

Representation (5-32) again shows four shared electrons, so the molecule still has a double bond. Now, however, there are five electrons with spin "up" and seven with spin "down." Thus the molecule can have a double bond and be paramagnetic as well.

Returning to nitrogen, we see that there is no analogous pair of structures. If we try to divide the ten electrons into sets of four and six, the spin set with six electrons will be one of those shown in (5-31), but there is no suitable way of filling the corners of two tetrahedra with only four electrons. So nitrogen must be triple bonded and diamagnetic, while oxygen can be double bonded and paramagnetic.

That leaves in question the two spin sets (5-31) for oxygen. What is wrong with these sets? They predict another double-bonded structure for oxygen, hence one of about the same energy as the paramagnetic oxygen structure (5-32). In fact oxygen does have such a state only a few kilocalories higher in energy than (5-32), in gratifying agreement with the spin-quartet prediction.

(c) SUMMARY ON ELECTRON-QUARTETS

There is some merit in the electron-quartet approach and it may enjoy increasing use. The tetrahedra seem cumbersome, but a shorthand notation has been developed to simplify the picture. For example, the representations for O_2 and NO can be written, in the form

O_2

$$\underset{\mathrm{x}}{\overset{\mathrm{x}}{-\!-}}\,\mathrm{O}\,\underset{\mathrm{o}}{\overset{\mathrm{o}}{-\!-}}\,\mathrm{O}\,\underset{\mathrm{x}}{\overset{\mathrm{x}}{-\!-}} \qquad (5\text{-}33)$$

NO

$$\overset{\mathrm{x}}{-\!-}\,\mathrm{N}\,\overset{\mathrm{o}}{=\!=}\,\mathrm{O}\,\overset{\mathrm{x}}{-\!-} \qquad (5\text{-}34)$$

to be compared with (5-32) and (5-30). In (5-33) and (5-34), the line indicates two electrons that are similarly placed near an atom or between two atoms but *not* in the same region of space. The x and o notations indicate spin. Hence (5-33) conveys the information that four electrons bind the atoms together to form a double bond, and that there are, net, two spins unpaired, to give a paramagnetic molecule. Nitric oxide has five electrons between the atoms to form a $2\frac{1}{2}$ order bond, and one net unpaired spin, to give, again, paramagnetism.

Even with this notational simplification, it is possible that the electron-quartet scheme was born twenty years too late to fall into everyday use. Like its electron-dot progenitor, the electron-quartet representation invokes the inert gas magic and ignores again the energy differences between the s and p orbitals. The quartets extricate the electron-dot scheme from some of its irritating failures in simple molecules (e.g., O_2, NO), but they are awkward to apply and sometimes require rather arbitrary choices. Meanwhile, the attention of most chemists has shifted toward very complex molecules and toward much more sophisticated understanding of simpler molecules. Nevertheless, the electron-quartet description points out a valuable quantum mechanical concept; that electron-spin relationships do furnish some important nuances of chemical bonding, even though the magnetic forces between the electron magnets are entirely negligible.

5-4 Pigeonholes and chemical bonding

Neither the Lewis electron pairs nor the Linnett electron quartets makes direct reference to the orbitals involved in the bonding. The molecular orbital scheme, however, indicates the importance of the orbital parentage of a bond. Discussion of the bonding in the transition elements requires explicit attention to the available orbitals. So it behooves us to be reminded of the pigeonhole scheme which portrays the valence orbitals and their occupancy. This scheme is of particular importance because it embodies the essence of the valence bond description of bonding and yet a clear connection to molecular orbitals can be shown.

(a) H_2 AND F_2

We begin by representing the valence orbitals as pigeonholes, spaced according to energy sequence, and by "filling" them with electrons as far as possible. The partially filled orbitals are then considered to interact to form bonds according to a ritual of rules. We will examine, first, two examples that present no difficulties, H_2 and F_2.

For H_2, each atom has one valence orbital and one valence electron. Higher orbitals are ignored, so the pigeonhole diagram is very simple. Each valence orbital is shown as a circle* (a pigeonhole) with a slash to indicate that it contains one electron —it is half-filled.

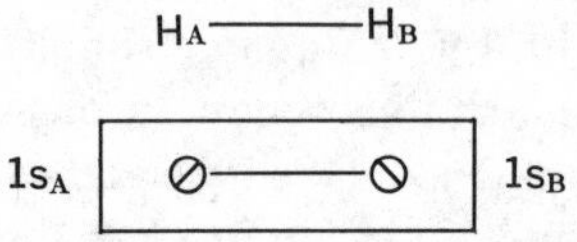

$$\begin{array}{ccc} & H_A\text{——}H_B & \\ 1s_A & \boxed{\oslash\text{——}\oslash} & 1s_B \end{array} \qquad (5\text{-}35)$$

According to this scheme, the situation portrayed in (5-35) has potentiality for bond formation. Each atom has a valence orbital containing a single electron which it can share with the other atom. This potentiality is designated by the rectangular box and the bond is signified by a line between the pigeonholes.

Fluorine is only a bit more complex. Its pigeonhole representation is as follows:

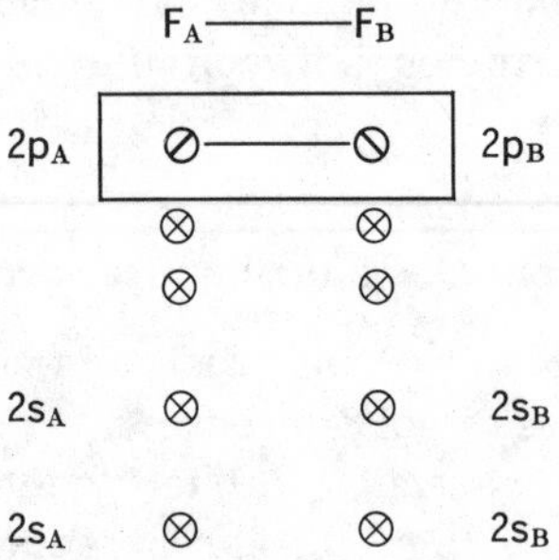

$$\begin{array}{cccc} & F_A & \text{——} & F_B & \\ 2p_A & \oslash & \text{——} & \oslash & 2p_B \\ & \otimes & & \otimes & \\ & \otimes & & \otimes & \\ 2s_A & \otimes & & \otimes & 2s_B \\ 2s_A & \otimes & & \otimes & 2s_B \end{array} \qquad (5\text{-}36)$$

(In the printed diagram the $2p_A$—$2p_B$ pair is enclosed in a rectangular box.)

Only the last 2p orbital of each fluorine atom satisfies the condition for bonding—a half-filled occupancy. All of the other orbitals are filled completely and play no role in the bonding. The box shows that one bond can be formed.

(b) CONNECTION TO VALENCE BOND AND MOLECULAR ORBITAL SCHEMES

The two examples, (5-35) and (5-36), are sufficient to show the conceptual connection and differences between valence

* Some pigeons prefer square pigeonholes and arrows up or down to indicate spin. Thus (5-35) could be written: $\boxed{\uparrow\text{—}\downarrow}$. This representation contains no new information but neither does it contain less, so it can be used by those who have square tendencies.

bond and molecular orbital schemes. In the former, a normal bond requires interaction between two half-filled valence orbitals. The picture suggests that atom A "possesses" one of the electrons which it shares with atom B in return for the privilege of sharing the valence electron "possessed" by B. This can be compared to the manner in which (5-35) would be altered to make it a molecular orbital approach:

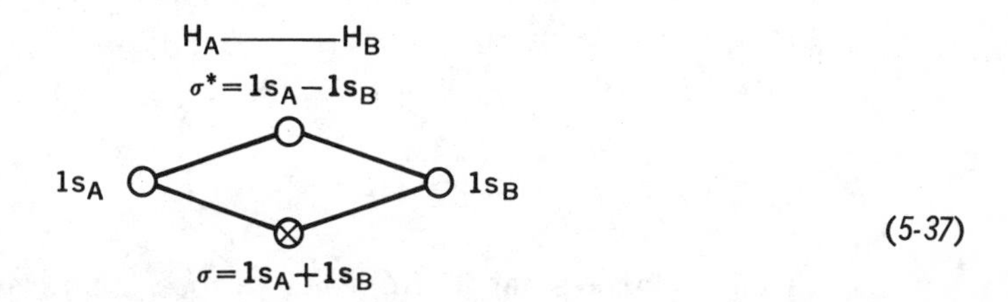

(5-37)

In (5-37), the orbitals $1s_A$ and $1s_B$ are considered to interact to form two molecular orbitals σ and σ^*. One is bonding and the other is antibonding, as suggested by the vertical spacing. *Now* the valence electrons are added, not one in $1s_A$ and one in $1s_B$, but *both in* $\sigma = 1s_A + 1s_B$*!*

The fluorine pigeonhole picture (5-36) also can be recast into the molecular orbital representation:

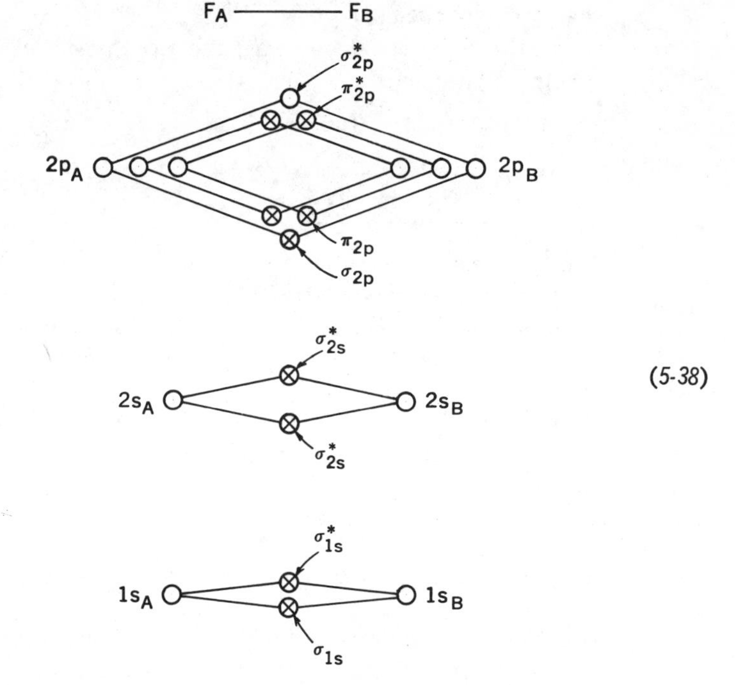

(5-38)

The diagram (5-38) shows that the M.O. "reason" that the filled $2s_A$ and $2s_B$ orbitals in (5-36) do not contribute to bonding is because filling them both is equivalent to filling both the bonding and antibonding σ_{2s} and σ^*_{2s} orbitals in (5-38).

(c) PIGEONHOLES AND HALF-ORDER BONDS

It is only a small intellectual leap to apply the diagram (5-35) to the molecule ion H_2^+.

$$(H_A \text{-------} H_B)^+$$

$$1s_A \boxed{\oslash \text{-------} \bigcirc} 1s_B \tag{5-39}$$

At this point, we depart from the valence bond view, since it requires that both $1s_A$ and $1s_B$ be half-occupied for bond formation. The molecular orbital interpretation of (5-39) is not discomfitted, however. If we think of (5-39) in connection with the M.O. diagram (5-37), we see what (5-39) should be interpreted to mean. The single electron should not be considered to be associated preferentially with H_A. Instead, it should be considered to half fill the bonding molecular orbital that can be formed from the orbitals within the rectangular box. The half-order bond that results is indicated by the dashed line connecting the two pigeonholes.

In an exactly analogous way, the He_2^+ molecule ion can be accommodated.

$$(He \text{------} He)^+$$

$$1s_A \boxed{\otimes \text{------} \oslash} 1s_B \tag{5-40}$$

The three electrons within the rectangular box are sufficient to occupy fully the bonding molecular orbital that can be formed from $1s_A$ and $1s_B$ and also to half occupy the antibonding orbital. The net bond order is $1 - \frac{1}{2} = \frac{1}{2}$, again indicated by a dashed line.

Thus we extend the pigeonhole scheme by admitting one- and three-electron occupancies with implied half-order bond contributions.

(d) PIGEONHOLES AND THE FIRST-ROW DIATOMICS

Now we are prepared to apply the pigeonhole representation to the diatomics, including those that are awkward to represent by traditional electron-pair and valence bond descriptions. The series C_2, CN, N_2, NO, and O_2 demonstrate the ease with which this simple scheme describes these molecules. (The 1s orbitals are omitted to simplify the diagrams.)

Nothing exciting turns up in the series until we reach oxygen. Then, with the option of half-order bonds, two possibilities appear. Both (5-45) and (5-46) have net double bonds so neither is to be favored on energy grounds. The traditional valence bond view would select (5-45) but Nature prefers (5-46). Our molecular orbital interpretation of the pigeonhole representations tells us why. In both (5-45) and (5-46), antibonding orbitals must be occupied but, as shown in Section 4-3, (5-46) has lower energy because electron repulsions are diminished.

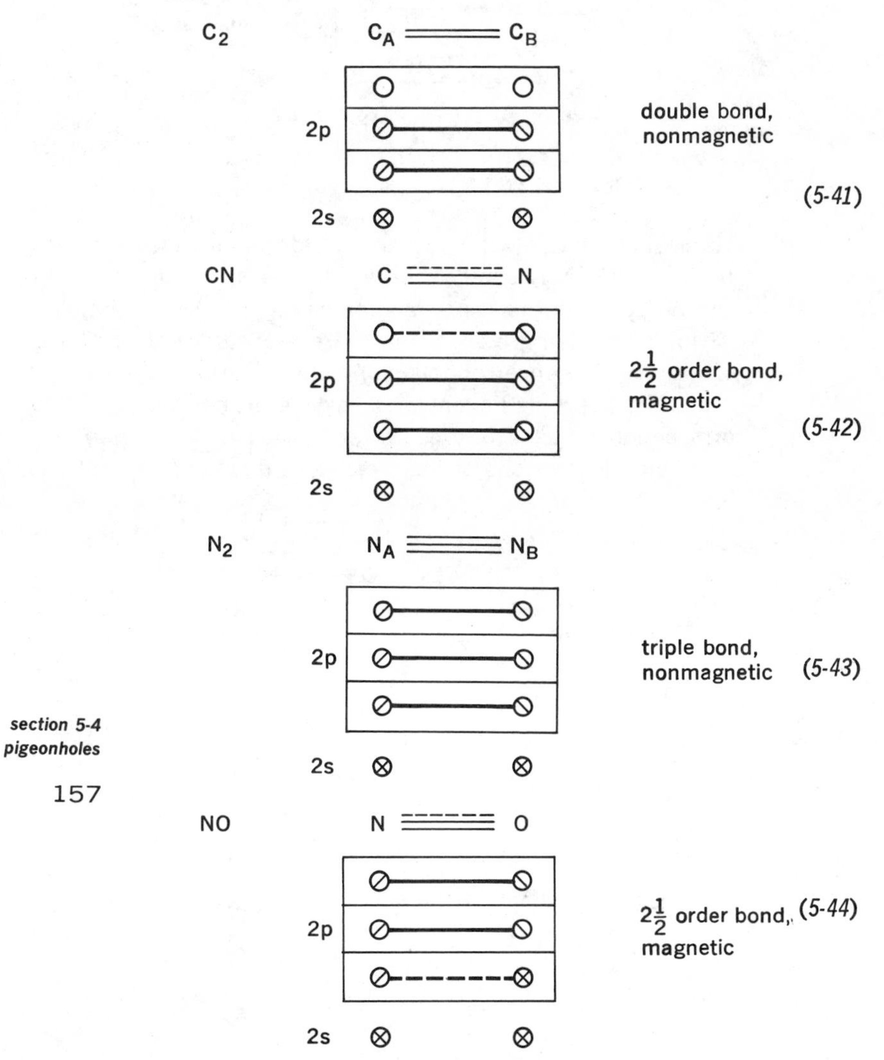

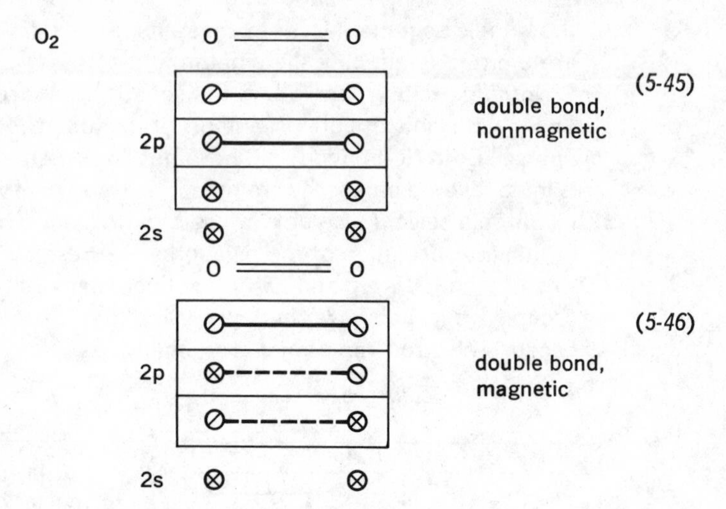

(e) THE HYDRIDES AND FLUORIDES OF THE FIRST-ROW ELEMENTS

It is clear how the pigeonhole diagrams for the hydride molecules HF, H_2O, and NH_3 should be written. Selecting the simplest possible scheme, the 2s orbitals can be regarded as noninfluential in the bonding as long as there are sufficient electrons to utilize the valence orbitals completely. The hydrogen atom, of course, has only its single 1s valence orbital with which to form bonds. In each of these three molecules, then, there are eight valence electrons at the molecule's disposal.

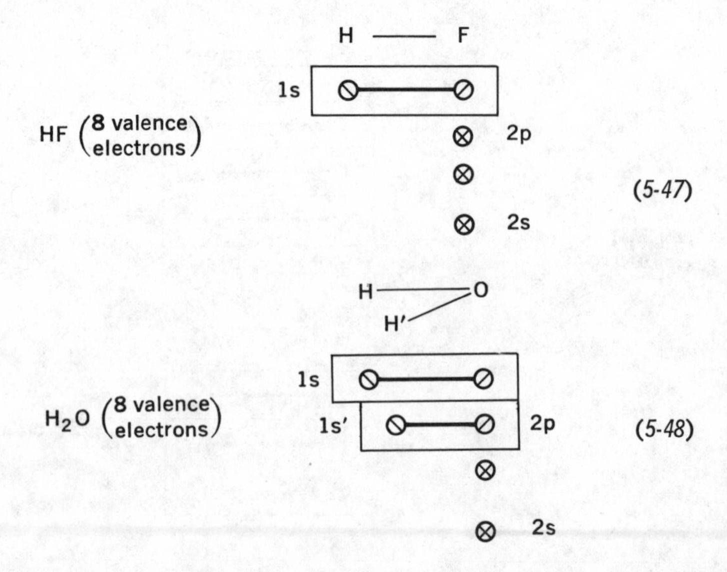

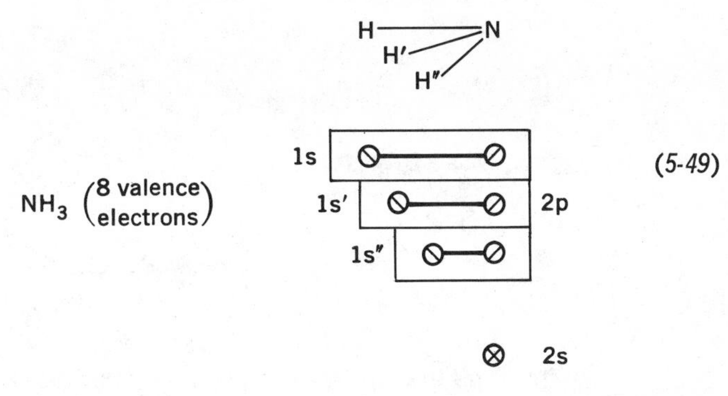

The representations (5-47), (5-48) and (5-49) are sufficient to explain the formulas of these three hydride molecules. The shapes of the boxes are not important, they merely remind us of the orbital parentage of the bonds. The disregard of the 2s orbitals does not mean that they make no contribution whatsoever; it merely means that they are not needed in the simplest scheme that explains the number of bonds.

The plot becomes more interesting when we reach the carbon atom. Now, if we consider the carbon 2s orbitals to be filled and unimportant in the bonding, there are two valence electrons remaining with which to form shared, electron-pair bonds. The methylene molecule CH_2 results.

H—C
H′
1s 2p
1s′
2s

CH_2 (6 valence electrons) (5-50)

Thus the methylene molecule should be a stable, diamagnetic molecule. However, the vacant $2p_z$ orbital implies that residual bonding capacity exists and the molecule will be quite reactive. The empty orbital also implies that CH_2 might have another state in which one of the 2s electrons is promoted to the $2p_z$ orbital if the energy price to be paid for this promotion is more than neutralized by removing some of the electron–electron repulsion that goes with having two electrons in the same orbital (the 2s orbital). In fact, this state of CH_2 is well known, and it is estimated to be a few kilocalories per mole lower in energy than the state shown above in (5-50). The lower state (5-51) is paramagnetic and, with two half-filled valence orbitals, it is extremely reactive.

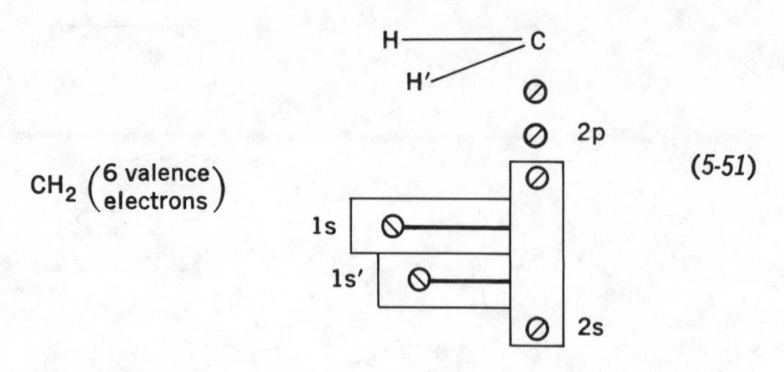

This time the box has been drawn to suggest that each hydrogen atom is held by a bond whose parentage is part 2p and part 2s—that is, to a "hybrid" orbital. The reason for this is that experimentally the two C—H bonds are found to be identical. More will be said about this in Chapter Six.

Experimentally, methylene is found to be stable enough to have a transient existence (its ultraviolet absorption spectrum has been recorded), but it is so reactive that it lasts only a few microseconds. Within that time, if it finds no other molecule with which to react (such as oxygen, nitrogen, carbon monoxide, any other hydrocarbon, etc.), two CH_2 molecules will react with each other to form ethylene, C_2H_4. More pertinent to our discussion is that successively, two more hydrogens might be added, stolen from some less reactive, hydrogen-containing molecule. If one is added, the resulting molecule, methyl, still has a residual bonding capacity:

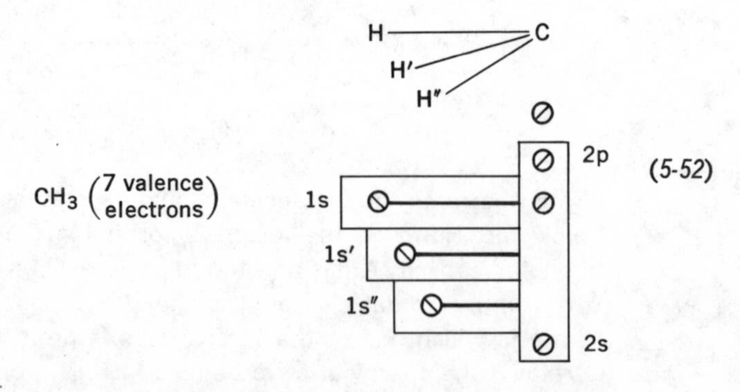

The single unpaired electron in the $2p_z$ valence orbital causes methyl to be a free radical and to be extremely reactive. Though not quite so reactive as CH_2, the methyl radical also has transient lifetimes less than a millisecond, depending upon the molecules it encounters. If it meets nothing with which to react, two methyls will combine to give ethane, C_2H_6.

Another possible fate for CH_3 is that it might gain a fourth hydrogen, stealing it from some other less powerful molecule. The result is methane, CH_4.

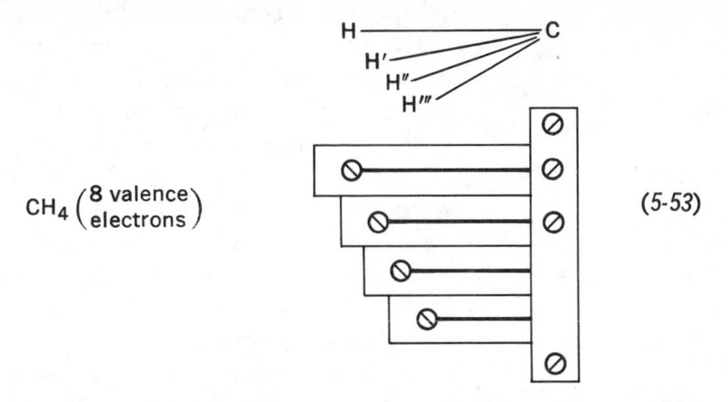

(5-53)

At last we have a carbon–hydrogen molecule with no residual bonding capacity: all the valence orbitals are fully occupied by shared electron pairs. This molecule, CH_4, is not as hyper-reactive as CH_2 and CH_3. In fact, one can purchase a tank car full of methane at a few cents a pound.

(f) SO WHY BOTHER THE PIGEONS?

There is an obvious difference in emphasis that distinguishes the pigeonhole representations of bonding from either of the electron-dot representations. The electron-dot approaches, both that of Lewis and of Linnett, treat the 2s and 2p orbitals as though they are identical. This greatly simplifies the business of predicting bonding capacity and residual bonding capability. However, there is no doubt that the 2s and 2p orbitals differ quite significantly in energy, as do the 2p σ and 2p π orbitals. The effects of these energy differences are lost in the electron-dot schemes.

The pigeonhole diagrams are justified, then, by the belief that the atomic orbital parentage of an M.O. is influential in determining the properties of that M.O. This belief is not shared by all chemists today, despite the successes displayed in Chapter Four. It behooves the budding chemist to cover all bets by learning to apply all the representations given in this chapter. That this is so is demonstrated once again by current views on molecular geometry. In explaining molecular shapes, two views are in vogue, one of which directly relates to the electron-pair point of view, and the other of which continues to attribute a role to the orbital parentage of the occupied M.O.'s. These two contrasting views are presented in Chapter Six.

six molecular geometry and molecular dipole moments

Molecular geometry proves to be almost as important as bond energies in fixing chemical properties. This seems to be particularly so in biological functions. The outcome of competing reactions can be entirely determined by spatial relationships connected with the molecular configurations of the competitors. A molecule with the wrong shape may not react at all, whereas a similar one, differently configured, may react in a trice. Many biological growth and reproductive processes depend upon "lock and key" shape relationships between reactants and products. Furthermore, many of our newer drugs are effective because of their similarity in molecular appearance to some misbehaving biological molecule. The look-alike drug takes the place of the offending molecule and blocks its action.

So chemists are vitally interested in molecular geometries. The experimentalist wants to be able to measure them and all of us want to generalize the results into a guiding theory. This theory should allow us to understand the bond angles and molecular shapes that are known and to predict with reasonable confidence those that are not.

In this chapter we'll examine two explanations of the bond angles observed among the elements in the first three rows of the periodic table. One model concentrates on the atomic-orbital parentage of the bonds—as does the pigeonhole representation of bonding. This model is called "Orbital hybridization." The second model ignores the orbital parentage and concentrates only on the number of pairs of valence electrons near each atom—like the electron-pair representation of bonding. This approach is called the "Electron repulsion" model. We'll examine these two theories and then consider the implications of molecular shape on the molecular dipole moment.

6-1 Orbital hybridization and bond angles

There is, of course, only one question to ask about the molecular geometry of a diatomic molecule: What is its bond length? We have already discovered that, for a given pair of atoms, the higher the bond order, the shorter the bond. This relationship is entirely consistent with our representations of bonding as presented in the dot diagrams (5-8) and (5-5) for H_2O_2 and O_2 (oxygen bond lengths 1.47 and 1.21 Å for single and double bonds, respectively), as presented in the spin-quartet diagrams (5-31) and (5-32) for O_2 and (5-29) for N_2, and as presented in the pigeonhole representations (5-41) to (5-46) for C_2, CN, N_2, NO and O_2. This bond length–bond order correlation (the higher the bond order, the shorter the bond) applies equally well to polyatomic molecules.

(a) THE MOLECULES H_2O AND NH_3

Molecular structures become more interesting when we pass on from diatomic molecules to polyatomics like H_2O and NH_3. The water molecule could be imagined to be linear or it could be bent. Its two oxygen—hydrogen bonds could be of equal or unequal length. Ammonia could be pyramidal or it could have all its atoms in the same plane; its NH bonds could all be of equal length or they could differ. We need some guiding principles to tell us which of these to expect.

Our pigeonhole picture gives us some guidance and it correlates the known facts. Examining representation (5-48) for H_2O, we see that each hydrogen is shown sharing electrons with an oxygen 2p orbital. There is no reason to expect one 2p orbital to be more effective than the other, so the picture implies two equivalent OH bonds (as they are). Furthermore, we might expect that the orbital geometry of the 2p orbitals would be represented in the bonding geometry. The 2p orbitals lie at right angles, so the picture implies a bent molecule with a bond angle near 90°. Figure 6-1 compares these expectations with the experimental facts: *the water molecule is a bent molecule and its angle is 104.5°.*

Turning to ammonia, (5-52), we see that each hydrogen is shown sharing electrons with a nitrogen 2p orbital. We expect, then, three equivalent bonds. The perpendicular geometry of p orbitals suggests a pyramidal structure for NH_3. *The ammonia molecule is a pyramid and its bonds have 107° angles between them.* Once again, the qualitative aspects of the molecular shape are represented, even if the bond angle expectations are 15–17° lower than observed.

(b) THE METHYLENE MOLECULE CH_2

We deduced two states for CH_2, a diamagnetic state (5-50) and a paramagnetic state (5-51). For spectroscopic reasons, these states are called "singlet" and "triplet" states, respectively.

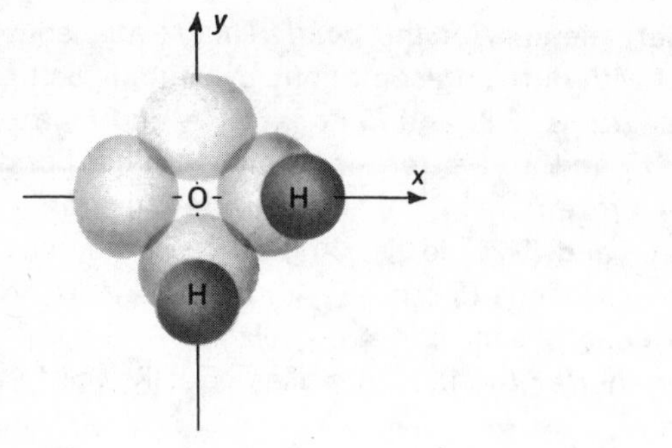

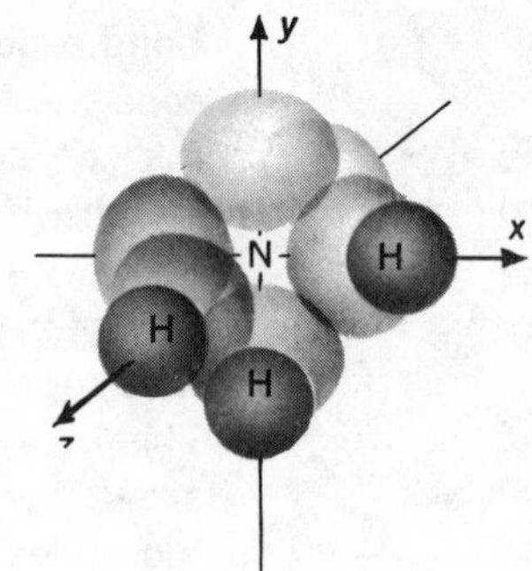

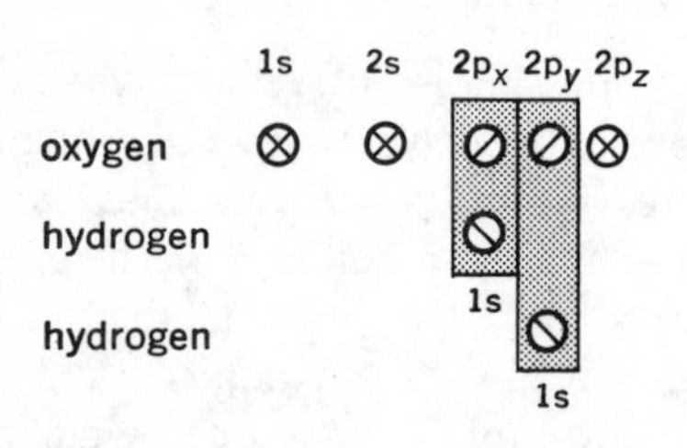

1s 2s $2p_x$ $2p_y$ $2p_z$
nitrogen
hydrogen
hydrogen
hydrogen
1s
1s
1s

H_2O: bent, 90° bond angles, equivalent bonds

NH_3: pyramidal, 90° bond angles, equivalent bonds

(*a*) Orbital prediction

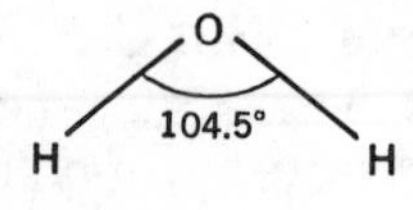

N
H
107°
H
H

H_2O: bent, 104.5° bond angle, equivalent bonds

NH_3: pyramidal, 107° bond angles, equivalent bonds

(*b*) Observed structures

Figure 6-1 The molecular shapes of water and ammonia.

The singlet or diamagnetic state of CH_2 is shown in (5-50) to involve a filled 2s orbital and bonding to the $2p_x$ and $2p_y$ orbitals. Accordingly, we should expect a bent molecule with two equivalent bonds, as in H_2O. This is the currently accepted structure of singlet CH_2: its bond angle is about 103° between two equivalent CH bonds.

The triplet or paramagnetic state (5-51) has been pictured to involve a bond to a 2s orbital and a bond to a $2p_x$ orbital. These are the lowest energy, unfilled orbitals; hence the best for bonding. Now we must deduce the angle between the bonds in accordance with our assumption that the bond angle is fixed by the directional features of the bonding orbitals of the central atom.

One of the central-atom bonding orbitals, the 2s orbital, is spherically symmetric and so it offers no immediate influence on the bond angle. The other carbon atom orbital, $2p_x$, is directed along the x coordinate axis and so it favors formation of bonds along that axis. Thus, the $2p_x$ orbital seems to dictate a linear molecule. Furthermore, it directs as much electron probability along the x axis to the right as it does to the left (as does the 2s orbital). Neither bond should be stronger than the other. We see that the carbon atom orbitals ought to lead to two equivalent bonds directed at 180° from each other. This view is consistent with our molecular orbital treatment of linear CH_2, presented in Chapter Four (Section 4-6). We learned from this three-center M.O. consideration that a carbon 2s and its axial 2p orbital form two bonding molecular orbitals with two equal bonds in a linear configuration. Thus the two C—H bonds are equivalent. We shall call these bonds *sp hybrids,* the word "hybrid" indicating that each bond is a mixture of the carbon 2s and $2p_x$ orbitals. To convey this hybridization in the pigeonhole diagram, we might represent CH_2 as follows:

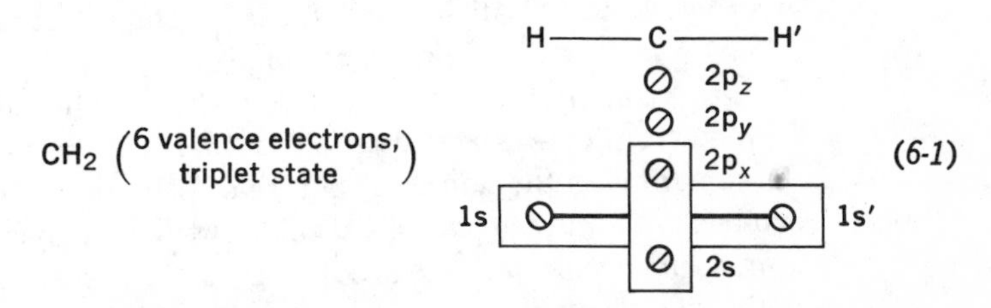

This pigeonhole analysis and hybridization idea prove to have general applicability, and they give us a predictive tool. *Whenever the pigeonhole diagram indicates that the central atom forms two bonds with an s and a p orbital, we can expect linear geometry and two equivalent (sp) bonds.*

(c) THE METHYL MOLECULE CH_3

The representation (5-52) for methyl can be interpreted in a similar way. If the lowest energy orbitals are used for bonding, the carbon 2s orbital and two of its 2p orbitals are the important ones. A molecular orbital approach like that for CH_2 is more complicated than we can cope with here, but it leads to results that can be derived intuitively in the light of the triplet CH_2 treatment. First, the two 2p orbitals of carbon define a plane in which bonding will take place. As in CH_2, the spherical symmetry of the 2s orbital implies no preferred direction. Hence the directional properties of the bonds are entirely fixed by the two 2p orbitals: *all three CH bonds will be in the same plane.*

Experimentally it is found that CH_3 is planar and *that all three C—H bonds are identical.* This last result proves to be easily accommodated in the molecular orbital development, though it is not so readily predicted from theory alone. In a symmetric, planar structure, with three C—H bonds formed at 120° angles to each other, each bond will involve a hybrid mixture formed from a 2s and two 2p orbitals. Extending the sp notation used for CH_2, we'll call these bonds sp^2 hybrids. Henceforth we shall expect that *whenever the central atom has at its disposal a half-filled s and two half-filled p orbitals, three equivalent sp^2 hybrid bonds will be formed at 120° angles in a plane.*

(d) THE METHANE MOLECULE CH_4

It is well to be reminded that while CH_2 and CH_3 are well-defined molecules with accurately known structures, they are also extremely reactive because of their unused valence orbital bonding capacity. Methyl has an average lifetime no longer than 10 to 100 microseconds under normal conditions. One of the ways in which it satisfies its unused bonding capacity is by adding a hydrogen to form methane, CH_4.

Again, a molecular orbital consideration of the bonds to be formed from one 2s and three 2p orbitals can be brought into agreement with the known structure of CH_4. Since the three 2p orbitals are mutually perpendicular, we can expect a three-dimensional molecule. As in both CH_2 and CH_3, the s orbital is uniformly divided among four exactly equivalent molecular orbitals. We designate these as sp^3 hybrids. To be equivalent, the bonds must point at the four vertices of a regular tetrahedron; 109°28′ bond angles are thus obtained.

Again we shall expect that *whenever the central atom has at its disposal a half-filled s and three half-filled p orbitals, four sp^3 bonds will be formed at tetrahedral angles.* According to the simple molecular orbital treatment, these angles will be obtained independent of the atom to be bonded. This is true to remarkable accuracy, as shown by the many halogen-substituted methanes whose bond angles are well known. Some of these are listed in Table 6-1. In every case the bond angles are within one or two degrees of the tetrahedral angle 109°28′.

Figure 6-2 shows the molecular geometry that accompanies sp, sp^2, and sp^3 hybrid bonds. We must remember, though, that neither the molecular orbital treatment nor the hybridization idea is sufficiently rigorous to prove that the structures of CH_2, CH_3, or CH_4 must be what they are observed to be. Rather,

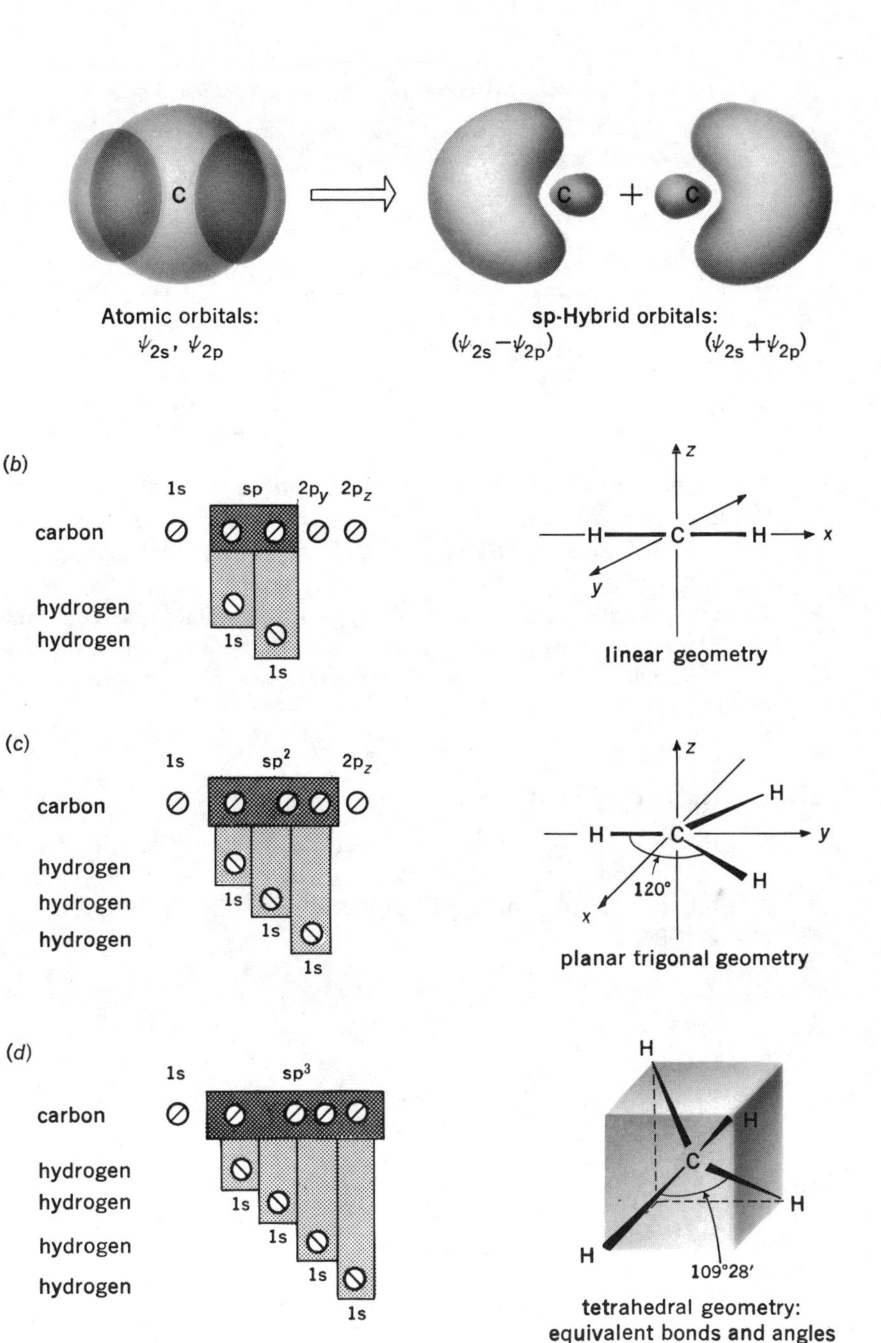

Figure 6-2 Orbital shape (a) and molecular geometry (b), (c) and (d) of hybrid bonds.

Table 6-1 Some Bond Angles in Halogen-Substituted Methanes

	Angle H—C—H	Angle H—C—X	Angle X—C—X
CH_4	109°28′	—	—
CH_3F	110°	108°56′	—
CH_2F_2	111°54′	109°9′	108°18′
CHF_3	—	110°38′	108°48′
CF_4	—	—	109°28′
CH_3Cl	110°30′	108°26′	—
CH_3Br	111°12′	107°44′	—
CH_3I	111°30′	107°26′	—

these elementary theories provide a framework that is consistent with the experimental facts such as those shown in Table 6-1. The value of the theory is that once tied to firm structures, it provides a basis for predicting the structures of other molecules from an examination of the orbitals available for bonding. Since the molecular structure is important in the chemistry of a molecule, this is a significant step towards understanding chemistry.

(e) THE MOLECULES BH_3 AND BF_3

In our earlier discussions, we did not mention the bonding of boron and beryllium. Considering boron first, a pigeonhole representation would initially predict stable molecules BH and BF.

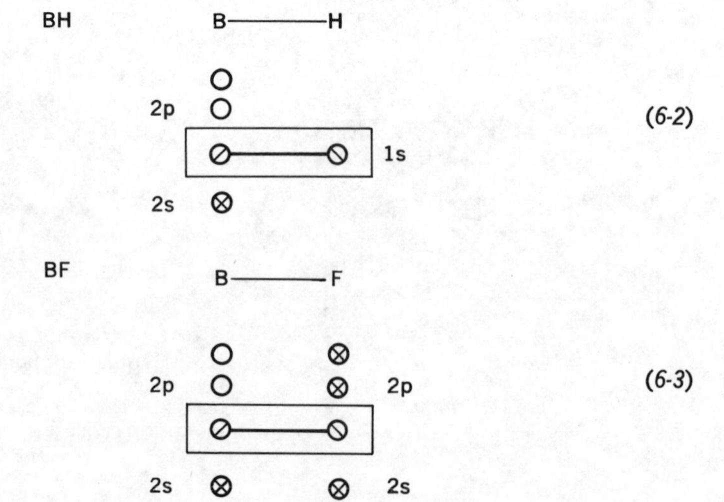

However, as in CH_2, it is possible to promote an electron to a vacant 2p orbital, at once reducing electron repulsions in the 2s orbital and increasing the number of electrons available for sharing in bonds. Now there are three half-filled orbitals, a 2s orbital and two 2p orbitals. Following the treatment for CH_3, we would expect three equivalent sp^2 hybrid molecular orbitals. The sp^2 hybrid orbitals are all in the same plane at angles of 120°. Hence we would expect planar, symmetric molecules BH_3 and BF_3.

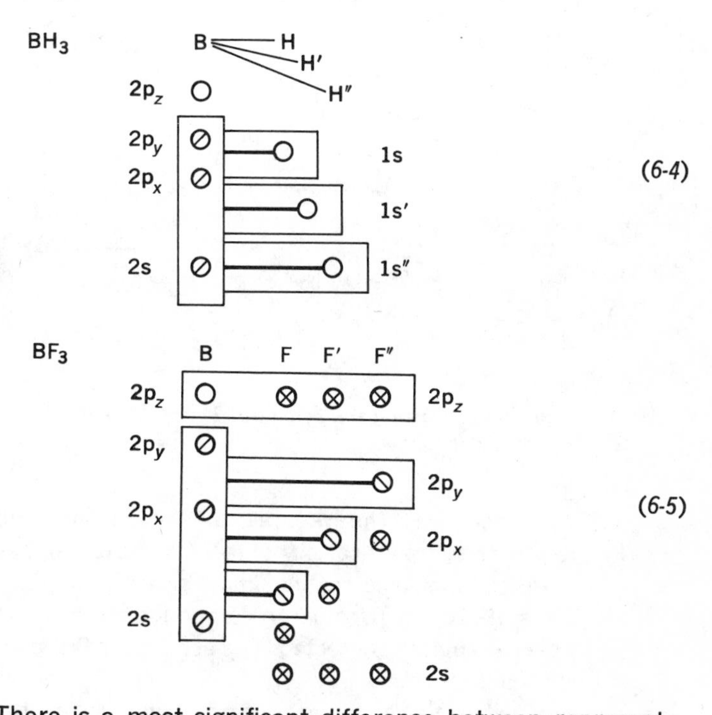

There is a most significant difference between representations (6-4) for BH_3 and (6-5) for BF_3. The borine (BH_3) representation displays a completely vacant $2p_z$ valence orbital on boron. This orbital, which is oriented perpendicular to the molecular plane, should cause borine to have residual bonding capacity. The molecule should be reactive. In contrast, the molecule BF_3 places the boron $2p_z$ orbital within a cluster of three fluorine atom $2p_z$ valence orbitals, all perpendicular to the molecular plane (see Fig. 6-3). We can expect the fluorine $2p_z$ electrons (the "pi" electrons) to occupy molecular orbitals that extend over and partially occupy the boron $2p_z$ orbital. Hence additional bonding can result and boron trifluoride, BF_3, should be more stable than borine, BH_3.

These expectations are all in accord with experience. While BF_3 shows some residual bonding capacity that can be attributed to its $2p_z$ orbital, the molecule is an inexpensive stock-

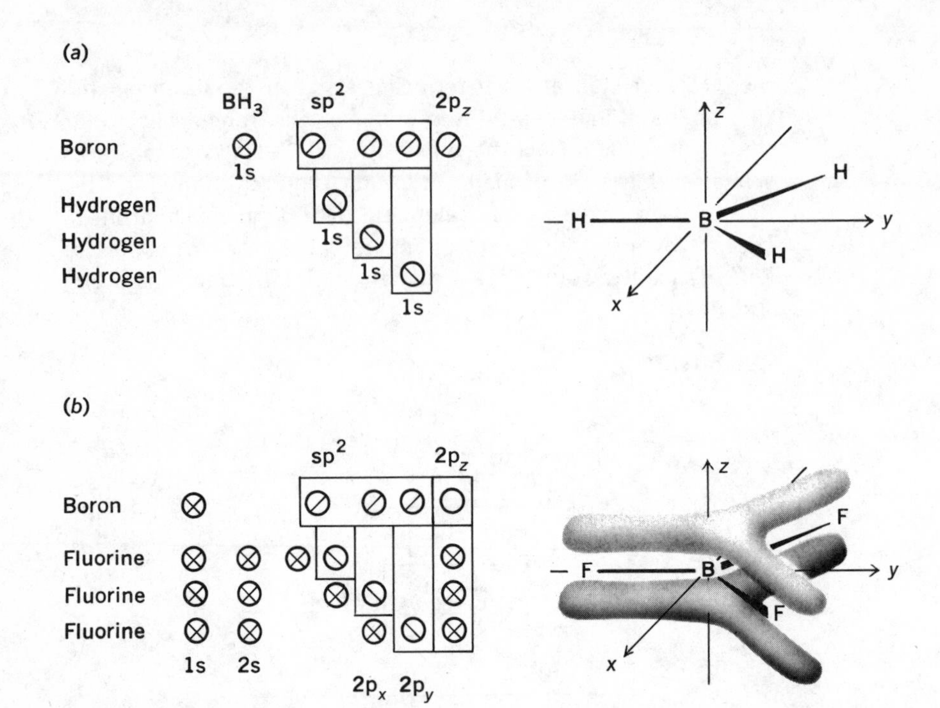

Figure 6-3 Borine (a) and boron trifluoride (b): sp^2 hybrids with and without pi-orbital stabilization.

room chemical. The molecules of this gaseous substance have a planar symmetrical structure. On the other hand, BH_3 is a transient molecule that is almost surely produced during reactions of a number of boron compounds, but it has never been spectroscopically detected. The search for BH_3 continues (using methods specially developed for such reactive species), and when it is finally found, the most interesting question to be answered will be whether it has the planar structure we have predicted.

(f) THE MOLECULES BeH_2 AND BeF_2

For BeH_2 or BeF_2 the pigeonhole representations are as follows, provided an electron is promoted to a vacant 2p orbital.

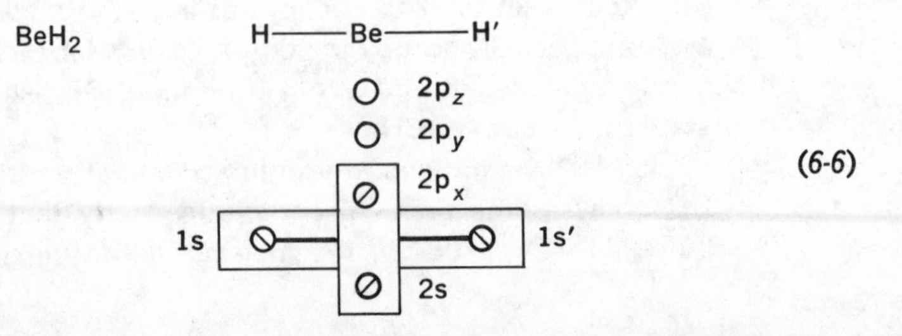

(6-6)

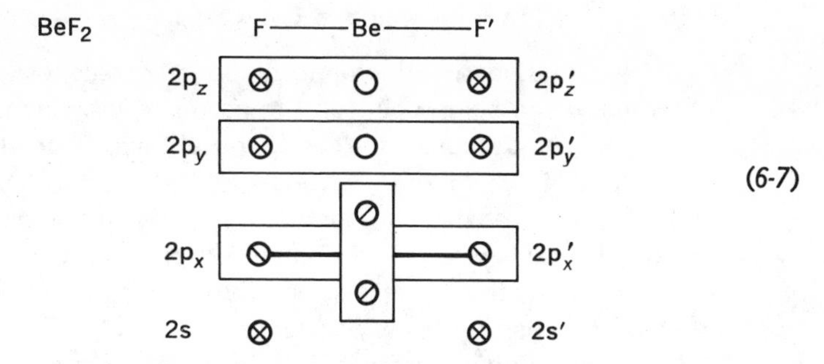

(6-7)

These two representations, (6-6) and (6-7), both predict linear structures with equivalent sp bonds. As in the case of BH_3, we can expect BeH_2 to be extremely reactive due to the completely vacant $2p_y$ and $2p_z$ valence orbitals of beryllium. This situation is less uncomfortable for BeF_2 because the fluorine $2p_y$ and $2p_z$ orbitals can form molecular orbitals with the boron orbitals and furnish electrons to add to the bonding.

Neither of the triatomic molecules BeH_2 or BeF_2 is known. The beryllium halides are, of course, easily prepared, but as solids in which the $2p_y$ and $2p_z$ valence orbitals participate in the bonding within the crystal. The closest we can come to testing our structural expectations for these two molecules is to compare the molecular geometries of other alkaline earth dihalides, a number of which have been detected in high temperature reactions. Table 6-2 shows that two experimental techniques have been used and that they are not in complete agreement. The evidence suggests that some alkaline earth dihalides are indeed linear ($CaCl_2$, $CaBr_2$, $SrBr_2$ and SrI_2) but that some are likely to be bent molecules (CaF_2, SrF_2, BaF_2 and $BaCl_2$). Thus, limitations of our simple theory based on sp hybrid molecular orbitals are in evidence here.

Table 6-2 Bond Angles in Gaseous Alkaline Earth Dihalides, MX_2: First Value, Electron Diffraction; Parenthetical Value, Molecular Beam, Electric Deflection†*

	F	Cl	Br	I
Mg	180 ± 30°	180 ± 10°	180 ± 10°	
Ca	180° (bent)	180 ± 10° (180°)	180 ± 10° (180°)	180 ± 10°
Sr	180° (~120°)	180 ± 30° (bent)	180 ± 10° (180°)	180 ± 10° (180°)
Ba	180° (~120°)	180 ± 40° (~120°)	180 ± 30° (bent)	180 ± 20° (bent)

* Electron diffraction; P. A. Akishin, V. P. Spiridonov, G. A. Sobolev, and V. A. Naumov, *Zhur. fiz. Khim.*, **31,** 1871 (1957); **32,** 58 (1958).
† Molecular beam, electric deflection; L. Wharton, R. A. Berg, and W. Klemperer, *J. Chem. Phys.*, **39,** 2023 (1963).

6-2 Electron repulsion and bond angles

The Linnett quartet scheme of bonding recognizes that electrons of the same spin tend to correlate their motion so as to stay apart because of electron repulsions. Four electrons are farthest apart when in tetrahedrally related positions.

There is a simple scheme for explaining molecular bond angles that is also based upon electron repulsions. The scheme is tuned to the Lewis electron-dot representation of bonding. It embodies the following premises:

(i) Each atom forms bonds until it fills its valence orbitals with electrons to reach the inert gas configuration.

(ii) These electrons are considered to distribute to act as pairs (of opposite spin).

(iii) The molecule takes that geometric structure that keeps these electron pairs as far apart as possible.

(a) ELECTRON REPULSIONS IN H_2O, NH_3 AND CH_4

This scheme is very easy to apply. For example, for H_2O, NH_3, and CH_4, each central atom has eight electrons in its valence orbitals (some electrons provided by the hydrogen atoms, of course). In each of these molecules there are, then, four pairs of electrons repelling each other and striving to stay apart. The four pairs will tend to be directed towards the vertices of a regular tetrahedron. Figure 6-4 shows the result. In each molecule the bond angle expected is the tetrahedral angle 109°28′.

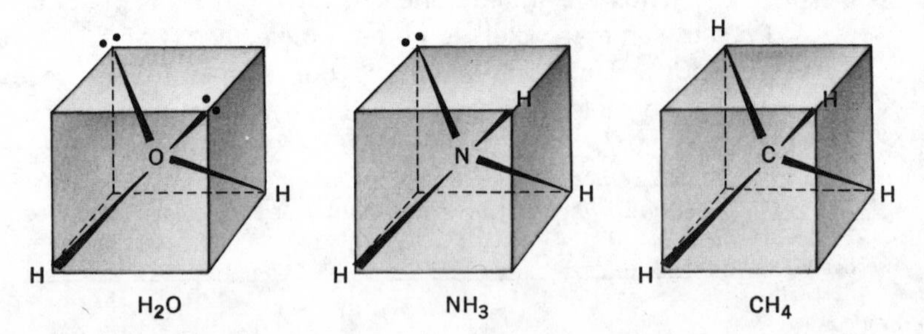

Figure 6-4 Electron repulsions and the structures of H_2O, NH_3 and CH_4.

For water, two of the electron pairs link H atoms to the oxygen atom (at the tetrahedral angle) while the other two pairs are merely "unused pairs." The molecule is expected to be bent. For ammonia, three electron pairs link H atoms and one pair is unused in bonding. The mutually tetrahedral bonding predicts a pyramidal molecule with 109°28′ bond angles. Finally, for methane, four bonds at the tetrahedral angles are expected.

These predictions can be contrasted with those based on the directional characteristics of the occupied valence orbitals. For water, the p^2 bonding predicts a bent molecule with a 90° bond angle, whereas electron repulsion predicts it to be 109°28′. Experimentally we find that the molecule is bent, with a measured angle of 104°30′. For ammonia, both the valence orbitals p^3 and the electron repulsion predict a pyramidal molecule, the former with 90° angles, the latter with 109°28′ angles. The molecule is pyramidal and its bonds have 107.3° angles between them. For methane, the electron repulsion predicts tetrahedral bonds in a three-dimensional molecule, exactly as observed. The same is true for the sp^3 hybrid orbital prediction, but only after a molecular orbital argument to see how one s and three 2p orbitals can form molecular orbitals that correspond to four equivalent, tetrahedral bonds.

The observed deviations from tetrahedral angles in NH_3 and H_2O can also be interpreted within the context of the electron repulsion scheme. Ammonia has one pair of electrons not engaged in bonding. Such a pair is called a "lone pair." In the first approximation it is considered to be exactly equivalent to the three "bonding pairs" that bind the protons in the molecule. However, we find empirically that the angles between bonds (hence, between bonding pairs) are 107.3°, less than tetrahedral. This can be attributed to a difference between bonding pair–bonding pair repulsions (of which there are three) and the lone pair–bonding pair repulsions (of which there are also three). Since the observed bond angles are less than tetrahedral, we conclude that a lone pair repels a bonding pair more than two bonding pairs repel each other. In shorthand notation, we write

$$\underset{\text{repulsions}}{\ell\text{p–bp}} > \underset{\text{repulsions}}{\text{bp–bp}} \tag{6-8}$$

Water takes us one more step. This molecule has two lone pairs and two bonding pairs. This gives one bonding pair–bonding pair repulsion, four lone pair–bonding pair repulsions, and, now, one lone pair–lone pair repulsion. It is a reasonable guess that lone pair–lone pair repulsions are even larger than lone pair–bonding pair repulsions, so the bond angle in H_2O should be even smaller than those in NH_3. This is what is observed: H_2O has a bond angle of 104.5°. We can expand (6-8) to include our guess about lone pair–lone pair repulsions.

$$\underset{\text{repulsions}}{\ell\text{p–}\ell\text{p}} > \underset{\text{repulsions}}{\ell\text{p–bp}} > \underset{\text{repulsions}}{\text{bp–bp}} \tag{6-9}$$

We shall find these empirical rules useful in more complex situations, but, before discussing them, we should examine the structures of molecules that do not have eight electrons around the central atom.

(b) ELECTRON REPULSIONS IN BH_3, BF_3, BeH_2 AND BeF_2

The pigeonhole representations for the boron compounds BH_3 and BF_3 show three electron pairs shared in each molecule. The fact that the $2p_z$ orbital is partially occupied in BF_3 is ignored. Hence, in each molecule, electron repulsions act to separate the three bonds as much as possible. This criterion dictates that the three bonds lie in a plane at angles of 120°. This is the same prediction as that deduced from the sp^2 orbital hybridization, in agreement with the known structure of BF_3. Beryllium compounds, with only two shared pairs, are even simpler. Clearly two bonds are farthest apart when they point in opposite directions, that is, with a bond angle of 180°. This also agrees with the sp hybridization picture, and again the expectation is only in partial agreement with a cloudy experimental picture (see Table 6-2).

(c) ELECTRON REPULSIONS IN PCl_5 AND SF_6

The compound PCl_5 must dispose ten valence electrons around the central phosphorus atom; five from phosphorus and one each from the five chlorine atoms. According to our electron repulsion scheme, the molecule will take that geometric structure that separates five electron pairs as much as possible. However, there is no way to distribute five bonds so that all of the angles are equal. The best we can do is to arrange them as shown in Figure 6-5(*a*). The bonds should point to the corners

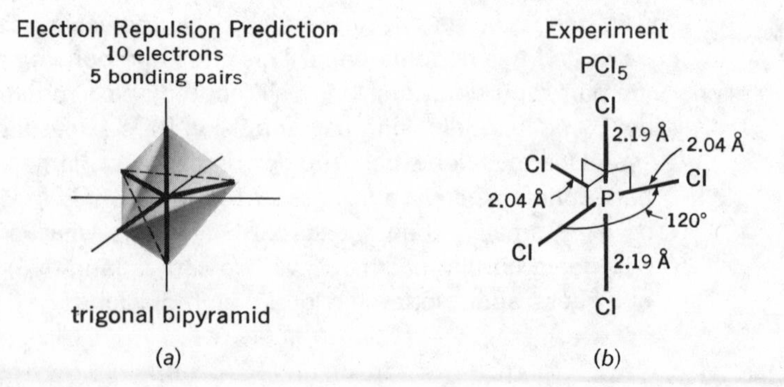

Figure 6-5 Electron repulsions and the structure of PCl_5.

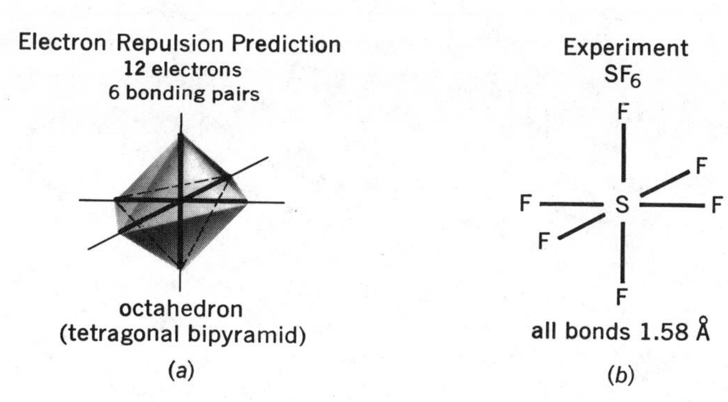

Figure 6-6 Electron repulsions and the structure of SF_6.

of the structure obtained when two three-sided pyramids are placed base-to-base (a *trigonal bipyramid*). In this structure there are two kinds of P—Cl bond, one kind that forms the base of the pyramids and the other kind that is directed axially. Indeed, PCl_5 does have this structure and the experimentally measured bond lengths differ noticeably, as shown in Figure 6-5(*b*).

Sulfur hexafluoride, SF_6, has twelve valence electrons around the central sulfur atom; six from sulfur and one each from the six fluorine atoms. This time, the bonds should point at the corners of the structure obtained when two four-sided pyramids are placed base-to-base. Such a structure can be called a *tetragonal bipyramid* but it is usually called an octahedron (it has eight sides). The angle between any pair of bonds is 90°, as pictured in Figure 6-6(*a*). The experimentally observed structure of SF_6 is in accord—all bond angles are 90° and all bond lengths are equal, 1.58 Å, in a perfect octahedral structure.

(d) ELECTRON REPULSIONS IN SF_4 AND XeF_4

Sulfur tetrafluoride, SF_4, has the same number of valence electrons as PCl_5 (10) so, according to the electron repulsion view, its structure should be similar. The bonds should be directed at the corners of a trigonal bipyramid. Figure 6-7 shows, however, that the molecule still has two options. In Figure 6-7(*a*), the lone pair is placed in the base of the pyramids whereas in Figure 6-7(*b*) it is placed in an axial position. The implication of (*a*) is that there are two ℓp–bp and four bp–bp interactions at the acute, 90° angle. In (*b*), these are equally divided, three ℓp–bp and three bp–bp interactions at the acute angle. Since ℓp–bp repulsions exceed bp–bp repulsions (from 6-8), structure 6-6(*a*) should be preferred. Experiment shows that it is, as displayed in Figure 6-7(*c*).

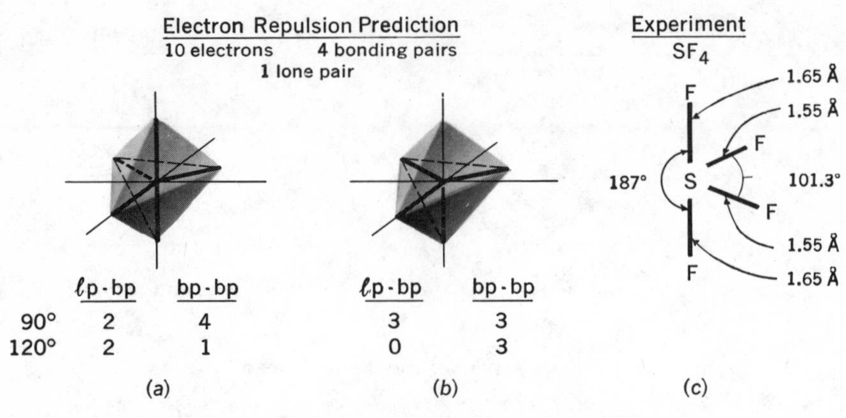

Figure 6-7 Electron repulsions and the structure of SF_4.

The inert gas compound XeF_4 has twelve valence electrons around the xenon atom, hence its structure should be related to SF_6. Again the molecule has two options. In the structure shown in Figure 6-8(*a*), the bonds are in a square, planar arrangement and there are eight ℓp–bp but no ℓp–ℓp repulsions. The structure shown in Figure 6-8(*b*) (which resembles the SF_4 structure) has one ℓp–ℓp but only six ℓp–bp repulsions. To explain the observed square, planar molecular geometry (Fig. 6-8(*c*)), we must postulate that ℓp–ℓp repulsions exceed ℓp–bp repulsions (as guessed in (6-9)) and by more than ℓp–bp repulsions exceed bp–bp repulsions.

6-3 Contrast of the hybridization and electron repulsion models

We have, then, two bases for explaining bond angles. The bond angles could reflect, and hence be predicted from, the orbital parentage of the bonds. On the other hand, the bond angles

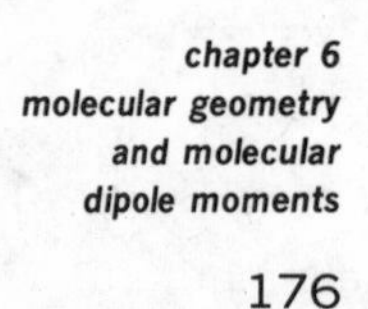

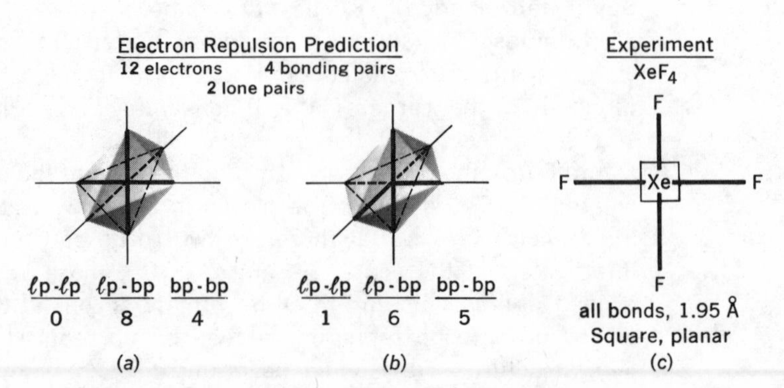

Figure 6-8 Electron repulsions and the structure of XeF_4.

Table 6-3 Contrast of Orbital Hybridization and Electron Repulsion Predictions

Orbital Hybridization			Electron Repulsion		
Number of bonds	Hybridization	Bond angles	Number of bonds	Number of electron pairs	Bond angles
2	sp	180°	2	2	180°
3	sp^2	120°	3	3	120°
2	p^2	90°	2	4	109°28′
3	p^3	90°	3	4	109°28′
4	sp^3	109°28′	4	4	109°28′
5	dp,sp^2	trigonal bipyramid	5	5	trigonal bipyramid
6	d^2sp^3	octahedron	6	6	octahedron

Table 6-4 Bond Angles in Fifth and Sixth Column Hydrides and Fluorides

H_2O	104.5°	F_2O	102°	Cl_2O	110°				
H_2S	92.2°	—		Cl_2S	102°				
H_2Se	91.0°	—		—					
H_2Te	88.5°	—		—		Br_2Te	98°		
NH_3	107.3°	NF_3	102°	—					
PH_3	93°	PF_3	104°	PCl_3	100°	PBr_3	101.5°	PI_3	98°
AsH_3	91.5°	AsF_3	102°	$AsCl_3$	98°	$AsBr_3$	101°	AsI_3	98.5°
SbH_3	91.3°	SbF_3	88°	$SbCl_3$	99.5°	$SbBr_3$	97°	SbI_3	99°
				$BiCl_3$	100°	$BiBr_3$	100°		

Bond Angle Predictions

	H_2M, X_2M	H_3M, X_3M
Hybrid Orbitals	90°	90°
Electron Repulsion	109°28′	109°28′

could be fixed merely by the repulsions between pairs of electrons in valence orbitals (ignoring energy differences between s and p orbitals, as well as their directional properties). The two theories qualitatively agree in all cases (see Table 6-3); they differ quantitatively in their bond angle predictions for H_2O and NH_3. In fact, the bond angles are closer, for these two molecules, to the electron repulsion expectation of 109°28′. There are, however, other opportunities to test these two theories lower in the periodic table. The other elements in the sixth column of the periodic table should, like oxygen, form dihydrides and difluorides, with the same bond angles predicted. In a similar way, the elements below nitrogen in the fifth column of the periodic table should form trihydrides and trihalides with angles like those in ammonia. The known facts are summarized in Table 6-4.

Among the hydrides we see that only H_2O and NH_3 are close to the electron repulsion prediction of a tetrahedral angle. All the others are remarkably close to the 90° angle expected for orbital bonds. For all the halide molecules the bond angles are found to be irritatingly close to halfway between the two predictions. Thus there is little basis for preferring either theory. Nevertheless, proponents of each theory seek rationalizations for the discrepancies. The orbital hybridization advocates speak of repulsions between the electrons in the bonding pairs and repulsions between the terminal atoms, both of which tend to open the bond angles from 90° (as observed in H_2O and NH_3). As the central atom becomes larger (as in the series O, S, Se, Te or N, P, Sb, As), the repulsion becomes less important and the angle approaches the orbital prediction. Thus, the fact that the H_2O and NH_3 angles are larger than predicted, while those analogues lower in the periodic table are quite close to the expected values, is comfortably explained. We have already seen that electron repulsion supporters argue that repulsions between nonbonding electron pairs must also be considered. Empirically it is found that nonbonding electrons repel more strongly than bonding electrons. Thus, the nonbonding electron repulsions open the angles involving unused (and unseen) electron pairs and, hence, close the angles between the bonds from the tetrahedral angle. Unfortunately, only a rather strained empirical argument can be mustered to account for the almost 90° angles in H_2S, H_2Se, and H_2Te. The same is true for PH_3, AsH_3, and SbH_3. On the other hand, the natural way in which tetrahedral angles in carbon compounds and the planar structure of BF_3 are rationalized is appealing.

Though there is frequent controversy over the relative merits of these two schemes, they have a generic likeness that is seldom expressed. The electron repulsion theory assumes wrongly that the 2s and 2p orbitals are exactly equivalent, that is, that the s and p orbitals are *completely* hybridized in the first approximation in all cases. The orbital hybridization view probably overemphasizes the energy difference between s and p orbitals and hence, tends to "under-hybridize." Both views, however, embody the concept of "mixing" s and p character to form bonds. If we knew where the valence electrons in H_2O spend most of their time, and if we attempted to describe the distribution in a series expansion based on hydrogen atom orbitals, the expansion would surely involve predominantly 2p orbitals, with some contribution from the 2s orbital. The electron repulsion and orbital hybridization schemes merely offer different first guesses as to the extent of this contribution.

The most important point to remember, though, is that both

Table 6-5 Molecular Dipole Moments and Molecular Properties

Molecule	Dipole Moment (Debye)	Mol. Wt.	M.P. (°K)	B.P. (°K)	$\Delta H_{vap.}$ (kcal/mole)	$\frac{\Delta H_{vap.}(HX)}{\Delta H_{vap.}(X_2)}$
F_2	0.00	38.0	50	85	0.80	
Cl_2	0.00	71.0	171	239	2.2	
Br_2	0.00	159.8	266	332	3.82	
I_2	0.00	253.8	387	458	7.45	
HF	1.91	20.0	190	293	1.85	2.31
HCl	1.07	36.5	158	188	3.85	1.75
HBr	0.79	80.9	186	206	4.21	1.10
HI	0.38	127.9	222	238	4.72	0.63

of these extremely simple theories correlate the bond angles and molecular structures of a multitude of molecules. A detailed quantum mechanical calculation would claim a major accomplishment if it were to predict within a few degrees (without prior knowledge of the experimental result) even a single bond angle in a polyatomic molecule. This sort of predictive accuracy can be assumed using either the electron repulsion or orbital hybridization models provided it is applied within bounds limited by the experimental data.

6-4 Molecular dipole moments

A variety of reasons for the chemist's great interest in molecular geometries were given in the introduction to this chapter. One of these, the geometric influence on molecular charge distributions, deserves more consideration. In Chapter Three we found that the charge distribution in a chemical bond could result in a charge separation, an electric dipole moment. This means that one end of the molecule appears to be negatively charged and the other end positively charged. Naturally this appearance is of interest to other molecules that come in contact with it. Hence, the presence of a molecular dipole moment has profound influence upon the chemistry of that molecule. These charge separations provide strong intermolecular forces that influence how close two molecules approach, their preferred orientations as they approach, and the energy needed to pull them apart. These factors are reflected in such properties as the boiling point, melting point, crystal structure, solvent properties, ease of reaction, and a host of other chemically important phenomena. Some examples can be seen in Table 6-5, which contrasts the halogens with their hydrohalides.

(a) DIPOLE MOMENT AND BOILING POINT

Consider, first, the elemental halogens. All have zero dipole moments, of course, since each of these diatomic molecules involves identical atoms. The rising trends from fluorine to iodine in melting point (50–387°K), boiling point (85–458°K), and heat of vaporization (0.8–7.5 kcal), are due entirely to the increasing forces between similar molecules as the number of electrons increases. Turning to the hydrogen halides, we see that the dipole moment is highest for HF and lowest for HI. The dipole forces *decrease* from HF to HI. This trend is opposite to that associated with the number of electrons, and causes the change in boiling point from HCl to HI to be only 50°K, whereas from Cl_2 to I_2 it is 219°K. Even more startling is that HF has the highest boiling point of all the hydrogen halides.

(b) DIPOLE MOMENT AND HEAT OF VAPORIZATION

The heat of vaporization permits us to distinguish the forces due to the number of electrons from those additional forces due to dipole attractions. The heat of vaporization of HX can be divided by the heat of vaporization of the element X_2 to determine the factor of change. These ratios, shown in the last column of Table 6-5, are plotted against dipole moment in Figure 6-9. We see that as the dipole moment rises, the heat of vaporization is larger by an increasing factor over its elemental counterpart.

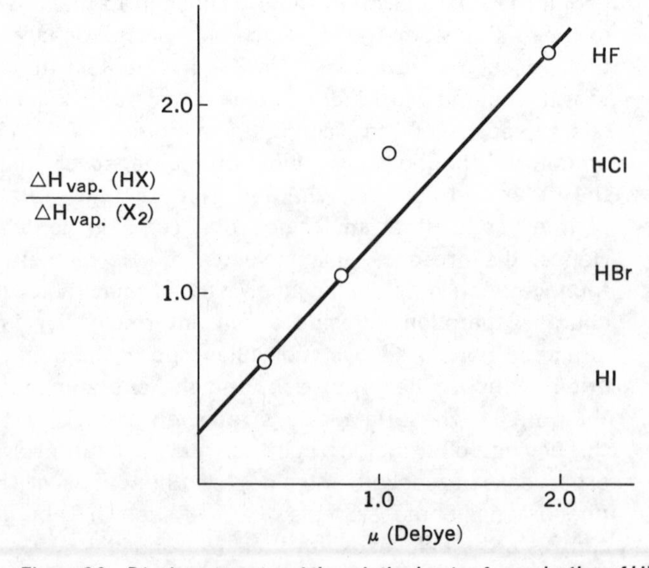

Figure 6-9 Dipole moment and the relative heats of vaporization of HX and X_2.

Table 6-6 Molecular Geometry and Dipole Moment

	Molecular Formula	Molecular Structure	Dipole Moment (D)
para-dichlorobenzene	$C_6H_4Cl_2$		0.00
meta-dichlorobenzene	$C_6H_4Cl_2$		1.49
ortho-dichlorobenzene	$C_6H_4Cl_2$		2.27

Because of these influences on intermolecular forces, chemists wish to be able to estimate and understand molecular dipole moments. Table 6-5 shows that the nature of the atoms bonded together is important. Table 6-6 shows that the molecular geometry matters as well.

(c) THE EFFECT OF MOLECULAR SHAPE ON DIPOLE MOMENT

Three molecules are shown in Table 6-6; each has the molecular formula $C_6H_4Cl_2$ and each is named "dichlorobenzene." They differ only in the geometrical arrangement of the atoms. All of the molecules are planar. In the first, the two chlorine atoms are on opposite sides of the molecule. This arrangement is identified by the prefix "*para*." If the chlorine atoms are next to each other, as in the last case, the arrangement is identified by the prefix "*ortho*." The remaining possibility, shown in the middle, is called "*meta*."

The last column shows that the molecular dipole moments vary greatly from *para* to *meta* to *ortho*. The zero dipole moment of *para*-dichlorobenzene is easiest to understand. The symmetry of the molecule is such that each bond has an equal counterpart oppositely directed on the other side. For example, if the electrons in one of the carbon—chlorine bonds are displaced toward the chlorine atom, they produce a local "bond dipole." The other carbon—chlorine bond will have a similar bond dipole but oriented to cancel exactly the first one. The C—H bonds may involve local bond dipoles in a similar way, but cancellation still leaves a net, molecular dipole moment of zero. These cancellations are evident in a vector diagram as shown in Figure 6-10(*a*).

The situation is quite different for *ortho*-dichlorobenzene. Figure 6-10(*b*) shows that this molecular geometry does not automatically cause perfect cancellations. In fact, the experimental dipole moment 2.27 D shows that a C—Cl bond working against a C—H bond does contribute a non-zero amount $(\mu_{CCl} - \mu_{CH})$ to the total molecular dipole moment μ_T. The amount contributed is such that the component along the bisector of the bond angle is one half the total value of 2.27 D. Using elementary trigonometry we can deduce the bond dipole moment $(\mu_{C-Cl} - \mu_{C-H})$ as follows:

$$2(\mu_{CCl} - \mu_{CH}) \cdot \cos 30° = \mu_T$$

$$(\mu_{CCl} - \mu_{CH}) \cdot 0.866 = \tfrac{1}{2} \cdot 2.27 \text{ D}$$

$$(\mu_{CCl} - \mu_{CH}) = 1.31 \text{ D} \qquad (6\text{-}10)$$

This view of the presence of bond dipoles in a molecule can now be tested through a calculation of the expected molecular dipole moment of *meta*-dichlorobenzene. Figure 6-10(*c*) shows that the situation resembles that of *ortho*-dichlorobenzene except for the different geometry. If we assume the same bond dipoles in the two molecules, the new vector sum would be

$$2(\mu_{CCl} - \mu_{CH}) \cdot \cos 60° = \mu_T$$

$$2 \cdot (1.31) \cdot \tfrac{1}{2} = \mu_{calc}$$

$$\mu_{calc} = 1.31 \text{ D} \qquad (6\text{-}11)$$

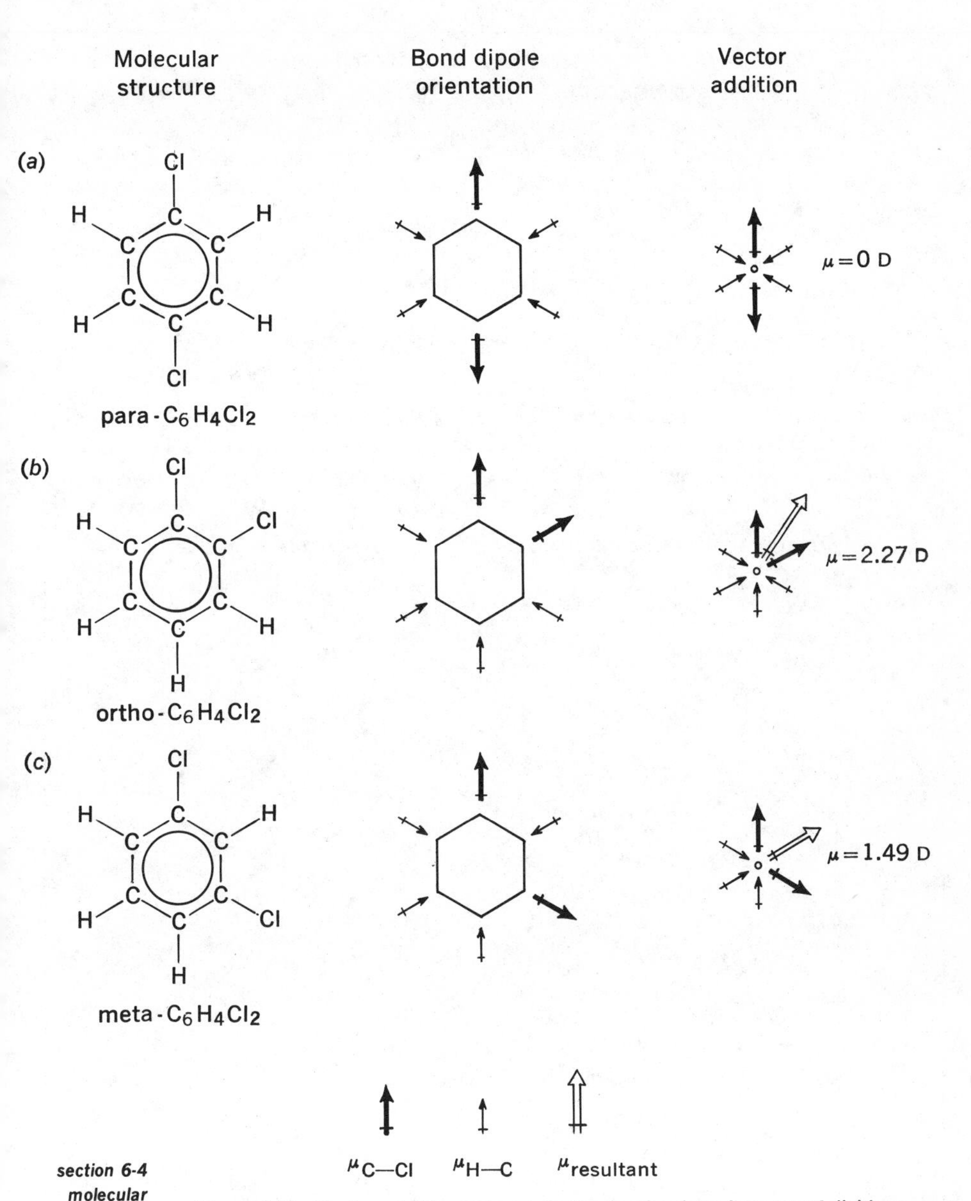

Figure 6-10 Vector addition of bond dipoles for the three isomers of dichlorobenzene (DCB): (a) para-DCB, $\mu_{\text{resultant}} = 0$ D; (b) ortho-DCB, $\mu_{\text{resultant}} = 2.27$ D; (c) meta-DCB, $\mu_{\text{resultant}} = 1.49$ D.

We see that the calculated value of μ_T is only 12 percent below the experimental value of 1.49 D. Evidently this bond dipole additivity deserves further exploration.

Table 6-7 shows the structure of two groups of molecules with similar bonds. Within each group it is possible to predict all the molecular dipole moments from the experimental measurement of one of the members. These calculated values and

Table 6-7 Effect of Molecular Geometry on Molecular Dipole Moments—Vector Addition Mode

		μ(expt) (D)	**μ(calc) (D)**	**Percent Discrepancy •**
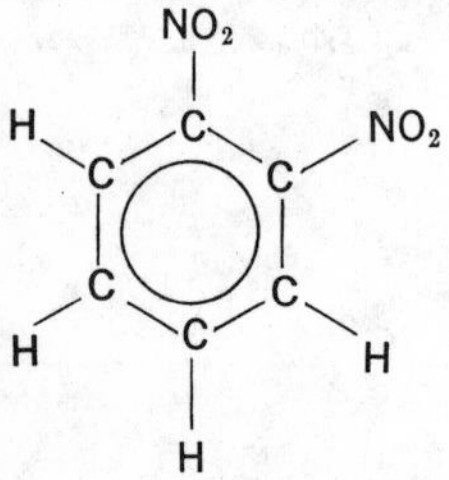	*ortho*-dinitrobenzene	6.5	(assumed, 6.5)	—
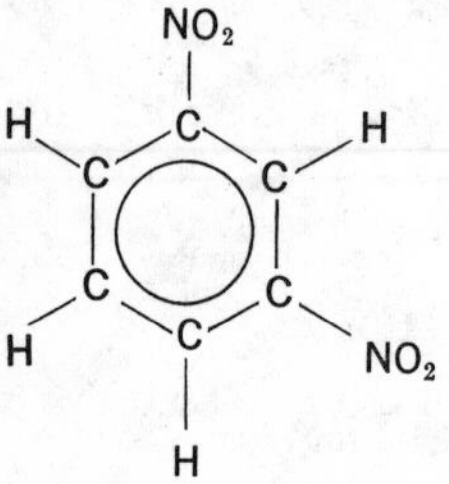	*meta*-dinitrobenzene	3.87	3.76	3
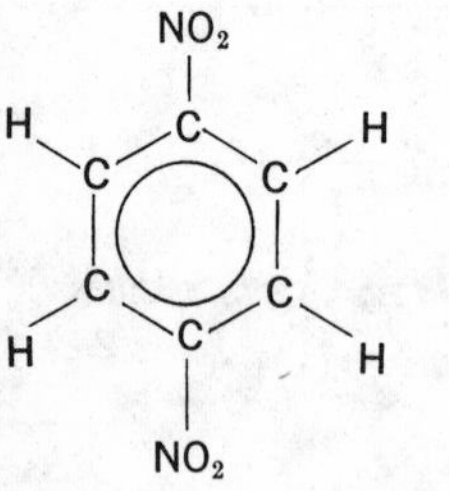	*para*-dinitrobenzene	0	0	0
CF_4	tetrafluoromethane	0	0	0
CHF_3	trifluoromethane	1.80	(assumed, 1.80)	—
CH_2F_2	difluoromethane	1.96	2.07	+6
CH_3F	monofluoromethane	1.60	1.80	+13
CH_4	methane	0	0	0

percent discrepancies are shown in the last two columns. The generally good agreement shows that the charge separations in individual bonds are reasonably additive if their spatial orientations are considered, at least in some cases.

The examples in Tables 6-6 and 6-7 leave no doubt that molecular dipole moments are reasonably attributed to charge distributions determined by the molecular geometry. These days chemists feel that a molecular formula or even a molecular structure is not well characterized until the entire three-dimensional perspective of the molecule is known. Entire fields of physics have been essentially taken over by chemists because of the importance of such data. These fields, which include infrared spectroscopy, nuclear magnetic resonance, X-ray diffraction, electron diffraction, and several others, make up the general subject of molecular spectroscopy.

Chemists still identify a variety of chemical bond types, despite the unifying influence of quantum mechanics. The value of these bond categories is undeniable—they enable the chemist to predict with reasonable confidence when a bond can form and what bond properties to expect. They are responsible for the ultimate measure of success—new compounds, never before seen on Earth, currently being produced at a rate of 300 per day, including Saturdays, Sundays, and the Fourth of July.

7-1 Covalent bonds

Most of the bond situations we have considered thus far involve two identical atoms, each with a valence orbital that contains a single electron. Such *a pair of electrons, equally shared between two atoms* (*one electron furnished by each atom*) *forms a* **covalent bond.** The bond forms because the two half-occupied valence orbitals give rise to a bonding and an antibonding M.O. Both electrons can occupy the bonding M.O. They lower the potential energy because both electrons are then simultaneously near two nuclei.

(a) COVALENCY: SHARE AND SHARE ALIKE

A large fraction of the known chemical compounds possess some bonds with the parentage just described. When the atoms bonded are identical, the electron probability distribution is symmetrically disposed, favoring neither atom.

We have represented in a very approximate way the M.O.'s of H_2^+ and H_2 in the form

seven chemical bond types

$$\sigma(1s) = \psi_A(1s) + \psi_B(1s) \quad (7\text{-}1)$$

$$\sigma^*(1s) = \psi_A(1s) - \psi_B(1s) \quad (7\text{-}2)$$

These approximations correctly describe the qualitative features of the probability distribution—particularly the nodal properties. They also make an explicit statement about where the electron is likely to be found. Hence the bond energy can be calculated for a hypothetical H_2 molecule, with an electron distribution in which one electron remains in the 1s orbital centered on proton A and the other remains in the 1s orbital centered on proton B. This calculation falls far short of a quantitatively accurate description; it predicts a bond energy only about 10 percent as large as the observed value. It is most revealing how this approximate electron distribution differs from an accurate portrayal. Figure 7-1 shows

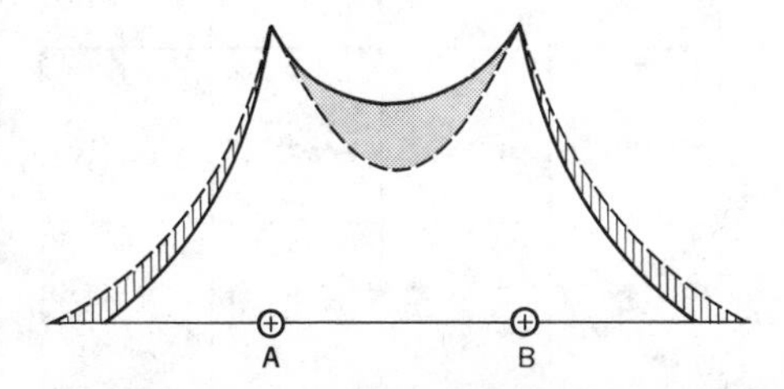

Figure 7-1 The electron probability distribution in H_2 (——) contrasted with the superposition of two 1s probability distributions (- - -). The simple combination of atomic orbitals underestimates the electron density between the nuclei (shaded portion) and overestimates the electron density on the outside of the nuclei in the antibinding regions (crosshatched portions).

the discrepancy, in terms of the probability density along the line connecting the nuclei. The actual distribution has a far higher density between the nuclei (see the extra, shaded area) than does the superposition of the 1s probabilities. This extra likelihood of finding the electrons between the nuclei (in the binding region) was stolen from the antibinding region. Hence this redistribution accounts for most of the energy-lowering. Crude approximations of bonding M.O.'s like (7-1) always underestimate the extent to which the electrons are placed between the nuclei in the binding region.

(b) COVALENT BOND ENERGIES IN HOMONUCLEAR DIATOMIC MOLECULES

The energy-lowering that actually results from the equal sharing of electrons in homonuclear diatomic molecules varies quite a bit across the periodic table. Figure 7-2 shows the bond energies presently known, 32 of them. They range from the very weak interactions of Hg_2, Cd_2 and Zn_2 up to the 225 kcal bond of nitrogen. There are obvious vertical trends. With only two exceptions, covalent bonds become weaker as we move down a particular column in the periodic table. One of the exceptions is fluorine, which has a much weaker bond than is consistent with the rest of the halogens. This 36 kcal bond accounts for the extreme reactivity of fluorine.

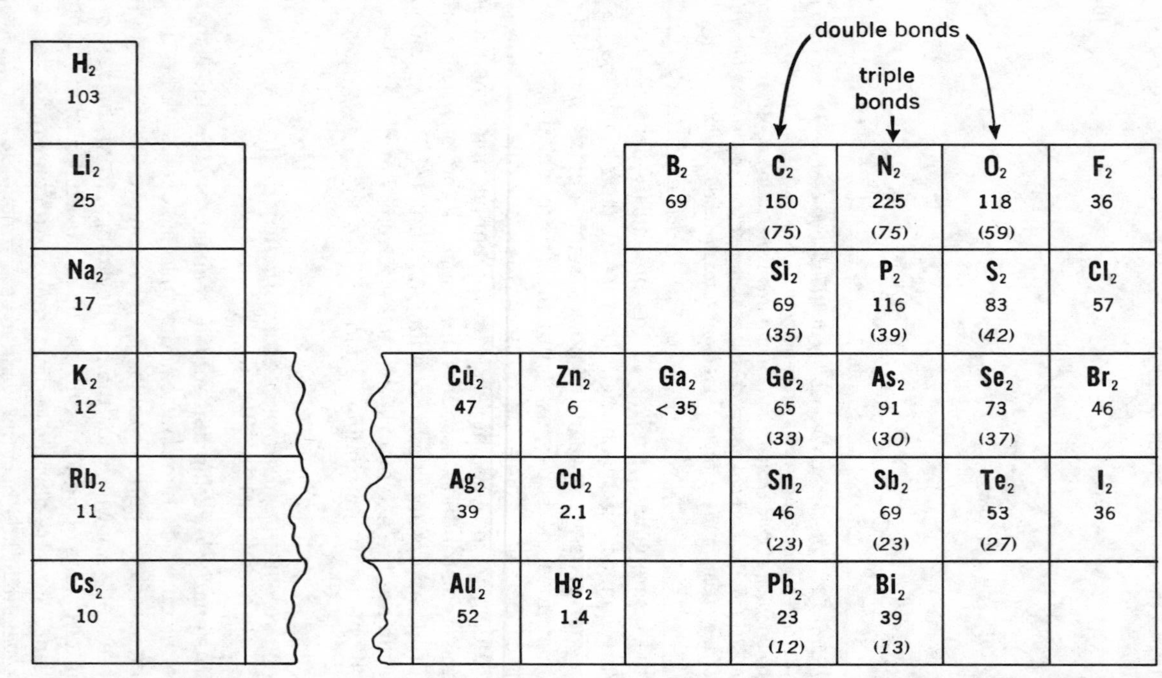

H_2 103								
Li_2 25				B_2 69	C_2 150 (75)	N_2 225 (75)	O_2 118 (59)	F_2 36
Na_2 17					Si_2 69 (35)	P_2 116 (39)	S_2 83 (42)	Cl_2 57
K_2 12		Cu_2 47	Zn_2 6	Ga_2 < 35	Ge_2 65 (33)	As_2 91 (30)	Se_2 73 (37)	Br_2 46
Rb_2 11		Ag_2 39	Cd_2 2.1		Sn_2 46 (23)	Sb_2 69 (23)	Te_2 53 (27)	I_2 36
Cs_2 10		Au_2 52	Hg_2 1.4		Pb_2 23 (12)	Bi_2 39 (13)		

Figure 7-2 Bond energies for homonuclear diatomics. Bond energy per bond order in parentheses (where applicable).

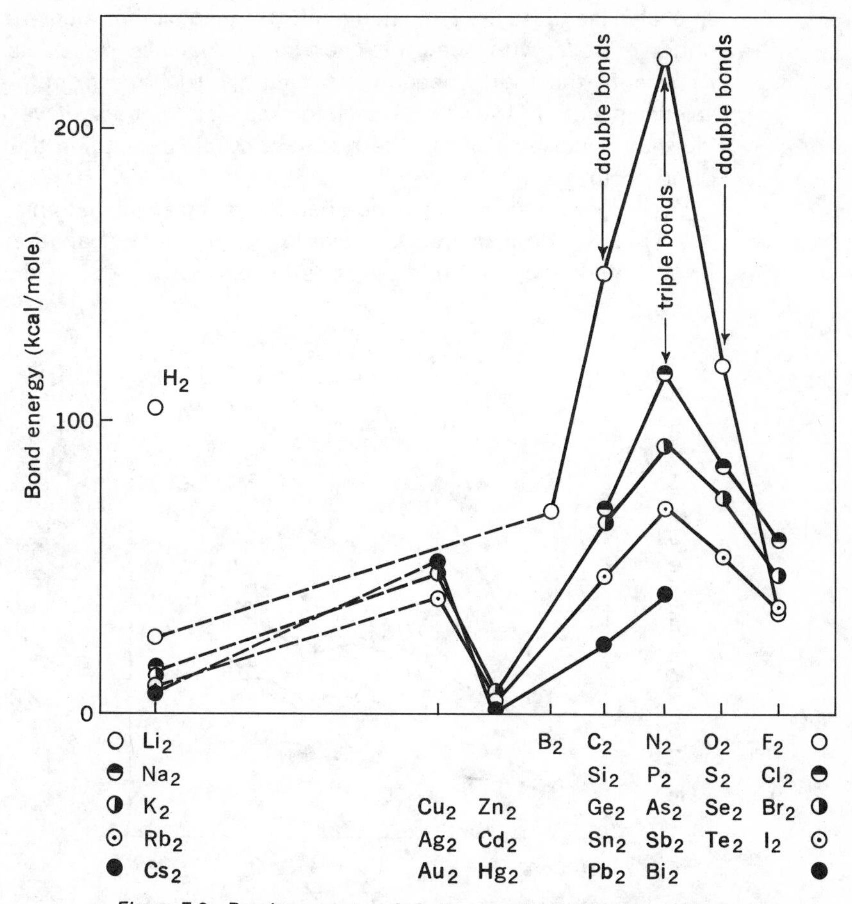

Figure 7-3 Bond energy trends in homonuclear diatomic molecules.

Figure 7-3 displays these energies graphically. The peaked shapes of the curves plainly reveal the triple bonds throughout the nitrogen family (N_2, P_2, As_2, Sb_2 and Bi_2). They also show that double bonds are obtained throughout the carbon family (C_2, Si_2, Ge_2, Sn_2, and Pb_2) and the oxygen family (O_2, S_2, Se_2, Te_2). Equally interesting is the Zn column with effectively zero bond energies (Zn_2, Cd_2, and Hg_2). This is appropriate to the orbital occupancies of these elements as gaseous atoms:

	3d	4s	4p	
Zn ·····	⊗⊗⊗⊗⊗	⊗	○○○	(7-3)

	4d	5s	5p	
Cd ····	⊗⊗⊗⊗⊗	⊗	○○○	(7-4)

	5d	6s	6p	
Hg ····	⊗⊗⊗⊗⊗	⊗	○○○	(7-5)

In each case there are just enough electrons to fill the valence d and s orbitals, with none left over for the valence p orbitals. Apparently the energy needed to promote an electron from the valence s state to the valence p state exceeds the energy-lowering that would result from the resultant covalent bonding that could occur.

Lastly, note the very low bond energies of the alkali diatomics (Li_2, Na_2, K_2, Rb_2, and Cs_2). These low values reflect another trend, a tendency for bond energies to rise as we move to the

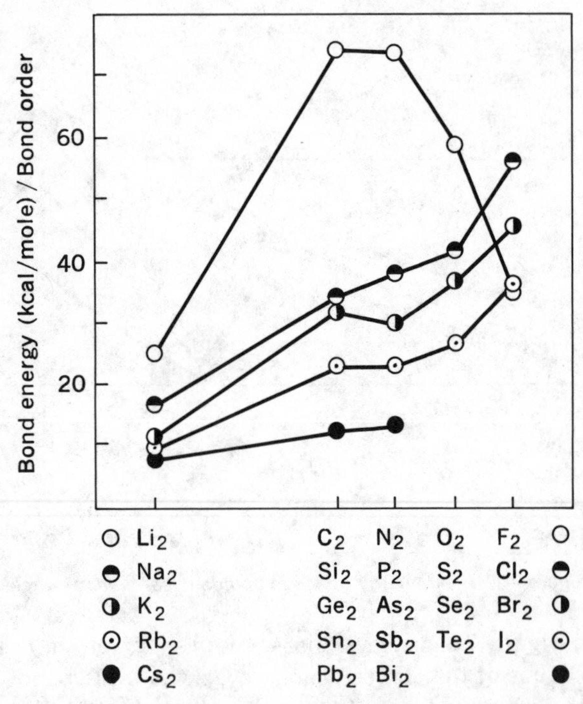

Figure 7-4 Bond energy per bond order trends in homonuclear diatomics.

right along a row. The effect is masked, at first glance, by the multiple bonds on the right-hand side of the periodic table. It can be revealed, however, by dividing each bond energy by its bond order to obtain the "bond energy per bond order." These values are shown parenthetically in Figure 7-2. The most uniform trend is obtained in the second row, as shown in Figure 7-4. The first row is unique in that its multiple bonds are especially stable and the single bond of fluorine is especially weak. All the other rows show the rising trend appropriate to the tendency for electron affinity to rise as one moves to the right in the periodic table (see Table 2-11).

(c) COVALENT HOMONUCLEAR BOND ENERGIES IN POLYATOMIC MOLECULES

The oxygen—oxygen bond in the O_2 molecule is a double bond—its bond energy is 118 kcal/mole. However, in hydrogen peroxide, H_2O_2, there are two oxygen atoms bonded to each other, but they share only one pair of electrons. This should be a single bond. Experimentally we find it is so. To break this bond (to give two OH molecules) requires 51 kcal, about half the energy needed to break the O_2 bond. The other measures of bond order—bond length and vibrational force constant (see Section 3-4d)—are also consistent. Table 7-1 shows the variations of bond properties with bond order for the four elements, carbon, nitrogen, oxygen and fluorine. As was found for diatomic molecules in Chapter Three, the three bond order criteria, bond energy, bond length and force constant, correlate and indicate the bond strength in polyatomic molecules. Any one of these measurements can be a useful indicator to the nature of the bonding in a polyatomic molecule.

Of the data in Table 7-1, the energy per bond order is of particular interest. Contrast the decreasing energy per bond order for the carbon compounds with the opposite trend for the nitrogen compounds. These trends undoubtedly contribute to the facts that acetylene, HC≡CH, is quite reactive, whereas elemental nitrogen, N≡N, is quite inert.

(d) COVALENT BOND LENGTHS

Table 7-1 lists covalent bond lengths for carbon and nitrogen as measured in compounds with single, double, and triple bonds. These trends are useful in the evaluation of the bonding in molecules with unorthodox structures. Figure 7-5 shows a plot of covalent bond length against bond order for carbon, nitrogen, and their second-row counterparts, silicon and phosphorus. The curve for the carbon—carbon bonds is extended to show that the bond length must approach infinity as the bond order approaches zero.

Figure 7-6 shows two important carbon compounds whose bonding is not readily explained with conventional bonding rules. Benzene, for example, can be written with two equally acceptable bond representations. Either structure implies alternate double and single bonds which are alternately long and short. The observed molecular structure shows six *equal* bond lengths of 1.40 Å! This length corresponds to a bond order near $1\frac{2}{3}$. The equal bond lengths force us to invoke the resonance argument that was used to discuss the bonding in sulfur dioxide, nitrate ion, carbonate ion, and so on, in Section 5-2e. Neither

Table 7-1 Bond Properties as a Function of Bond Order—Homonuclear Bonds

Compounds	**C**	**N**	**O**	**F**
single bond	$H_3C—CH_3$ ethane	$H_2N—NH_2$ hydrazine	HO—OH hydrogen peroxide	F—F fluorine
double bond	$H_2C{=}CH_2$ ethylene	HN=NH diimide	O=O oxygen	—
triple bond	HC≡CH acetylene	N≡N nitrogen	—	—
Bond energies (kcal)	**C**	**N**	**O**	**F**
single	83	60	51	36
double	143	?	118	—
triple	194	225	—	—
Bond energy/ bond order (kcal)	**C**	**N**	**O**	**F**
single	83	60	51	36
double	72	?	59	—
triple	65	75	—	—
Bond length (Å)	**C**	**N**	**O**	**F**
single	1.54	1.47	1.49	1.44
double	1.34	1.25*	1.21	—
triple	1.20	1.10	—	—
Force constants (mdyne/Å)	**C**	**N**	**O**	**F**
single	4	4	4	4.4
double	11	12	11.8	—
triple	15	23.0	—	—

* For FN=NF.

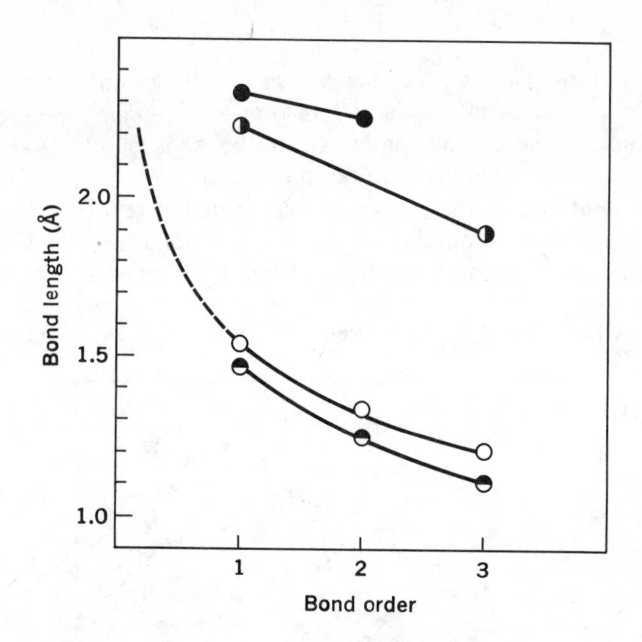

Figure 7-5 Change of bond length with bond order: ⊙, C—C bonds; ◓, N—N bonds; ●, Si—Si bonds; ◑, P—P bonds.

of the individual benzene structures shown in Figure 7-6 accurately describes the molecule—we need a superposition of the two. However, the bond order is about $1\frac{2}{3}$, a stronger bond than would be suggested by the simple average of a single and a double bond. This is generally the case: when the resonance

	Bond representation	Observed: C—C bond length	Observed: C—C bond order
Benzene	(two Kekulé structures of C_6H_6)	1.40 Å (all identical)	1.67 (all identical)
1, 3-Butadiene	$H_2C{=}CH{-}CH{=}CH_2$	C==C 1.35 Å; C—C 1.46 Å; C==C 1.35 Å	C==C 1.9; C—C 1.4; C==C 1.9

Figure 7-6 Molecules with more than one double bond: benzene and 1,3-butadiene. The observed bond orders are obtained from the graph of Figure 7-5 at the observed bond lengths.

argument is needed, the bonds are somewhat stronger than indicated by averaging. (Note that a molecular orbital description of the pi contribution to the benzene bonds does predict equal bond lengths and bond orders.)

Butadiene, the second molecule in Figure 7-6 has only one conventional representation, but the bond lengths show that the center bond has a bond order of 1.4, much stronger than a

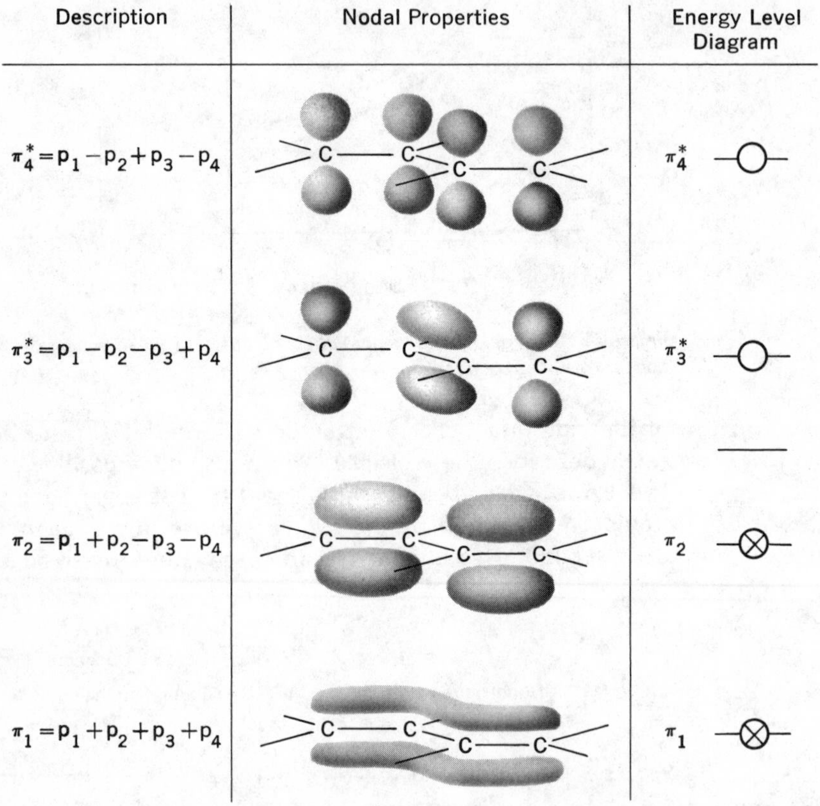

Description	Nodal Properties	Energy Level Diagram
$\pi_4^* = p_1 - p_2 + p_3 - p_4$	C—C C—C	π_4^*
$\pi_3^* = p_1 - p_2 - p_3 + p_4$	C—C C—C	π_3^*
$\pi_2 = p_1 + p_2 - p_3 - p_4$	C—C C—C	π_2
$\pi_1 = p_1 + p_2 + p_3 + p_4$	C—C C—C	π_1

Figure 7-7 Pi molecular orbitals for 1,3-butadiene, $CH_2CHCHCH_2$.

normal single bond, while the end bonds have a bond order near 1.8. Again this failing of the line representation is corrected in the molecular orbital description of the pi contribution to the bonding. Figure 7-7 shows the nodal properties of the pi M.O.'s and the resultant energy level diagram. With four electrons in these M.O.'s, the lowest two M.O.'s are occupied. The lowest M.O. contributes to the bonding of the center as well as to the end C—C bonds. A single electron pair divided among three bonds will contribute to each bond about one third of a bond order. The next M.O. strengthens only the end bonds and hence contributes one half of a bond order to each. These pi contributions must be added to the single bond contributions of the sigma orbitals to each bond. We are led to expect the end bond orders to be $1 + \frac{1}{3} + \frac{1}{2} = 1.8$ and the center bond order to be $1 + \frac{1}{3} = 1.3$.

Returning to Figure 7-5, we see that N—N covalent bonds are shorter than C—C bonds. On the other hand, the figure shows that phosphorus and silicon bonds are much longer than their nitrogen and carbon counterparts. We want to understand the factors that determine these trends.

We already have the answer at hand from our consideration of the hydrogen atom. For a one-electron atom, an increase in nuclear charge pulls the electron in closer—the average radius depends on $1/Z$. On the other hand, the radius increases with the square of n, the principal quantum number.

These two factors can be seen to be at work in many-electron atoms. In fact, we have already used a one-electron approximation of atomic size in discussing the observed bond lengths in H_2 and Li_2. To recapitulate that scheme, the nuclear charge Z^* felt by the bonding electrons is estimated from the first ionization energy (2-3).

$$E_1 = -313.6 \frac{Z^{*2}}{n^2}$$

This value of Z^* fixes the average radius of a hypothetical one-electron atom having the same quantum numbers (n and ℓ) and the same ionization energy as the valence electron (2-4).

$$\bar{r}(\text{Å}) = \frac{0.529 n^2}{Z^*}\left\{\frac{3}{2} - \frac{\ell(\ell+1)}{2n^2}\right\}$$

Figure 7-8 shows the calculated trends in $\bar{r}$ for the first 18 elements. Also shown, as crosses, are 11 covalent bond lengths (single bonds only). The observed bond lengths clearly reflect the sawtooth systematics attributable to the trend in atomic size, as indicated by $\bar{r}$. The periodic spikes are associated with a change in n, and the general shortening of bonds across a row is caused by an increase in nuclear charge with constant principal quantum number.

7-2 Heteronuclear bonds: ionic bonds

A bond between atoms of different elements is called a heteronuclear bond. We have already encountered them in earlier chapters. In Section 4-5, the bonding in first-row heteronuclear

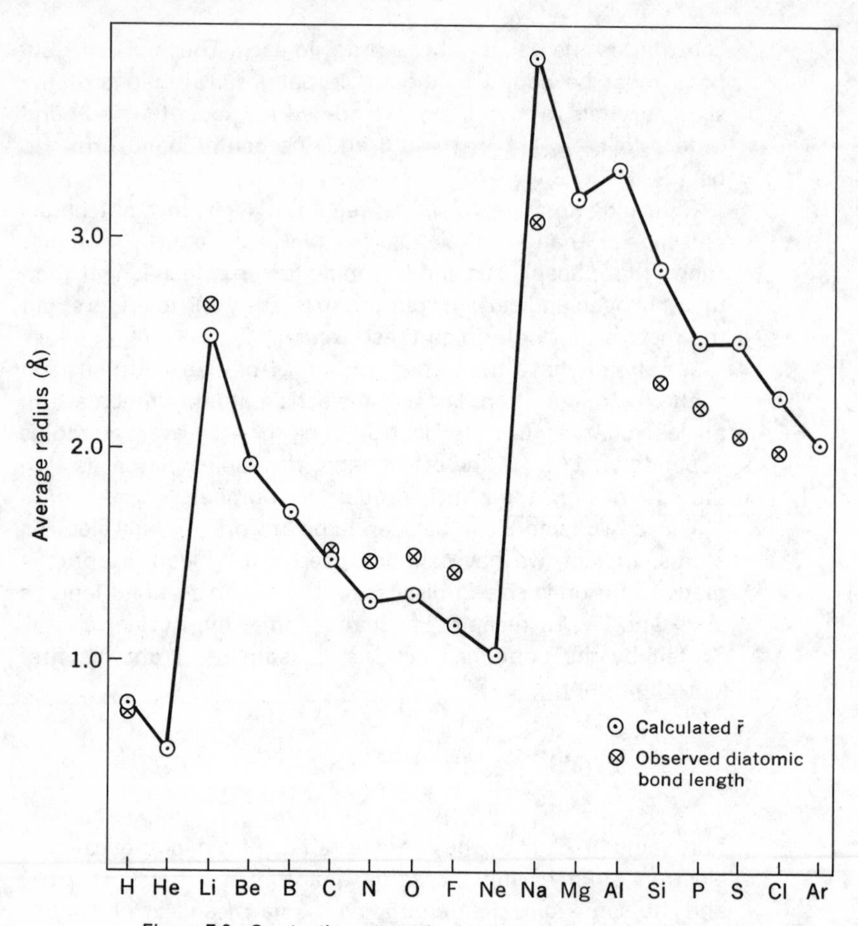

Figure 7-8 Sawtooth systematics in covalent bond lengths; ⊙, calculated $\bar{r}$; ⊗, observed diatomic bond length.

diatomic molecules was adequately considered in terms of the molecular orbitals of homonuclear diatomics. Earlier, though, in Section 3-4, three heteronuclear molecules were found to differ from homonuclear counterparts. In each of the molecules LiH, HF, and LiF, the electron distribution is skewed toward one of the atoms, producing a dipole moment. The electron movement takes place because the energy is lowered even more than the energy-lowering obtained from a perfect-sharing (or, covalent) distribution. These are important consequences—a dipole moment and a specially strong bond can make a molecule's fortune!

(a) IONIC BOND CHARACTER AND EXCESS BOND ENERGY

A bond in which the electrons concentrate nearer to one of the atoms is said to have *ionic character.* This ionic character is manifested in a dipole moment μ and in bond energy D_0 or, rather, in the excess bond energy above that to be expected from a perfect sharing distribution, $\bar{D}_{AB}$. The reference bond energy between two atoms A and B, $\bar{D}_{AB}$, is best taken to be the geometric mean of the bond energies of D_{A_2} and D_{B_2}, the bond energies of the covalent molecules A_2 and B_2. Either μ or $D_0 - \bar{D}_{AB}$ could be adopted as a quantitative measure of ionic character (indeed, both are used). Unfortunately they do not always agree, as we saw in Table 3-6. The molecule HF has a value of $D_0 - \bar{D}$ equal to 73 kcal/mole, midway between those of LiH and LiF, but it has the lowest dipole moment of the three.

Bond energies are generally more influential in determining chemistry than are dipole moments. Hence chemists have placed more emphasis on this feature in their attempts to assess ionic character. The magnitudes of these effects and their trends are shown in Table 7-2 for some molecules that involve only single bonds.

Table 7-2 Excess Bond Energies in Heteronuclear Single Bonds (kcal/mole)

MX	D_0(MX)	$D(M_2)$	$D(X_2)$	$\bar{D}$	$D_0 - \bar{D}$
LiF	137	25	36	30	107
LiCl	115	25	57	38	77
LiBr	101	25	46	34	67
LiF	81	25	36	30	51
NaF	107	17	36	25	82
NaCl	98	17	57	31	67
NaBr	88	17	46	29	59
NaI	71	17	36	25	46
HF	134	103	36	61	73
HCl	102	103	57	77	25
HBr	87	103	46	69	18
HI	71	103	36	61	10

The first molecule listed, LiF, is composed of two elements that form very weak covalent bonds. The diatomic molecules Li_2 and F_2 have bond energies of only 25 and 36 kcal/mole, respectively. Yet the gaseous diatomic molecule LiF has a bond energy of 137 kcal/mole! As was discussed in Section 3-4e, this extra energy must be attributed to an electron distribution skewed toward the fluorine atom, but it cannot be caused by "removal" of an electron from the lithium atom. The formation of gaseous ions Li^+ and F^- from neutral atoms requires an input of energy of 41 kcal/mole. The extra bond energy results from the fact that the movement of the lithium atom's electron over to the fluorine atom localizes that electron and holds it even closer to the lithium nucleus than in the neutral lithium atom. (The bond length in LiF is 1.56 Å, whereas the average radius of the lithium 2s orbital is 2.52 Å.) The importance of this effect is emphasized by the observation that the electron affinity of fluorine, 81 kcal, is 60 percent of the entire bond energy. The remaining 56 kcal must be attributed to the interaction of the bonding electrons with the lithium nucleus. This is more than double the Li_2 bond energy!

Turning to the next lithium halide entry, we see that the value of $D_0 - \bar{D}$ decreases to 77 kcal/mole in NaCl, despite the fact that both the bond energy and the electron affinity of chlorine are above those of fluorine. The larger size of the chlorine atom implies that localization of the bonding electrons on the chlorine is less beneficial to the lithium atom. The bond length in LiCl, 2.02 Å, is still less than the average radius of the lithium 2s orbital but almost 0.5 Å larger than the LiF bond length.

The remaining lithium halides continue the trend. As we move down in the periodic table, the excess bond energy attributable to ionic character decreases. The same trend is exhibited by the sodium halides and the hydrogen halides. We conclude that in each of these series, the ionic character decreases as the halogen atom becomes larger.

(b) EXCESS BOND ENERGY AND ELECTRONEGATIVITY

Over three decades ago, Linus Pauling attempted to use these excess bond energies to define a scale that encompassed what was then known about ionic character. Chemists had earlier attributed to each element a quality called "electronegativity." Pauling decided to establish a quantitative scale of electronegativity based on the excess bond energy. Each element would be assigned a number x, such that when two elements A and B form a bond, the difference between x_A and x_B would determine the excess bond energy. Well, that's the reverse of what Pauling did. He took the known excess bond energies and hunted around for suitable values of x so that the differences $(x_A - x_B)$ would be consistent with the facts. The functional form he worked with was

$$D_0 = \bar{D}_{AB} + 23(x_A - x_B)^2 \qquad (7\text{-}6)$$

$$\bar{D}_{AB} = \sqrt{D_{A_2} D_{B_2}} \qquad (7\text{-}7)$$

The difference $(x_A - x_B)$ appears squared to express the expectation that a difference between the x's is what is important,

Table 7-3 Pauling's Electronegativity Scale Based upon Excess Bond Energies. Elements in the Same Column of the Periodic Table Are Joined By Heavy Lines

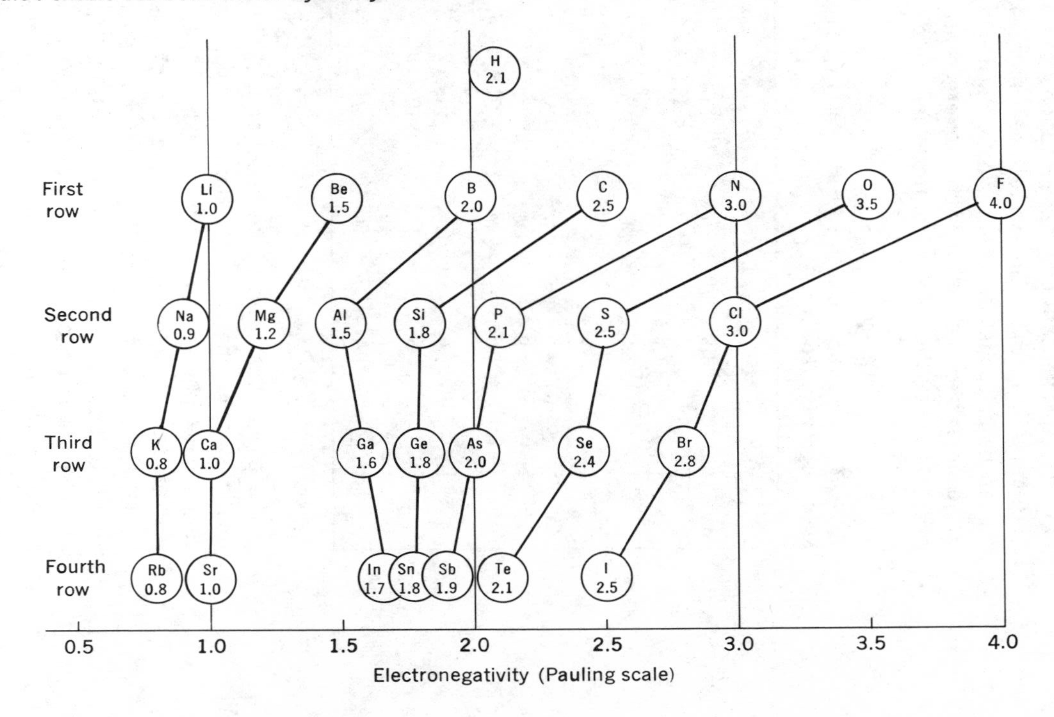

not its sign. The number 23 is an accident of history—it merely adjusts energy units to kilocalories. With the bond energies available to him, Pauling selected the electronegativity values shown in Table 7-3. This scale guides the qualitative thinking of most chemists. A bond between two elements that differ quite a bit in electronegativity can be expected to be strong and to involve a significant charge separation.

Quantitatively, things are not too keen. From expression (7-6), D_0 can be calculated from the electronegativities in Table 7-3. However, Table 7-4 shows that bond energies are not accurately calculated if electronegativity differences are too large.

Table 7-4 Bond Energies Calculated from Pauling's Electronegativities (kcal/mole)

AB	$x_A - x_B$	D_0 (calc.)	D_0 (expt.)	Discrepancy
LiF	3.0	238	137	101
LiCl	2.0	130	115	15
LiBr	1.8	109	101	8
LiI	1.5	82	81	1
NaF	3.1	247	107	140
NaCl	2.1	133	98	35
NaBr	1.9	111	88	23
NaI	1.6	84	71	13
HF	1.9	144	134	10
HCl	0.9	95	102	−7
HBr	0.7	80	87	−7
HI	0.4	64	71	−7

The other dimension of the electronegativity concept is its relationship to charge separation, or dipole moment. Once again, large charge movement should correlate with a large electronegativity difference. The extent of charge movement is best measured by the quantity $\delta = \mu/re$, in which r is the bond length and e is the charge on an electron. This hypothetical quantity δ is a fraction of an electron charge. It has a magnitude such that if $+\delta$ were to be placed on one atom and $-\delta$ on the other, the molecule would have the observed dipole moment μ. Figure 7-9 shows how δ and $(x_A - x_B)$ correlate for the molecules listed in Table 7-4. The hydrogen halides depend smoothly enough upon $(x_A - x_B)$, but they do not link up well with the alkali halides.

Many alternative electronegativity scales have been drawn

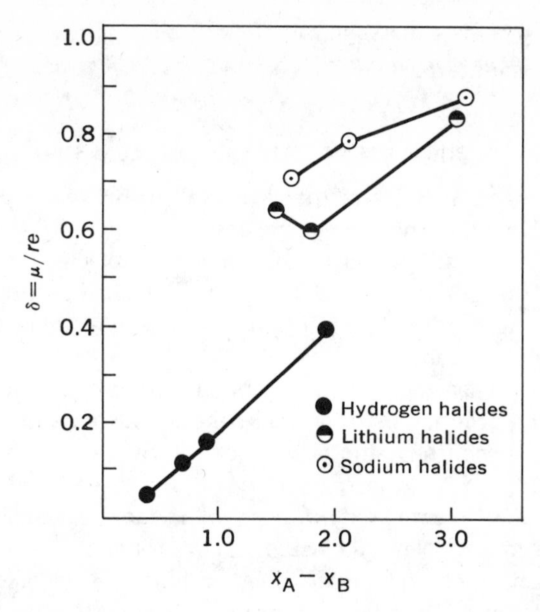

Figure 7-9 Charge separation and its relation to electronegativity difference: ●, hydrogen halides; ◓, lithium halides; ⊙, sodium halides.

up to polish the defects of the Pauling approach. None of these really reaches quantitative predictive value and Pauling's scale remains a useful guide to a chemist's intuition concerning the extent of ionic character.

7-3 Electron donor—acceptor bonds

Covalent bonds both with and without ionic character are bonds resulting from the proximity of two atoms each of which possesses a half-filled valence orbital (an orbital occupied by only one electron). Chemists call such an orbital "half-filled" since two electrons can occupy each orbital, be it an atomic orbital or a molecular orbital.

It is observed that even after all the "half-filled" valence orbitals of an atom have become engaged in bonding, the resultant molecule will still be reactive if it has other, completely vacant valence orbitals. We have already encountered some examples: CH_2 which reacts on almost every collision, BH_3 which is so reactive it has never been detected spectroscopically, and BeF_2 which can be produced in the gas phase only at very high temperatures because of the strong chemical bonds between BeF_2 groups in the solid state. The bonds formed by molecules containing atoms with vacant valence orbitals are very important

—we'll begin investigating them with the aid of BH_3 and its fluorine counterpart BF_3.

(a) BF_3 AND BH_3: VACANT VALENCE ORBITALS FOR RENT

The bonding in BF_3 and that expected in BH_3 were described in Section 6-1e. The pigeonhole representations (6-4) and (6-5) display three sp^2 sigma bonds for each molecule. In each of these bonds, each of the two atoms linked contributes one electron to make up an electron pair. The remaining p_z orbital of boron is vacant.

In each case the energy can be lowered still further if electrons can occupy this p_z orbital, because such electrons would thus be placed near the positive charge of the boron nucleus. In the case of BF_3, the $2p_z$ electrons of the three fluorine atoms do just that to some extent, accounting for the stability of BF_3 relative to BH_3. Nevertheless, this $2p_z$ orbital is not used to its full effectiveness in bonding. The boron atom would like to form a fourth bond but it doesn't have the one electron to contribute to a normal bond.

What is needed is another molecule with valence electrons to spare. Ammonia is such a molecule. After forming three bonds with hydrogen atoms, the nitrogen has a remaining pair of electrons in a valence orbital. This pair is called a "lone pair" or an "unused pair" and, with nitrogen's permission, it can be donated by the nitrogen atom and shared with a BH_3 molecule in the boron atom's vacant valence orbital. As is usual in chemical bonding, the electrons donated by the nitrogen are not "lost" to it—rather, they move so as to be simultaneously near both the nitrogen and the boron nuclei. An electron-dot representation shows the situation:

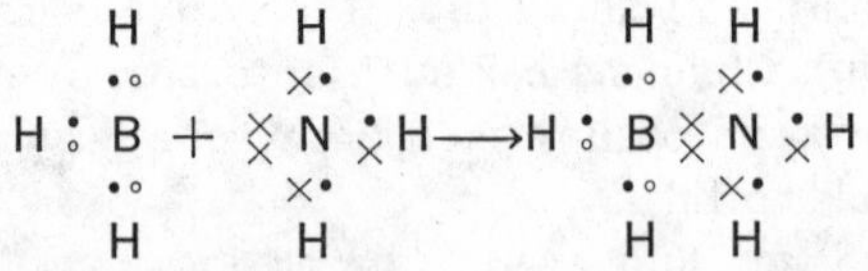

$$\begin{array}{ccccccccccccccc}
 & & H & & & & H & & & & & H & & H & \\
 & & \bullet\circ & & & & \times\bullet & & & & & \bullet\circ & & \times\bullet & \\
H & \overset{\bullet}{\circ} & B & + & \overset{\times}{\times} & N & \overset{\bullet}{\times} & H & \longrightarrow & H & \overset{\bullet}{\circ} & B & \overset{\times}{\times} & N & \overset{\bullet}{\times}\, H \\
 & & \bullet\circ & & & & \times\bullet & & & & & \bullet\circ & & \times\bullet & \\
 & & H & & & & H & & & & & H & & H &
\end{array} \qquad (7\text{-}8)$$

In (7-8) we see that boron, with its vacant valence orbital, acts as an *electron acceptor* and the nitrogen atom, with an unused valence pair of electrons, acts as an *electron donor.* The bond that can be formed is called an *electron donor–acceptor bond* (or, by some chemists, a "dative" bond).

The properties of some of the donor–acceptor compounds formed by BH_3 are well known, despite the illusory nature of BH_3 itself. For example, the bond energy of BH_3 with the ammonia-like compound trimethylamine $N(CH_3)_3$ has been measured to be 31.5 kcal/mole.

$$BH_3(g) + N(CH_3)_3(g) \rightarrow H_3B{-}N(CH_3)_3(g) \qquad (7\text{-}9)$$
$$\Delta H = -31.5 \text{ kcal/mole}$$

In this case, the electron donor–acceptor bond energy is similar in magnitude to the weakest covalent bonds known, those of the alkali metal diatomic molecules (in the range 10–25 kcal/mole) and those of F_2 and I_2 (each about 36 kcal/mole).

Despite the extra stability of BF_3 (because of the fluorine $2p_z$ electrons), it forms quite stable donor–acceptor bonds like those of BH_3.

$$BF_3(g) + NH_3(g) \rightarrow F_3B{-}NH_3(g) \qquad (7\text{-}10)$$

$$BF_3(g) + N(CH_3)_3(g) \rightarrow F_3B{-}N(CH_3)_3(g) \qquad (7\text{-}11)$$
$$\Delta H = -26.5 \text{ kcal/mole}$$

Again the bond energy is low compared with most chemical bonds. This is generally true; electron donor–acceptor bond energies are generally in the range 10–50 kcal/mole. To designate the weakness of this bond, the molecule is often written with a dot rather than a line—as in $BF_3 \cdot NH_3$ or $BF_3 \cdot N(CH_3)_3$—and called a "molecular complex." These notations imply that the molecular integrities of the reactants BF_3 and NH_3 are fairly well preserved in the new molecule. This is only partially true, as can be seen in the $F_3B{-}NH_3$ compound, whose molecular geometry is reasonably well known.

Figure 7-10 shows the structure of $F_3B{-}NH_3$. The hydrogen atoms are not very "visible" in the experimental technique that was used (which was electron diffraction), but all the heavy atoms are well located. Notice the dramatic change inflicted upon BF_3 as the bond forms. The parent molecule BF_3 is planar, with three equal BF bonds at 120° angles to each other. The BF bond lengths are 1.295 Å. In the donor–acceptor compound F_3BNH_3 the three BF bonds are folded back into a pyramid. The F—B—F bond angles are now 111°, only 1.5° from the perfect tetrahedral angle. The bond lengths, too, are significantly changed. The extension to 1.38 Å shows that the BF bonds are weakened, presumably because of the sacrifice of the extra pi bonding that stabilizes BF_3. The new bond formed, that between boron and nitrogen, is 1.60 Å in length. This is relatively long and suggests a bond order of about $\frac{1}{2}$ to $\frac{3}{4}$.

Oxygen compounds can also act as electron donors. For example, consider the reaction between BF_3 and dimethyl ether,

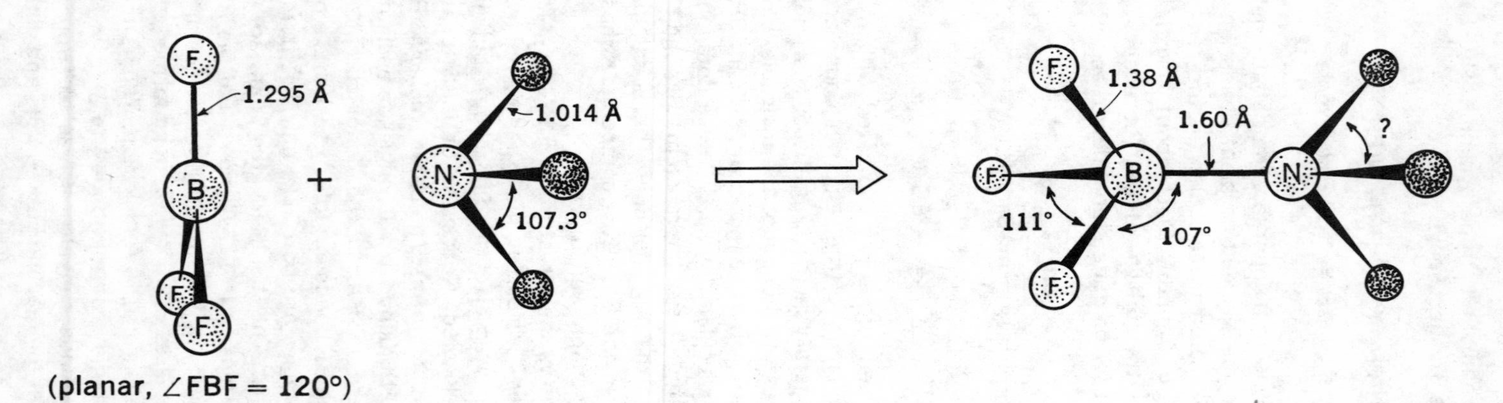

Figure 7-10 $BF_3 + NH_3 \rightarrow BF_3 : NH_3$ *Planar boron trifluoride folds into a pyramidal shape when it forms a donor–acceptor bond with the lone pair on ammonia. Note that the B—F bond lengthens in the complex since pi-bonding is not possible.*

CH_3OCH_3. Either the electron-dot or the pigeonhole representation displays the donor–acceptor bond possibility.

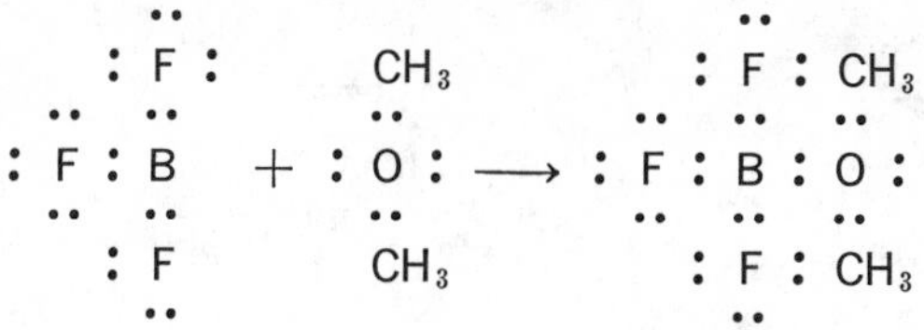

$$
\begin{array}{c}
\ddot{\text{:F:}} \\
\text{:}\ddot{\text{F}}\text{:}\ddot{\text{B}} \\
\text{:}\ddot{\text{F}}
\end{array}
\;+\;
\begin{array}{c}
CH_3 \\
\text{:}\ddot{\text{O}}\text{:} \\
CH_3
\end{array}
\longrightarrow
\begin{array}{c}
\ddot{\text{:F:}}\;CH_3 \\
\text{:}\ddot{\text{F}}\text{:}\ddot{\text{B}}\text{:}\ddot{\text{O}}\text{:} \\
\text{:}\ddot{\text{F}}\text{:}\;CH_3
\end{array}
\qquad (7\text{-}12)
$$

$$\Delta H = -13.9 \text{ kcal/mole}$$

or

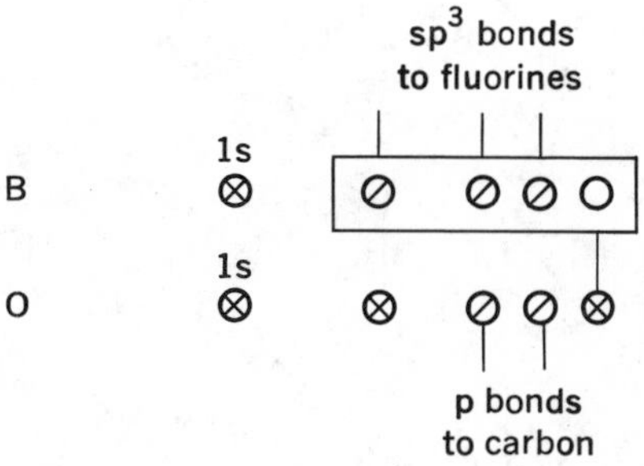

sp^3 bonds to fluorines

	1s				
B	⊗	⊘	⊘	⊘	○
O	⊗	⊗	⊘	⊘	⊗

(7-13)

p bonds to carbon

In the compound, the boron—oxygen bond length is 1.50 Å, to be compared with a normal single bond length of about 1.38 Å. This length and the low bond energy, 13.9 kcal/mole, indicate a bond order of $\frac{1}{2}$ or less.

Table 7-5 collects the bond energy and bond length data for the presently known BH_3 and BF_3 electron donor–acceptor compounds. Two distinctive features should be noted. First, as carbon monoxide forms the compound H_3B—CO, borine carbonyl, the C—O bond length remains at 1.13 Å, virtually unchanged from the triple bond length of the parent CO molecule. This is not atypical. The geometrical and bond length changes tend to concentrate in the electron acceptor, with less impact on the electron donor. The other feature is that the *nitrogen electron donors form stronger bonds than the oxygen donors.* In fact, pyridine, the last compound in Table 7-5, forms a donor–acceptor bond to boron with a 50.6 kcal/mole bond energy—a husky chemical bond!

(b) BERYLLIUM AND ALUMINUM—NOT TO BE OUTDONE

If vacant orbitals have capacity for bonding, beryllium should be twice as able as boron in this respect. A compound like BeF_2 requires that only two of the valence orbitals be employed—

Table 7-5 Electron Donor–Acceptor Compounds of BH_3 and BF_3

Electron Acceptor	Electron Donor	Compound	Bond Energy (kcal/mole)	Donor–Acceptor Bond Length (Å)
BH_3 borine	NH_3 ammonia	$H_3B—NH_3$	—	—
BH_3	$N(CH_3)_3$ trimethyl amine	$H_3B—N(CH_3)_3$	31.5	1.62
BH_3	C≡O carbon monoxide	$H_3B—C≡O$	—	1.54
BF_3 boron tri-fluoride	NH_3 ammonia	$F_3B—NH_3$	—	1.60
BF_3	$N(CH_3)_3$ trimethyl amine	$F_3B—N(CH_3)_3$	26.6	1.58
BF_3	$O(CH_3)_2$ dimethyl ether	$F_3B—O(CH_3)_2$	13.9	1.50
BF_3	H H C—C O—C C—H CH_3 C=C H H anisole	H H C—C F_3B—O—C C—H CH_3 C=C H H	12	—
BF_3	H H C—C N C—H C=C H H pyridine	H H C—C F_3B—N C—H C=C H H	50.6	—

there will be two vacant valence orbitals. Hence, we might expect to find BeF_2 (or $BeCl_2$) forming donor–acceptor bonds to two electron donors. Two examples are sufficient to validate this expectation.

Beryllium chloride forms a compound with diethyl ether, $O(C_2H_5)_2$, analogous to the BF_3 compound cited in (7-12). Because of the presence of two vacant orbitals, each $BeCl_2$ molecule combines with two ether molecules.

$$BeCl_2 + 2O(C_2H_5)_2 \rightarrow Cl_2Be\begin{matrix} \diagup O(C_2H_5)_2 \\ \diagdown O(C_2H_5)_2 \end{matrix} \qquad (7\text{-}14)$$

The second example is a bit different—hence more interesting. The substance $BeCl_2$ readily crystallizes, and its crystal structure arranges the molecules into infinitely long, bridged chains, as pictured in Figure 7-11. Each chlorine atom is equivalently bonded to two beryllium atoms, so that each beryllium

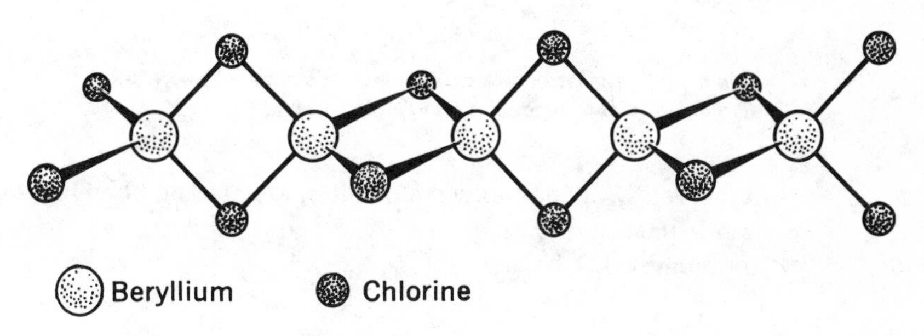

Figure 7-11 The structure of solid beryllium chloride, $BeCl_2(s)$. Each beryllium atom is surrounded by four chlorine atoms occupying the corners of a tetrahedron elongated along the chain axis. Every Be—Cl bond length is 2.02 Å.

is bonded to four chlorines. This can be seen to be the ultimate result of a combination of one bond in which a chlorine electron is shared along with a beryllium electron in a "normal" bond and another bond in which the chlorine donates an electron pair into a vacant beryllium valence orbital. With this parentage, the bonding is better described in terms of molecular orbitals. These M.O.'s give, in now familiar fashion, two equivalent bonds to each chlorine atom instead of a strong one and a weak one. The observed bond lengths, 2.02 Å, are rather long, indicating a bond order less than unity. Thus, the crystal structure of $BeCl_2$ is readily understood on the basis of electron donor–acceptor bonding. The electronegativity difference between beryllium and chlorine is 1.5, so the bonds are only moderately ionic in character. However, we see the interesting feature that chlorine is equivalently bonded to two atoms rather than to only one.

Aluminum is, of course, a member of the boron family. Hence we should expect donor-acceptor reactions like those shown in Table 7-5. The following is an example:

$$AlCl_3(g) + NH_3(g) \rightarrow Cl_3Al—NH_3 \qquad \Delta H = 40 \text{ kcal} \qquad (7\text{-}15)$$

Even more interesting is that aluminum chloride shows a fondness for the bridged structures like those in crystalline $BeCl_2$. However, a single such bridge expends the possibilities, so $AlCl_3$ merely dimerizes, as shown in Figure 7-12, instead of forming an infinite chain. Again we find the central chlorine atoms

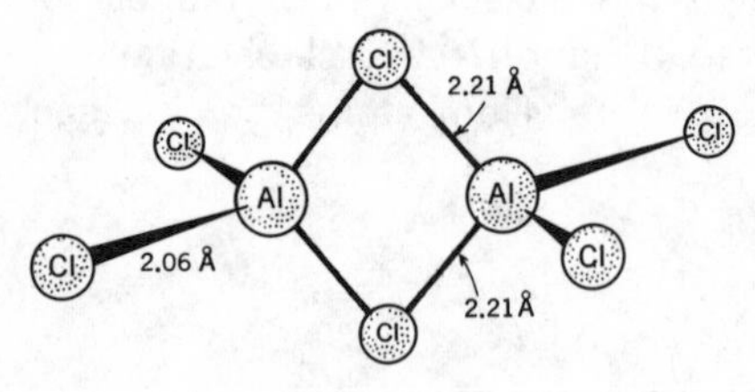

Figure 7-12 The structure of aluminum chloride dimer, $(AlCl_3)_2(g)$. Each aluminum atom is bonded to four chlorine atoms.

equivalently bonded to two aluminum atoms. The bond lengths show that these bridge Al—Cl—Al bonds are weaker than the terminal Al—Cl bonds.

(c) POSITIVE IONS (CATIONS) IN SOLUTION

One of the most striking aspects of aqueous chemistry is the appearance, in solution, of ions as separate entities, despite the enormous energies needed to form such ions in the gas phase. Consider, for example, the dissolving of $CuCl_2(s)$ in water to form $Cu^{+2}(aq)$ and $2Cl^-(aq)$. Without even worrying about the vaporization of $CuCl_2$ and the breaking of the copper—chlorine bonds to form atoms, we can see that large energies are involved in forming gaseous Cu^{+2} and Cl^- ions.

$$\begin{aligned} Cu(g) &\rightarrow Cu^+(g) + e^- && \Delta H = +178 \text{ kcal} \\ Cu^+(g) &\rightarrow Cu^{+2}(g) + e^- && \Delta H = +468 \\ 2e^- + 2Cl(g) &\rightarrow 2Cl^- && \Delta H = -172 \\ \hline Cu(g) + 2Cl(g) &\rightarrow Cu^{+2}(g) + 2Cl^-(g) && \Delta H = +474 \text{ kcal} \end{aligned}$$

Yet solid copper chloride blissfully and readily dissolves in water to form aqueous Cu^{+2} and Cl^- ions.

$$CuCl_2(s) \rightarrow Cu^{+2} + 2Cl^-(aq) \qquad \Delta H = -12 \text{ kcal} \qquad (7\text{-}16)$$

Nor does the formation of aqueous ions require the absorption of any 474 kcal of energy—exactly the opposite, a modest amount of energy (12 kcal) is *released* as heat. The only conclusion we can reach is that both Cu^{+2} and Cl^- involve strong interactions with the solvent.

In the classical view of this interaction the water molecules are pictured as dipoles, and the favorable orientation of these dipoles around the ions is considered to account for the stability of $Cu^{+2}(aq)$. We can, however, view this aquation process rather differently now, if we choose. The orbital occupancy of Cu^{+2} leaves four vacant valence orbitals:

	3d	4s	4p	
Cu^{+2}	⊗⊗⊗⊗⊘	○	○○○	*(7-17)*

In keeping with our current discussion, Cu^{+2} can act as an electron acceptor, making use of its vacant 4s and 4p orbitals. It should be a particularly effective acceptor because of the net charge on the ion. Any electron donor nearby should cause significant energy-lowering—for example, an oxygen atom from a water molecule. We might picture an electron donor–acceptor interaction involving four water molecules:

$$Cu^{+2} + 4H_2O \rightarrow Cu(OH_2)_4^{+2} \qquad (7\text{-}18)$$

This type of reasoning would also lead us to expect that other electron donors would be able to replace the H_2O molecules. For example, ammonia might do so—nitrogen is supposed to be a better electron donor than oxygen. We might expect a sequence of reactions such as

$$Cu(OH_2)_4^{+2} + NH_3 \rightarrow CuNH_3(OH_2)_3^{+2} + H_2O \qquad (7\text{-}19a)$$

$$CuNH_3(OH_2)_3^{+2} + NH_3 \rightarrow Cu(NH_3)_2(OH_2)_2^{+2} + H_2O \qquad (7\text{-}19b)$$

$$Cu(NH_3)_2(OH_2)_2^{+2} + NH_3 \rightarrow Cu(NH_3)_3(OH_2)^{+2} + H_2O \qquad (7\text{-}19c)$$

$$Cu(NH_3)_3(OH_2)^{+2} + NH_3 \rightarrow Cu(NH_3)_4^{+2} + H_2O \qquad (7\text{-}19d)$$

The equilibrium constants for all the reactions (7-19a–d) are known, and it is easy to calculate the effect of adding ammonia to an aqueous solution of Cu^{+2}. For example, suppose we add 0.15 moles/liter of ammonia to a 0.010 M Cu^{+2} solution. Calculations show that, at equilibrium, 92 percent of the $Cu^{+2}(aq)$ will have been converted to $Cu(NH_3)_4^{+2}$, and most of the rest will be present as $Cu(NH_3)_3(OH_2)^{+2}$. About one quarter of the ammonia has been consumed in forming the copper–ammonia complexes. Yet, in a 0.15 M NH_3 solution there are 370 times more water molecules competing for the opportunity of donating electrons to the Cu^{+2} electron acceptor! The NH_3 is obviously a better competitor—again the nitrogen compound is a better electron donor than the oxygen compound. This is contrary to expectations based upon dipole moments, since NH_3 has a molecular dipole moment smaller than that of water (1.47 D and 1.80 D, respectively).

(d) OXYGEN ATOM AS AN ELECTRON ACCEPTOR

There is one more important electron acceptor we should consider. The gaseous oxygen atom has, in its ground state, its orbital occupancy represented by

	1s	2s	2p		
O	⊗	⊗	⊗⊘⊘	^3P	*(7-20)*

This state, called the "triplet P" state, is lower in energy by 45 kcal than the next higher state because of decreased electron repulsion. This higher state, called the "singlet D" state, has the following pigeonhole diagram:

	1s	2s	2p		
O	⊗	⊗	⊗⊗○	^1D	*(7-21)*

The representations (7-20) and (7-21) suggest two possible bonding situations for the oxygen atom. The ^3P state could bond with any other pair of atoms, each with a half-filled valence orbital, giving such normal compounds as HOH, FOF, H_3COCH_3, HOCl, etc. The ^1D state, however, is an electron acceptor and it suggests that oxygen atoms should be able to form donor–acceptor bonds with atoms holding unused pairs. The energy requirement is that the bond energy exceeds the $^1D - {}^3P$ energy difference, 45 kcal.

A chloride ion, Cl^-, may be taken as the simplest example. This ion has four unused electron pairs and one might expect that one, two, three, or four oxygen atoms could bond to it through donor–acceptor bonding. The species thus predicted

Table 7-6 *¹D Oxygen as an Electron Acceptor*

Ion	Name	Structure
Cl^-	chloride ion	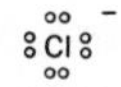
ClO^-	hypochlorite ion	Cl—O ?
ClO_2^-	chlorite ion	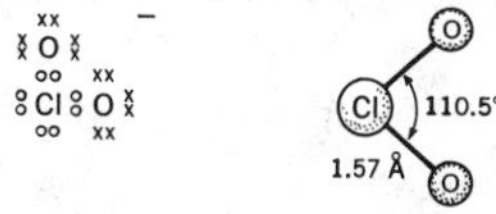 1.57 Å, 110.5°
ClO_3^-	chlorate ion	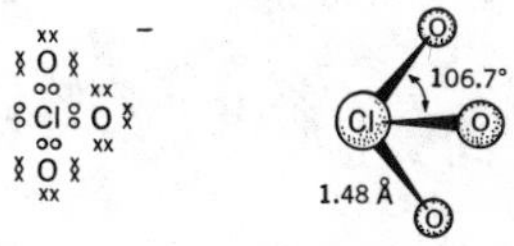 1.48 Å, 106.7°
ClO_4^-	perchlorate ion	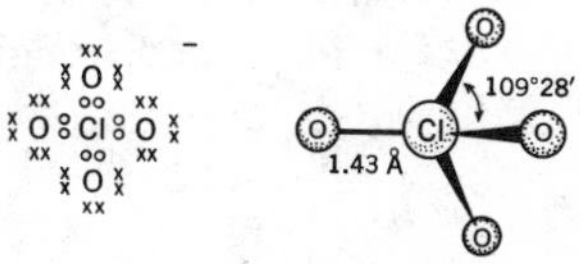 1.43 Å, 109°28′

are listed in Table 7-6 and they are all well known, both in salts and in aqueous solutions.

Table 7-6 also shows the electron-dot formulas and molecular structures for these ions. Concentrating on electron parentage, we see that the dot formula intrinsically describes the donor–acceptor nature of the bonding. The structures of these reveal a decreasing bond length as oxygen atoms are added. For contrast, the normal Cl—O bond length is 1.69 Å (as it is in Cl_2O). This shortening of the bonds as oxygen atoms are added is generally attributed to increasing ionic character of the Cl—O bonds. Since oxygen has the higher electronegativity, there will tend to be electron charge movement from chlorine to oxygen, leaving the chlorine slightly positive. As successive oxygen atoms bind to the chlorine, each pulls a little more negative charge off the chlorine, so its positive charge increases. This effect, in turn, pulls in the negatively charged oxygen atoms closer and closer.

Similar ions are formed by the other halogens to varying extents. Bromate BrO_3^-, iodate IO_3^-, periodate IO_4^- and, most recently discovered, perbromate BrO_4^-, are well known. No fluorine counterparts have been discovered.

From the bonding in these molecules, it is possible to predict some of their chemistry. As we invoke the ^1D description of the oxygen atom to explain the bond formation, it is implied that these compounds should be a potent source of oxygen atoms. These expectations are consistent with the observed chemistry —each of the species is a good oxidizing agent, as shown by the large positive reduction potentials displayed.

$$HClO_2 + 2H^+ + 2e^- \rightarrow HClO + H_2O \qquad \mathcal{E}^\circ = +1.64 \text{ volts}$$
$$ClO_3^- + 3H^+ + 2e^- \rightarrow HClO_2 + H_2O \qquad \mathcal{E}^\circ = +1.21 \text{ volts}$$
$$ClO_4^- + 2H^+ + 2e^- \rightarrow ClO_3^- + H_2O \qquad \mathcal{E}^\circ = +1.19 \text{ volts} \qquad (7\text{-}22)$$

Another way of comparing these compounds as oxidizing agents is to examine the exothermicity of some of their reactions. For example, we can compare the heats of reaction with $H_2(g)$ to form liquid water, using the similar reaction with oxygen itself as a reference.

$$H_2(g) + \tfrac{1}{2}O_2(g) = H_2O(\ell) \qquad \Delta H = -68.3 \text{ kcal/mole} \qquad (7\text{-}23)$$

$$3H_2(g) + HClO_4(aq) = 3H_2O(\ell) + HClO(aq) \qquad \Delta H = 3(-68.3) + 3.6 \text{ kcal/mole} \qquad (7\text{-}24)$$

$$2H_2(g) + HClO_3(aq) = 2H_2O(\ell) + HClO(aq) \qquad \Delta H = 2(-68.3) - 4.3 \text{ kcal/mole} \qquad (7\text{-}25)$$

$$H_2(g) + HClO_2(aq) = H_2O(\ell) + HClO(aq) \qquad \Delta H = -68.3 - 14.2 \text{ kcal/mole} \qquad (7\text{-}26)$$

These exothermicities show that, from an energy point of view, the molecules $HClO_4$, $HClO_3$, and $HClO_2$ are as good oxygen atom sources as oxygen itself and, in the case of $HClO_2$, substantially better.

A most exciting implication of this type of bonding is connected with the similarity between the halide ions and their adjacent neutral inert gas atoms. Thus, both Cl^- and Ar have eight electrons in their 3s and 3p orbitals. They are called "isoelectronic." Similarly, Br^- and Kr are alike, as are I^- and Xe as well. We are led to expect the existence of molecular analogues such as ArO_3, KrO_3, and XeO_3, to ions like ClO_3^-, BrO_3^-, and IO_3^-. In fact, one of these, XeO_3, was discovered in 1963. Its structure is extremely close to that of IO_3^-, as shown in Figure 7-13. In some respects, even its chemistry is similar! Like chlorate ion, iodate ion is a good oxidizing agent: it readily accepts electrons by giving up its electron-accepting oxygen atoms. Xenon trioxide is an even better oxidizing agent! It gives up its electron-accepting oxygen atoms so enthusiastically that it detonates when merely touched with oxidizable material such as a piece of tissue.

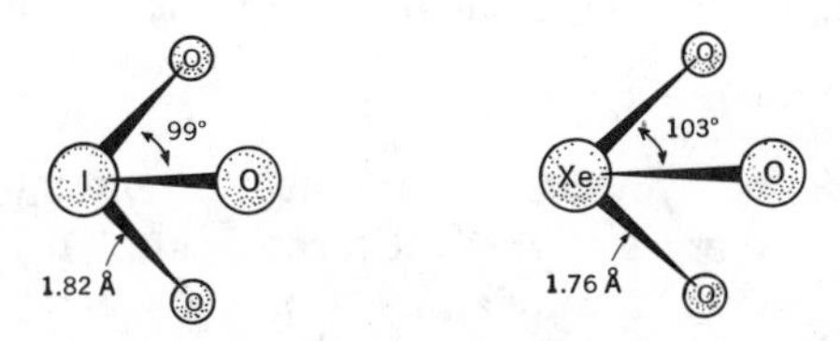

Figure 7-13 The structures of iodate ion, IO_3^-, and xenon trioxide, XeO_3.

(e) SUMMARY

Thus, the electron donor–acceptor bonding view provides a basis for discussion of bonding in boron trihalide—ammonia complexes, in crystalline $BeCl_2$, aquated and ammoniated cations in water, the oxides of chlorine and even in the oxide of xenon. We see that electron donor–acceptor bonding can be regarded as a link between the bonding in a wide variety of molecular types. Most important, it emphasizes once again that all bonding depends ultimately upon placing electrons simultaneously near two nuclei. A vacant valence orbital signals that possibility just as much as does a half-filled valence orbital. An atom (or molecule) with a vacant valence orbital needs to find a partner with valence electrons to burn—an electron donor. An atom (or molecule) with a half-filled valence orbital needs to find a partner with the same type of orbital occupancy—another half-filled valence orbital. An atom or molecule will continue to display residual bonding capacity until it has used all its valence orbital space as effectively as possible. We recognize this as merely a paraphrase of the classical bonding rule that atoms seek the inert gas electron configuration.

7-4 Electron-excess compounds

There are two more classes of compounds to add to our catalogue of bond types. Both classes challenge the classical representations of bonding presented in Chapter Five. Some compounds with laboratory-shelf stability have more electrons than the older bonding rules seem to allow—these we will call *electron-excess* compounds. Other easily stocked substances have too few electrons—these mavericks are called *electron-deficient* compounds. We'll deal with the electron-excess class first.

(a) TRIHALIDE IONS: I_3^- AND ITS FAMILY FRIENDS

The electron-excess compounds are well typified by the trihalide ion I_3^-. This ion forms in aqueous solution by reaction (7-27) and is found in ionic crystals as a linear, symmetric molecule.

$$I^-(aq) + I_2(aq) \rightarrow I_3^-(aq) \tag{7-27}$$

The trihalide ion long mystified theorists of chemical bonding. There is no convenient electron-dot formulation. Furthermore, neither I^- nor I_2 is apparently an electron acceptor, since both I^- and I_2 have achieved filled valence orbitals. From the classical bonding point of view, there are apparently too many electrons.

The molecular orbital picture saves the day for these electron-excess compounds. Following the pattern used in discussing the bonding of CH_2 (Section 4-6), we can examine the nodal surfaces of the molecular orbitals formed from the three axial (sigma) p orbitals. As in the case of CH_2, we must consider the nodal pattern of the central p orbital and link it with orbital combinations of the terminal atoms that possess the same nodal behavior. Figure 7-14 shows the molecular orbitals so formed. One, $p_x + (p_x' + p_x'')$, is a bonding orbital because it concentrates electrons between the atoms. The other, $p_x - (p_x' + p_x'')$, is antibonding since it does the opposite.

There is a third molecular orbital, $p_x' - p_x''$. If we regard all the iodine atom 5s orbitals to be ineffective in bonding, then

Description	Nodal Properties	Energy Level Diagram
$\sigma_2^* = p_x - (p_x' + p_x'')$	I I I	σ_2^*
$\sigma_N = p_x' - p_x''$	I I I	σ_N
$\sigma_1 = p_x + (p_x' + p_x'')$	I I I	σ_1

Figure 7-14 Axial molecular orbitals for triiodide ion, I_3^-.

there is no axial valence orbital on the central atom that doesn't have a nodal surface perpendicular to the molecular axis. Consequently the $p_x' - p_x''$ molecular orbital does not involve the central atom at all. Since p_x' and p_x'' are geographically distant, this M.O. is a *nonbonding* orbital. It neither helps nor hinders the bond as it forms.

We must now determine the orbital occupancy of the axial molecular orbitals shown in Figure 7-14. There are $3 \times 7 = 21$ valence electrons available from the 5s and 5p electrons of the three iodine atoms. Then there is one extra electron to give the ion its negative charge—so our total is 22. First, we place six electrons in the three low-energy 5s orbitals. Then 12 more go into the three $5p_y$ and three $5p_z$ orbitals, filling them completely. We are left with $22 - 6 - 12 = 4$ electrons with which to occupy the axial molecular orbitals. Two will go into the bonding M.O., $\sigma_1 = p_x + (p_x' + p_x'')$, and two into the nonbonding M.O., $\sigma_N = (p_x' - p_x'')$. The occupied bonding M.O. indicates that there will be a bond, hence there may be a stable species. Of course, one pair of electrons must occupy a three-center molecular orbital, so it will give only $\frac{1}{2}$ order bonds. But, energy-wise, we are about as well off with the I_3^- product as we were with the reactants I_2 and I^-. The I_2 molecule involves one single bond, and I_3^- involves two $\frac{1}{2}$ order bonds.

Turning to the other halogens, we might expect there to be other trihalide ions, such as Br_3^-, Cl_3^-, F_3^-, IBr_2^-, $BrCl_2^-$, and so on. However, our theory predicts very small energy effects as these ions form, so some may be easily prepared while others may be non-existent. That is the case. For example, no trihalide ion has been discovered in which a fluorine atom occupies the central position. Table 7-7 lists some of the well-known trihalide

Table 7-7 Some of the Known Trihalide Ions

	Bond Lengths		Reference Bond Lengths		
(X—Y—Z)⁻	X—Y	Y—Z	X—Y	Y—Z	Notes
(I—I—I)⁻	2.90	2.90	2.66	2.66	a
	2.83	3.04	2.66	2.66	b
(Br—Br—Br)⁻	2.53	2.53	2.28	2.28	
(Cl—I—Cl)⁻	2.36	2.36	2.32	2.32	c
(Cl—I—Br)⁻	2.38	2.50	2.32	2.47	
(Cl—Cl—Cl)⁻	—	—	2.01	2.01	d
(Cl—Br—Cl)⁻	—	—	2.14	2.14	e

(*a*) In crystalline compounds $N(C_2H_5)_4 \cdot 2I_2 \cdot I_3^-$ and $(C_6H_5)_4AsI_3$; (*b*) in the crystalline compound CsI_3; (*c*) known also through its vibrational spectrum: ICl force constant in ICl_2^- is 0.46 that of ICl; (*d*) known through its vibrational spectrum: Cl—Cl force constant in Cl_3^- is 0.3 that of Cl_2; (*e*) known through its vibrational spectrum: BrCl force constant in $BrCl_2^-$ is 0.3 that of BrCl.

ions and contrasts the bond lengths with those of a suitable parent halogen. All of them display weakened bonds—one pair of bonding electrons is trying to keep a three-atom molecule's body and soul together. Note that some of the trihalide ions are not symmetric. This will somewhat modify the molecular orbital description needed, but not in a way that contradicts the description. We conclude that there is nothing unique about I_3^- except its common occurrence—in dealing with standardized aqueous I_2 solutions, every freshman chemistry student (knowingly or un-) depends upon I_3^- to reduce I_2 volatility.

(b) OTHER POLYHALOGEN MOLECULES

We have explained the stability of such molecules as I_3^- and ICl_2^- with a molecular orbital description that places two electrons in a bonding M.O. and two in a nonbonding M.O. This raises a question. Why would an iodine atom be limited to one such three-center arrangement? Only one of the valence p orbitals of the central halogen is involved—there are two more such orbitals (both filled) perpendicular to the first one. Either one of these could contribute its pair of electrons to another three-center M.O. system with another pair of lucky halogen atoms. We are led to expect reactions such as:

$$I_3^- + I_2 \rightarrow I_5^- \qquad (7\text{-}28)$$

$$ICl_2^- + Cl_2 \rightarrow ICl_4^- \qquad (7\text{-}29)$$

$$ICl_4^- + Cl_2 \rightarrow ICl_6^- \qquad (7\text{-}30)$$

For ICl_4^- it is natural to picture a second Cl—I—Cl bond formation whose orientation is perpendicular to the first Cl—I—Cl group. All the atoms should be in the same plane and the bond lengths should be close to those of ICl_2^-. This is the observed structure as shown in Figure 7-15.

A similar arrangement for I_5^- is not observed. Rather, the I_5^- structure places the common atom in the end position, while preserving the two three-center bonds at about a 90° angle. This difference from ICl_4^- is easy to rationalize. The ICl_4^- structure for I_5^- would have to crowd four large iodine atoms around a central atom of the same size. The actual structure places the terminal iodine atoms in a less crowded region. The M.O. description does not demand that the iodine atom furnishing two electrons be the center atom—it can, as well, be on the end.

The third ion mentioned, ICl_6^-, is not yet known. It will turn up, no doubt, one of these days. Fluorine counterparts, IF_6^- and BrF_6^- have been prepared.

With this success we can expect another kind of polyhalogen molecule. If one of the filled valence p orbitals in I_3^- or ICl_2^- can form its own three-center M.O. bond, why can't the same happen with I_2 or ICl? This would lead to neutral polyhalogen molecules with structures reminiscent of those for the ions shown in Figure 7-15. Several of these are known and some are displayed in Figure 7-16. We will discuss only ClF_3, but the bond relationships in the others are clearly related.

In ClF_3, the chlorine atom forms a normal covalent bond with one fluorine atom. This is shown by the 1.60 Å bond length which is quite close to the 1.63 Å bond length in the diatomic ClF. Then using one of its perpendicular and filled p orbitals, the chlorine can form a pair of three-center bonds to two other fluorine atoms that should resemble those of ClF_2^-. The 1.70 Å

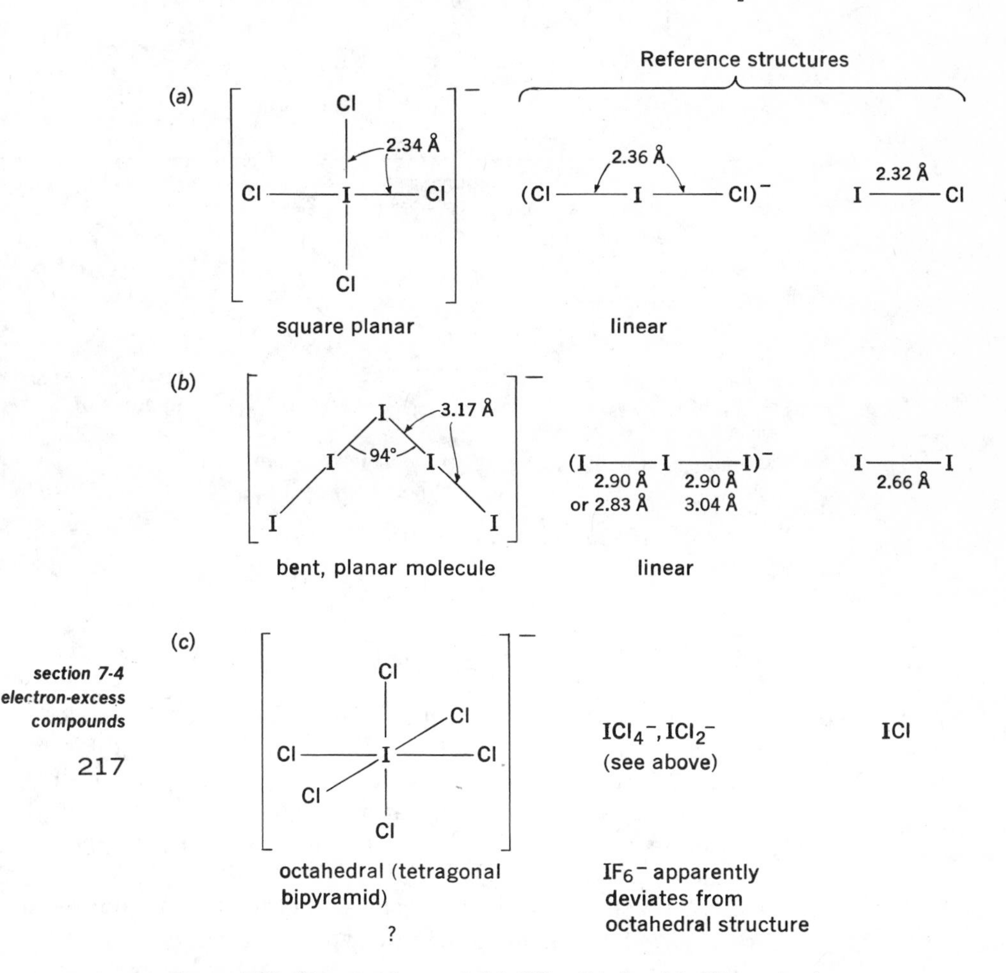

Figure 7-15 The structures of (a) ICl_4^-, (b) I_5^-, (c) ICl_6^- and reference structures.

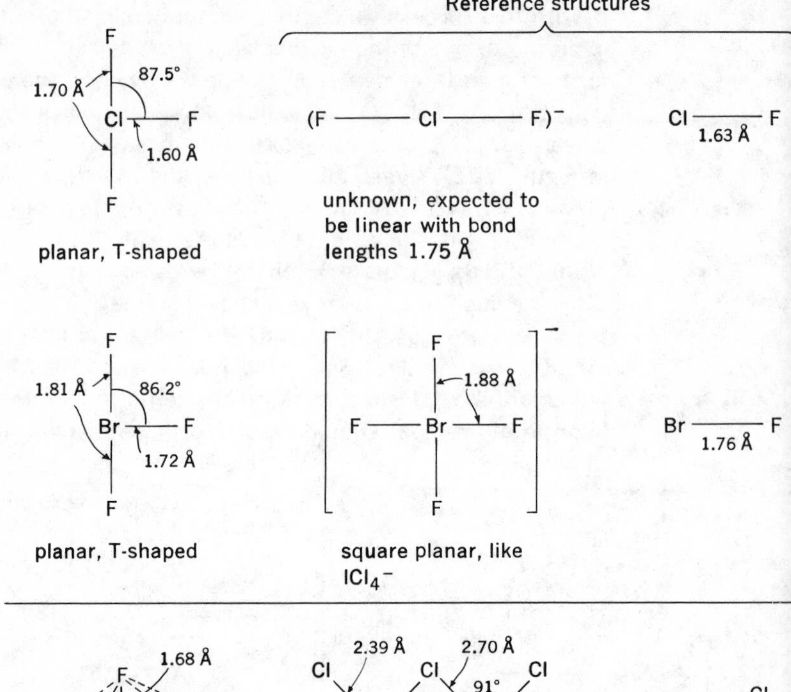

Figure 7-16 The structures of some neutral polyhalogens.

bonds, oriented almost perfectly in line and perpendicular to the short bond, fill this bill.

The other molecules shown in Figure 7-16 demonstrate the variety that exists, all understandable with the simple three-center M.O. argument developed for I_3^-. There are a few known polyhalogens with structures that may prove to be more complicated (e.g., IF_7 and I_8). However, even if a few polyhalogen structures prove to be unlike those pictured in Figures 7-15 and 7-16, we must be struck by the similarity of the others and the adequacy of a simple, three-center M.O. description of the bond lengths and bond angles for these electron-excess molecules.

(c) INERT GAS COMPOUNDS

As in the case of the IO_3^-—XeO_3 similarity, there are exciting possible analogues between the interhalogen compounds and their inert gas counterparts. Consider, for example, the reactions (7-31) and (7-32):

$$I^-(aq) + Cl_2 \rightarrow (Cl—I—Cl)^- \qquad (7\text{-}31)$$

$$Xe(g) + Cl_2 \rightarrow Cl—Xe—Cl \qquad (7\text{-}32)$$

Again, the identical orbital occupancies of I^- and Xe imply that as we explain the existence of ICl_2^-, we are predicting the possible existence of $XeCl_2$. In fact, such compounds were predicted on just this basis in 1951, long before they were discovered. The first such inert gas compounds were XeF_2 and XeF_4, prepared ten years after their being predicted from the molecular orbital description of I_3^-. Today, five simple inert gas–halogen compounds are known. Their structures, shown in Figure 7-17, are obviously related to those for the polyhalogens shown in Figures 7-15 and 7-16. It is an interesting sidelight that only two of the isoelectronic polyhalogens are actually known, ICl_2^- and IF_6^-. It seems significant that XeF_2 and XeF_4 are relatively easy to prepare but that no one has yet reported the preparation of either an IF_2^- or an IF_4^- salt.

The bond energies in XeF_2, XeF_4, and XeF_6 are about 30 kcal/mole. These are comparable to the bond energies in F_2 and I_2, and, in fact, all the xenon fluorides are energetically stable with respect to decomposition into the elements. The opposite is true for KrF_2 in which the bond energies are only 12 kcal/mole.

$$XeF_2(g) \rightarrow Xe(g) + F_2(g) \qquad \Delta H = +25.9 \text{ kcal/mole} \qquad (7\text{-}33a)$$

$$XeF_4(g) \rightarrow Xe(g) + 2F_2(g) \qquad \Delta H = +51.5 \text{ kcal/mole} \qquad (7\text{-}33b)$$

$$XeF_6(g) \rightarrow Xe(g) + 3F_2(g) \qquad \Delta H = +70.4 \text{ kcal/mole} \qquad (7\text{-}33c)$$

$$KrF_2(g) \rightarrow Kr(g) + F_2(g) \qquad \Delta H = -14.4 \text{ kcal/mole} \qquad (7\text{-}34)$$

On the other hand, these bond energies are sufficiently low that the margin of stability can be expected to be lost in many other inert gas possibilities, particularly those with Cl_2 or Br_2 for which the X_2 bond energy is much higher than that of F_2. For example, concerted efforts to prepare ArF_2, $KrCl_2$, and $XeBr_2$ have as yet been unsuccessful.

Inert gas compound	Discovered by (year)	Isoelectronic wit
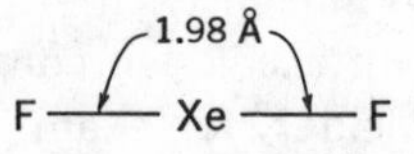 F — Xe — F (1.98 Å) linear, symmetric	Weeks, Chernick and Matheson (1962)	IF_2^- (not known
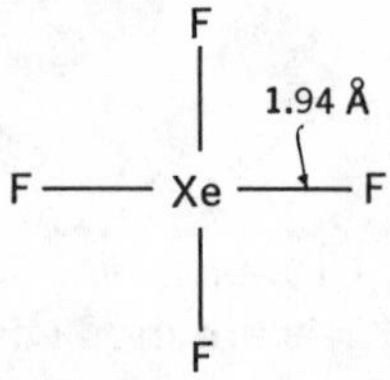 XeF_4 (1.94 Å) square planar	Claassen, Selig and Malm (1962)	IF_4^- (not known (BrF_4^-: See Fig. 7-
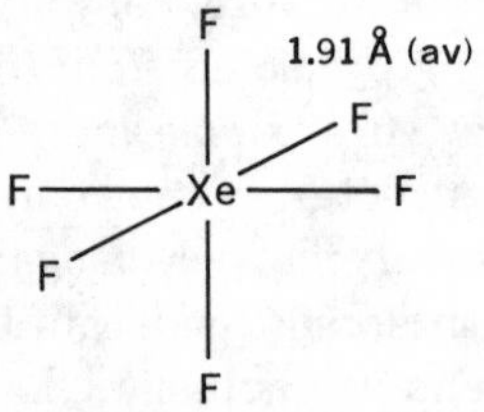 XeF_6 (1.91 Å (av)) Gas phase structure uncertain: *not* perfectly octahedral	Slivnick (1962)	IF_6^-: structure ma not be perfectly octahedral
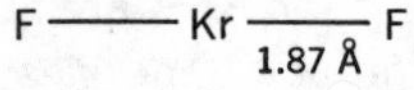 F — Kr — F (1.87 Å) linear, symmetric	Turner and Pimentel (1963)	BrF_2^- (not known (BrF_4^-: See Fig. 7- (BrF_3: See Fig. 7-1
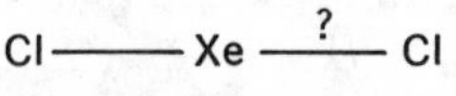 Cl — Xe — Cl (?) linear, symmetric	Pimentel and Nelson (1967)	(ICl_2^-: See Fig. 7-1

Figure 7-17 The structures of some inert gas compounds.

(d) HYDROGEN BIFLUORIDE ION, HF_2^-

Perhaps the most important electron-excess compound known is the hydrogen bifluoride ion HF_2^-. As for I_3^-, this ion is formed from two species, F^- and HF, both of which have filled valence orbitals.

$$F^-(aq) + HF(aq) \rightarrow FHF^-(aq) \quad (7\text{-}35)$$

Yet, the HF_2^- ion is present in every aqueous HF solution, since there is some F^- formed through dissociation of HF itself. Of course the concentration is raised if a fluoride salt is added

(NaF, KF, etc.). Solid bifluoride salts are also known, $NaHF_2$ and KHF_2 being the commonest.

The structure of the HF_2^- ion was first deduced to be linear and symmetric (as shown in Fig. 7-18) through entropy measurements. This conclusion has been firmly corroborated through infrared, neutron diffraction, and nuclear magnetic resonance spectra. The hydrogen atom is centrally placed between the fluorine atoms.

As in the trihalide ions, the molecular orbital point of view readily explains the bonding in HF_2^-. The M.O. energy level diagram is exactly the same as that for I_3^- (Fig. 7-14), although the bonding and antibonding orbitals have no nodal plane through the hydrogen atom since they involve its 1s orbital.

Just as in the trihalides, the last two electrons of HF_2^- are placed in the nonbonding orbital σ_N. Note that in both molecular types, the nonbonding orbital concentrates the extra charge on the terminal atoms. Consequently, these molecules will be most stable if the end atoms have high electronegativity, fluorine being the optimum. This expectation is most strikingly seen in the other possible bihalide ions. Until a few years ago it was felt that *only* fluorine formed such species. Since 1960, however, many bihalide ions have been discovered. Table 7-8 lists a number of these with their bond energies, all much

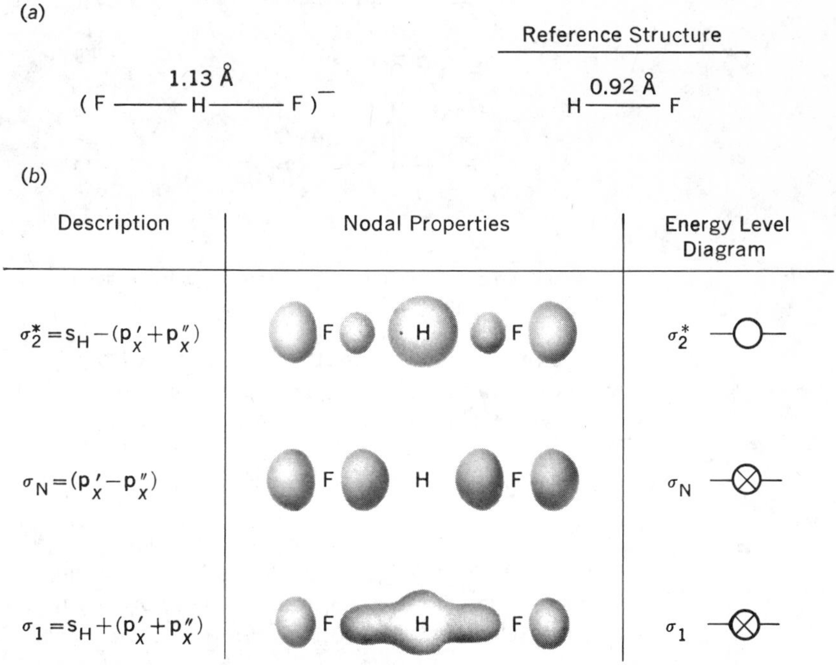

(*a*)

Reference Structure

$(F \overset{1.13\ Å}{—} H — F)^-$

$H \overset{0.92\ Å}{—} F$

(*b*)

Description	Nodal Properties	Energy Level Diagram
$\sigma_2^* = s_H - (p_x' + p_x'')$	F · H F	σ_2^*
$\sigma_N = (p_x' - p_x'')$	F H F	σ_N
$\sigma_1 = s_H + (p_x' + p_x'')$	F H F	σ_1

Figure 7-18 The structure (a) and molecular orbitals (b) of bifluoride ion, HF_2^-.

weaker than that of HF_2^-. Furthermore, infrared spectra have shown that HCl_2^- is linear but that its two bond lengths are different. Its hydrogen atom lies between the two chlorine atoms, bonded to both, but not quite equivalently. Again the situation is reminiscent of the polyhalogens. In some cases the three-center bond places the center atom equidistant from the end atoms, but in other cases it is closer to one than the other.

(e) THE HYDROGEN BOND—AN ELECTRON-EXCESS BOND

All the ions listed in Table 7-8 involve a hydrogen atom bonded simultaneously to two other atoms. A hydrogen atom so placed is said to form a *hydrogen bond*. This name is intended to distinguish these situations in Table 7-8 from normal bonds to hydrogen, as in H_2 or HF. The importance of HF_2^- stems from its prototype relationship to hydrogen bonds. Hydrogen bonds crucially influence the structure and chemistry of most biologically active molecules.

Table 7-8 Bond Energies of Some Hydrogen Bihalide Ions

$$HX + Y^- \rightarrow X—H—Y^-$$

X—H—Y⁻	ΔH(kcal/mole)
(F ——— H ——— F)⁻	−37
Cl ——— H ········· Cl⁻	−14
Br ——— H ········· Br⁻	−13
I ——— H ········· I⁻	−12
Cl ——— H ········· Br⁻	−9
F ——— H ········· Cl⁻	*a*
F ——— H ········· Br⁻	*a*
F ——— H ········· I⁻	*a*
Cl ——— H ········· I⁻	*a*

(*a*) Bifluoride ions known through infrared spectrum, bond lengths and bond energies not yet measured.

Experience tells us that whenever a hydrogen atom that is bonded to an atom A displays acidic properties, that hydrogen atom can form hydrogen bonds.

$$A—H + B \rightarrow A—H \cdots B \qquad (7\text{-}36)$$

The types of molecules B that can react with A—H are those we identified earlier (in Section 7-3) as good electron donors. For example, HCl forms a reasonably strong hydrogen bond to diethyl ether—ΔH is probably around 6 kcal/mole. Carboxylic acids, alcohols, phenols, and water—all of which are weak acids in aqueous solution—form hydrogen bonds to electron donors such as ethers R—O—R, ketones R—CO—R, ammonia NH_3, or amines RNH_2. The acidic proton acts as the electron acceptor.

Thus we have two explanations for the existence of the hydrogen bond. It can be regarded as an electron-excess compound with the bonding described in the molecular orbital framework as done here for the bihalide ions. Because the nonbonding electrons are placed on the terminal atoms these atoms should have high electronegativity. However, we can also regard the hydrogen atom as an electron acceptor because of charge displacement in its bond toward its highly electronegative partner atom (F, O, or N). The dipole moment of HF, 1.82 D, indicates that charge is moved away from the proton somewhat, leaving its valence orbital region partially vacant. This permits an electron acceptor–donor interaction with an electron donor. The smaller dipole moment of HCl, 1.07 D, suggests that it should form somewhat weaker interactions of this sort. In water, the molecular dipole moment, 1.82 D, implies a bond dipole moment of 1.49 D (see Section 6-4). Its hydrogen bonds should be intermediate to those of HCl and HF. If so, we can expect to find strong hydrogen bonds between water and negative ions like F^-, Cl^-, Br^-, or I^-. Table 7-8 shows the $Cl—H\cdots Cl^-$ bond energy to be 14 kcal and that of $Cl—H\cdots Br^-$ to be 9 kcal. The water—Cl^- bond energy ought to be about 20–25 kcal and that of water—Br^- about 15–20 kcal. Furthermore, in aqueous solution the abundance of water molecules implies that each halide ion will form several such hydrogen bonds. As in the chlorate ion, the four electron pairs of Cl^- might form four donor–acceptor bonds. Hence a gaseous Cl^- ion ought to release a large amount of energy when it dissolves in water—the bond energy of about four $H—O—H\cdots Cl^-$ hydrogen bonds.

$$Cl^-(g) + 4H_2O \rightarrow Cl^- \cdot 4H_2O(aq)$$
$$\Delta H \cong 4(-20 \text{ to } -25) \cong -(90) \text{ kcal} \qquad (7\text{-}37)$$

$$Br^-(g) + 4H_2O \rightarrow Br^- \cdot 4H_2O(aq)$$
$$\Delta H \cong 4(-15 \text{ to } -20) \cong -(70) \text{ kcal} \qquad (7\text{-}38)$$

The estimates, based upon a hydrogen bond view of the aquation of anions, compare favorably with the accepted aquation energies of Cl^- and Br^- (based upon heats of solution of HCl and HBr), 87 and 80 kcal/mole, respectively.

(f) HYDROGEN BONDING IN WATER

Now we see why water is such a remarkable solvent. As a substance like NaCl dissolves, Na^+ ions find that water molecules furnish an excellent environment because of their electron-donor capabilities, and Cl^- ions feel the same comfort because of water's electron-acceptor capabilities.

This schizophrenic personality of water is evident in practically every measurable property. Since H_2O can act both as an electron acceptor and as an electron donor, hydrogen bonds between water molecules interlink most of the molecules in liquid water. The most obvious result is that the boiling point of water is very much higher than those of similar molecules that do not possess the same hydrogen bonding capability. This is evident in Figure 7-19, which shows the distinctive boiling

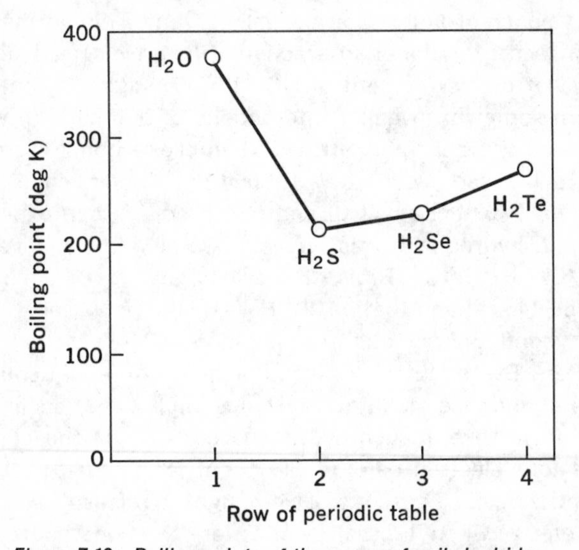

Figure 7-19 Boiling points of the oxygen family hydrides.

point of water among its periodic table counterparts. The viscosity of water is also abnormally high—it is five times the viscosity of diethyl ether and three times that of hexane. A most important manifestation of this intermolecular hydrogen bonding is found in the entropy of vaporization of water. The presence of special molecular orientations (to preserve approximately linear hydrogen bonds) implies a high degree of order in the liquid. Hence the disorder created on vaporization is much larger than that of the usual liquid $\Delta S_{\text{vap.}} = \Delta H/T =$ 26 cal/mole degK for water, whereas the norm (as expressed in Trouton's Rule) is 21 cal/mole degK. This entropy shows up in the solvent properties of water. A solute entering an aqueous environment must elbow into the interlinked liquid, breaking hydrogen bonds, which raises both the energy and the entropy. Whether the net effect works in favor of high solubility depends upon the extent to which solute–solvent interactions lower the energy again or create new, highly ordered orientations.

Finally, it is appropriate to note the crystal structure of ice. In this structure, each water molecule finds itself perfectly placed to form the maximum number of hydrogen bonds. It has four, tetrahedrally oriented neighbor oxygen atoms. To two of these the central water molecule forms hydrogen bonds by acting as the electron acceptor. To the other two neighbors it acts as the electron donor, with the neighbors furnishing the hydrogen atoms. The water–water hydrogen bond energy is 5–6 kcal/mole.

(g) MORE HYDROGEN BONDS

As we remarked earlier, there are many substances acidic enough to form hydrogen bonds. Most of these are both hydrogen bonding electron acceptors as well as donors. As for water molecules, their hydrogen bonding capability reflects into almost every measurable property. For example, it is a firm generalization that any molecule that can form hydrogen bonds to other molecules like itself, will surely crystallize in a lattice linked by such bonds. It is also generally true that hydrogen bonding compounds tend to be soluble in water. Their molecular configurations can be critically influenced by intramolecular hydrogen bonds, as is the case for many biologically important molecules. The protein molecule is held in its helical shape by hydrogen bonds and the double strands of DNA are held together by this same bond type.

Table 7-9 displays some hydrogen bond structures and their associated hydrogen bond energies. Typically they fall in the range 3–7 kcal/mole per hydrogen bond. Whenever an A—H···B hydrogen bond forms, the geometry is close to linear, and the A···B distance is a few tenths of an angstrom shorter than the A···B van der Waals separation expected with no hydrogen between. The A—H stretching frequency is substantially lowered by the hydrogen bond formation. For similar hydrogen bonds, there tends to be an inverse monotonic relationship between the bond energy and the bond length and a direct monotonic relationship between the bond energy and the downward shift in the A—H stretching frequency.

7-5 Electron-deficient compounds

The last class of compounds we'll consider confounds the classical valence rules as do the electron-excess compounds, but for the opposite reason: they seem to have too few electrons. Boron–hydrogen compounds furnish good examples.

Table 7-9 Some Hydrogen Bonds and Their Bond Energies $A—H\cdots B$

Alcohols	**ΔH(kcal/mole)**	**Acids**	**ΔH(kcal/mole)**
$CH_3O—H\cdots O(C_2H_5)_2$	2.5	$H—C(=O\cdots H—O)—$ / $—(O—H\cdots O=)C—H$ (cyclic dimer)	14.0(=2·7)
$CH_3O—H\cdots N(C_2H_5)_3$	3.0		
Phenols		**Chloroform**	
$C_6H_5O—H\cdots O(C_2H_5)_2$	3.7	$Cl_3C—H\cdots O{=}C(CH_3)_2$	2.5
$C_6H_5O—H\cdots N(CH_3)_3$	5.8	$Cl_3C—H\cdots N(C_2H_5)_3$	4.0
$C_6H_5O—H\cdots O{=}C(OCH_3)—C_2H_5$	5.7	**Inorganics**	
Amines		$N{\equiv}C—H\cdots N{\equiv}C—H$	3.3
$(C_6H_5)_2N—H\cdots O(C_2H_5)_2O$	2.3	$H—O—H\cdots OH_2$	5.0
Amides		$F—H\cdots F—H$	7.0
$C_2H_5C(=O)—N(CH_3)—H\cdots O{=}C(C_2H_5)—N(H)—CH_3$	3.6	$Cl—H\cdots Cl^-$	14
		$(F—H—F)^-$	37

(a) DIBORANE, B_2H_6

As we have said several times, the simplest borohydride, BH_3, has only transient existence. This reactivity arises, we explain, from the completely vacant valence orbital yearning for some electron occupancy. If an electron donor comes by (such as ammonia), a donor–acceptor bond is formed *toute suite*. If no such well-matched reaction partner appears, however, two BH_3 molecules react to form the simplest stable borohydride, diborane, B_2H_6.

The formula B_2H_6 is immediately reminiscent of ethane, C_2H_6.

However, the ethane structure requires 14 electrons to form the necessary bonds, shown in (7-39).

$$
\begin{array}{ccccc}
 & H & & H & \\
 & \circ\circ & & \circ\circ & \\
H \,\overset{\circ}{\underset{\circ}{}}\, & C & \overset{\circ}{\underset{\circ}{}} & C & \overset{\circ}{\underset{\circ}{}}\, H \\
 & \circ\circ & & \circ\circ & \\
 & H & & H &
\end{array}
\tag{7-39}
$$

If diborane were to adopt the ethane structure, some of the bonds would be short of electrons. Instead, it takes an entirely different structure in which two hydrogen atoms bridge between the two boron atoms (as pictured in Fig. 7-20). One is immediately struck by the similarity to hydrogen bonding—the central

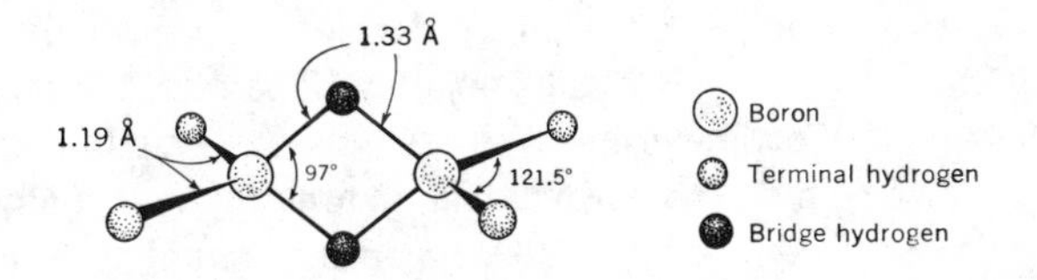

Figure 7-20 The structure of diborane, B_2H_6, an electron-deficient compound.

hydrogen atoms, in defiance of classical bonding rules, are each equally bonded to two atoms. The bond lengths contrast, too, as in hydrogen bonds—the four terminal B—H bonds, which are more normal, have bond lengths of 1.19 Å, whereas the B—H distances in the hydrogen bridges are longer, 1.33 Å.

In fact, the analogy to the hydrogen bond is a close one. Recalling the molecular orbital view of the linear FHF^- ion, we find the last two electrons lodged in a nonbonding orbital that localizes them on the highly electronegative fluorine atoms. This last pair of electrons is missing in the B—H—B bridge, as might be considered to be appropriate since the boron atoms have rather low electronegativity. More important, however, is the fact that the orbital so vacated is *nonbonding*, so the bond is not weakened.

In noting the geometry of the four bonds around the boron atom, we see that the boron orbitals are close to sp^3, tetrahedral hybridization. This can account for the non-linearity of the B—H—B bridge, again in contrast to the linear hydrogen bond. Nevertheless, the simplest description of each arm of the bridge is a three-center molecular orbital description like the hydrogen bond description, except that it involves sp^3 hybrids from the terminal atoms, a bent geometry, and only two electrons instead of four. Thus diborane solves its electron deficiency problem.

(b) SOME HIGHER BORANES: B_5H_9 AND $B_{10}H_{14}$

Diborane can react with itself to form higher boranes. As it does so, it loses hydrogen.

$$2\tfrac{1}{2}B_2H_6 \rightarrow B_5H_9 + 3H_2 \qquad (7\text{-}40)$$

$$5B_2H_6 \rightarrow B_{10}H_{14} + 8H_2 \qquad (7\text{-}41)$$

Both these structures involve some hydrogen bridges and some terminal B—H bonds like those that hold diborane together. The loss of hydrogen (in reactions (7-40) and (7-41)), however, accentuates the electron-deficiency problem. Consider pentaborane, B_5H_9. Each of the five boron atoms furnishes four valence orbitals (a 2s and three 2p), and the nine hydrogen atoms supply nine more. These 29 valence orbitals must make do with $5 \cdot 3 + 9 = 24$ electrons, only 12 pairs! Such an electron deficiency (or orbital excess) always causes "clustered" structures for which molecular orbitals, extending over many atoms, furnish the bonding. Thus B_5H_9 has a tetragonal pyramid structure, one boron atom "sitting" on top of four borons in a plane. The pyramid base is linked by four single hydrogen bridges, as shown in Figure 7-21(*a*). If we assign one pair of electrons to each B—H bond and one pair to each B—H—B bridge, 18 electrons are thus consumed. The remaining six electrons must occupy molecular orbitals extending over the whole boron pyramid,

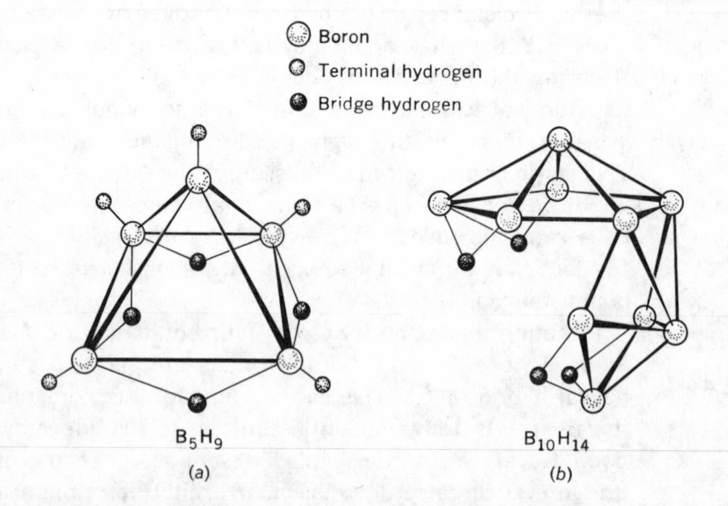

Figure 7-21 The cluster structures of (a) pentaborane, B_5H_9, and (b) decaborane, $B_{10}H_{14}$. The decaborane structure is drawn without the ten terminal hydrogen atoms—one on each boron. The boron framework is very close to a regular icosohedron with two vacant vertices. (An icosohedron is a figure with twelve vertices and twenty identical triangular faces.)

binding the apex boron atom to the base. Thus we can say that these last six electrons are "delocalized;" they occupy molecular orbitals that extend over the whole boron cluster.

Decaborane has an even more clustered, boat-like structure, as shown in Figure 7-21(*b*). There are ten normal B—H bonds, to each of which can be assigned an electron pair, and four B—H—B bridges that consume four more pairs. Twenty-eight electrons are thus used up. The entire molecule has $10 \cdot 3 + 14 = 44$ electrons, so 16 now remain. These eight pairs must link 19 boron—boron bonds (not counting the hydrogen-bridged B—B bonds). Again the electrons must occupy molecular orbitals that extend over the entire boron molecular skeleton.

We will not pursue these electron-deficient, orbital-excess compounds any further at this time. It is sufficient to note that *all metals are made up of electron-deficient, orbital-excess atoms.* Invariably, this type of atom condenses to a clustered or tightly packed solid and the electrons occupy molecular orbitals extending over the whole crystal. The result is that all the atoms share all the valence electrons, thereby occupying as effectively as possible the many vacant orbitals with the few bonding electrons available. Naturally an electron in an orbital that extends over an entire crystal has great mobility, which is the origin of the high electrical conductivity and reflectivity of metals.

7-6 New frontiers

Despite the emerging coherence in our views on chemical bonding, there is much to learn. Among the molecules that fall within our normal bonding rules there are many whose chemistry seems unnaturally reactive. For example, hypochlorous acid, HOCl, is a household chemical, but hypofluorous acid, HOF, was prepared for the first time in micromole quantities only a few years ago. The carbon compound ethylene, $H_2C{=}CH_2$, is a chemical of commerce, but the nitrogen counterpart, diimide, HN═NH, decomposes spontaneously at room temperature. We have already mentioned that sulfur readily forms a cyclic S_8 molecule, whereas there is no evidence that O_8 can exist, and the existence of O_4 is quite controversial. Turning to the electron-deficient and electron-excess compounds, we find our predictive powers to be quite limited. From the existence of the trihalide ions, we can predict the existence of inert gas halides, but we seem to have no firm basis for deciding which ones to seek. Then, the most stable of such compounds, XeF_2 and XeF_4, prove to be isoelectronic with polyhalide ions that are not known!

These dilemmas probe the limits now existing on our applications of bonding theory. We can predict the existence of molecules, picture their structures, and usually infer their chemistry. These predictions are correct often enough to be an extremely effective aid, but surprises occur now and then to keep life interesting and to warrant a becoming humility in the prophet. We will take up a few of these special cases that point in new directions.

(a) DIMERIZATION OF TWO FREE RADICALS: CN AND NO

The cyanide radical, CN, and nitric oxide, NO, are both free radicals. The bonding in each of these two molecules is helpfully discussed in terms of molecular orbitals. Their orbital occupancies, given in Figure 4-15, are repeated in Figure 7-22. Each

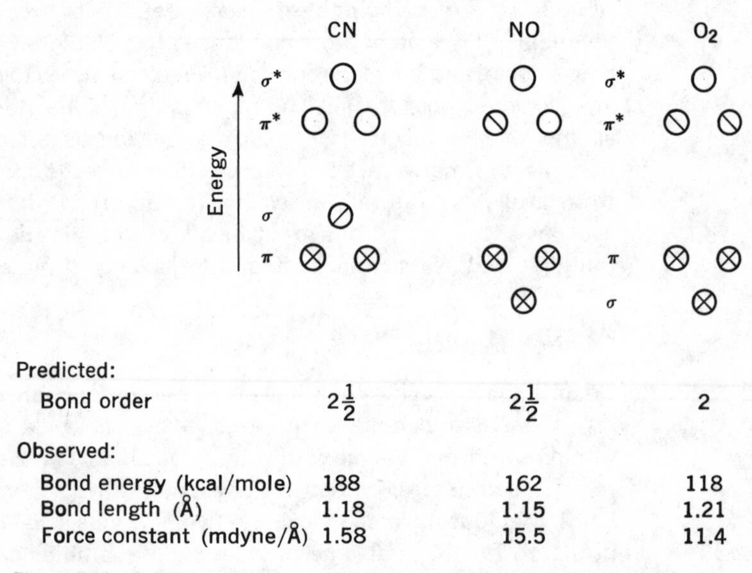

	CN	NO	O_2
Predicted:			
Bond order	$2\frac{1}{2}$	$2\frac{1}{2}$	2
Observed:			
Bond energy (kcal/mole)	188	162	118
Bond length (Å)	1.18	1.15	1.21
Force constant (mdyne/Å)	1.58	15.5	11.4

Figure 7-22 Orbital occupancy and bond properties of CN, NO, and O_2.

molecule has a $2\frac{1}{2}$ order bond—as is corroborated by the operational criteria of bond strength, energy, length, and force constant.

Each of these molecules could dimerize to a molecule with quite conventional bonding.

$$\text{C}\overset{\cdots}{=}\text{N} + \text{C}{=}\text{N} \rightarrow \text{N}{\equiv}\text{C}{-}\text{C}{\equiv}\text{N} \qquad \Delta H = -112\ \text{kcal} \qquad (7\text{-}42)$$

$$\text{N}\overset{\cdots}{=}\text{O} + \text{N}{=}\text{O} \rightarrow \text{O}{=}\text{N}{\diagdown}\text{N}{=}\text{O} \qquad \Delta H = -4\ \text{kcal} \qquad (7\text{-}43)$$

Both reactions occur, but the products differ greatly. The discrepancy in reaction heats is the first clue. The formation of cyanogen, C_2N_2, is accompanied by a generous release of energy, whereas nitric oxide dimer, N_2O_2, is formed in an almost thermoneutral reaction. However, this difference is readily rationalized. Counting bonds, there are seven in C_2N_2 but only $2 \cdot 2\frac{1}{2} = 5$ in the CN reactants. The formation of two extra bonds releases 112 kcal—nothing wrong there. The nitric oxide reaction involves only five bonds in the product—the same number as in the reactants. As the number of bonds is not increased, little energy is released.

Where then is the problem? The problem is that nitric oxide does *not* have the structure represented by the line drawing shown in (7-43). Instead it has a structure qualitatively described by the line drawing

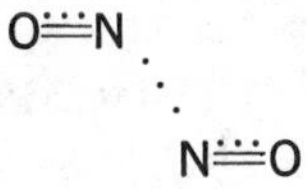

$$\begin{array}{l} O\overset{\cdots}{=}N \\ \quad\quad \cdot \\ \quad\quad\quad \cdot \\ \quad\quad\quad N\overset{\cdots}{=}O \end{array} \qquad (7\text{-}44)$$

The NO bond lengths and vibrational force constants show that the NO bonds are virtually unaffected by the dimerization. The N—N linkage is an extremely weak bond—not even as strong as a $\frac{1}{2}$ order bond. Thus we could have predicted that NO would dimerize in a thermoneutral reaction, but no one could have anticipated that Nature would pass up the "normal" structure (7-43) in favor of the dimer structure (7-44).

(b) NITRIC OXIDE AND THE NITROSYLS

Other simple compounds formed by nitric oxide add to the picture. Nitric oxide, a free radical, can react with a variety of atomic free radicals, such as hydrogen atoms, halogen atoms, lithium atoms, and so on.

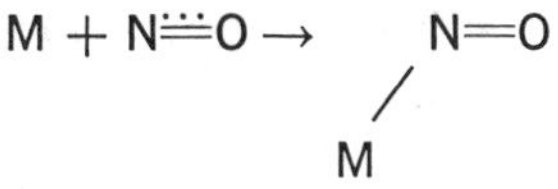

$$M + N\overset{\cdots}{=}O \rightarrow \begin{array}{l} \quad N{=}O \\ \ / \\ M \end{array} \qquad (7\text{-}45)$$

or?

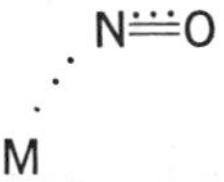

$$\begin{array}{l} \quad\ \ N\overset{\cdots}{=}O \\ \quad \cdot \\ \ \ \cdot \\ M \end{array} \qquad (7\text{-}45a)$$

This time we seem to be gaining. The three bonds shown in the product in (7-45) ought to be better than the $2\frac{1}{2}$ order bonds in nitric oxide. Yet the experience with N_2O_2 might make us cautious in our prediction. The facts are given in Table 7-10.

Table 7-10 The NO Bond Lengths and Force Constants in MNO Molecules

MNO Molecule	NO Bond Length (Å)	NO Force Constant (mdyne/Å)	M Electro-negativity
NO	1.15	15.5	—
FNO	1.13	14.7	4.0
ClNO	1.14	14.1	3.1
BrNO	1.15	14.1	2.8
HNO	1.21	10.5	2.1
LiON	—	8.0	1.0

The halo nitrosyls, XNO, are like the nitric oxide dimer. The NO bond length and force constant show that the $2\frac{1}{2}$ bond order is retained as represented in (7-45*a*). (The force constants show that the X—N bond is about a $\frac{1}{2}$ order bond.) In HNO, however, the NO bond length and force constant are quite appropriate to a double bond, as pictured in (7-45). The N—H bond is also quite normal—a single bond, as it should be. Then along comes lithium nitroxide! Not only does the NO bond weaken still further but now the lithium atom attaches itself at the other end of the NO group!

These experimental facts have been known only a few years and chemists are still trying to rationalize them. One view (proposed by the authors) builds on the observation that nitric oxide has a half-occupied antibonding orbital (see Fig. 7-22). Perhaps such a half-occupied π^* orbital can share electrons with a halogen atom (or another nitric oxide π^* orbital) to form a new and weak bond without disturbing the bonding in the rest of the molecule. Whether it does or not depends on the electronegativity of M. If χ_M is too low, the bonding changes. This is, of course, entirely an ad hoc proposal—but many theories start that way. The progress of science depends upon framing such models and then seeking a challenging experiment. In this case, oxygen and the peroxides furnish a test.

(c) OXYGEN AND THE PEROXIDES

Figure 7-22 shows that O_2 has half-occupied π^* orbitals. Perhaps it, also, should form weak bonds to its π^* orbitals. Table 7-11 shows how the situation looks at present. It parallels closely that displayed by nitric oxide. The oxygen molecule can add either one or two fluorine atoms without substantially reducing the bond order of the O—O bond. With hydrogen, however, the O—O bond order is reduced. The bond length and force constant of H_2O_2 define the conventional single bond and we deduce that HO_2 has a bond order of about $1\frac{1}{4}$.

One further agreeable test is provided by the absence of an easily identified O_4 species. Like NO dimer, the O_2 dimer should be approximately thermoneutral, with the molecular integrity of the O_2 retained. This is the case. If O_4 exists, its bond energy is only about 250 calories.

Finally, lithium again heightens the interest. The compound LiO_2 has recently been detected and the evidence strongly points to an isosceles triangular structure. The lithium atom can be pictured to be "embedded" in the oxygen—oxygen bond in a clustered arrangement. Could it be vacant valence orbitals at work?

Table 7-11 OO Bond Lengths and Force Constants in MO_2 and M_2O_2 Molecules

MO_2 or M_2O_2 Molecule	OO Bond Length (Å)	OO Force Constant (mdyne/Å)
O_2	1.21	11.4
FOO	—	10.5
FOOF	1.22	—
HOO	—	6.1
HOOH	1.48	4.6
LiO_2	—	~5

Table 7-11 does not mention the Cl_2O_2 peroxide. Its structure has only recently been deduced, and a line representation of its bonding matches *neither* F_2O_2 or H_2O_2. It is found to be

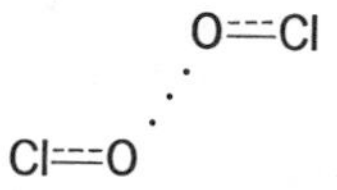

$$\begin{array}{cc} & O{=\!=}Cl \\ & \cdot^{\cdot} \\ Cl{=\!=}O & \end{array} \qquad (7\text{-}46)$$

This structure gives us only a momentary setback, however. This could be called ClO dimer, and its structure is like the NO dimer (7-44). Now we are reminded that ClO has a $1\frac{1}{2}$ order bond and a half-occupied antibonding orbital. So this structure is consistent with the π^* weak bonding possibility. In fact, we see that both F_2O_2 and Cl_2O_2 have two alternative structures, each involving the weak π^* bonds.

$$\begin{array}{ccc} & O{=}X & \quad O{=}O \\ & \cdot^{\cdot} & \quad \cdot^{\cdot} \quad \quad {}^{\cdot}\cdot \\ X{=}O & & X \qquad\qquad X \end{array} \tag{7-47}$$

The relative stability of OX and O_2 will determine which structure is found. For OCl the first structure exists and for OF, the second.

These attempts to verify the π^* explanation are reassuring, but by no means definitive. The value in these examples is that they indicate that we still have more to learn about chemical bonding. The frontiers are expanding rapidly enough to keep the field dynamic and exciting.

7-7 Reprise on chemical bonding

It has only recently become possible to sense any real coherence in the subject of chemical bonding. For many years, chemists had many more or less unrelated classes of chemical bonds: covalent bonds, ionic bonds, dative, metallic, one-electron, three-electron, bridge, donor–acceptor, chelate, coordinate, and hydrogen bonds all were used. Now with the guidance of quantum mechanics and with the use of the Pauli Principle, all of these types can be interrelated. The older terminology persists and will be useful for some time to come, but gradually the knowledge of one kind of bond is being correlated with that of others. Some useful generalities have emerged.

1. All bonds form because the potential energy is lowered as the bonding atoms approach.
2. The energy is lowered as a bond forms because electrons are then able to move simultaneously near the positive charges of two or more nuclei.
3. Electrons of one atom can move simultaneously near the positive charge of a second atom when that atom has valence orbitals that are either half-filled or vacant. Even when an atom has formed one or more bonds, it retains reactivity as long as there remain additional half-filled or vacant valence orbitals.
4. When a chemical bond forms between two atoms, each with a single, half-filled valence orbital, the bond energy will be in the range 30–130 kcal/mole. The bond energy is higher if one atom attracts electrons more strongly than the other, giving rise to the concepts of effective nuclear charge and electronegativity.
5. When a chemical bond forms between an atom with a

vacant valence orbital and an atom with an unused pair, the bond energy will usually be in the range 10–50 kcal/mole.

6. The bonding electrons in a molecule move in the field of all of the nuclei, hence in molecular orbitals. When the number of electron pairs engaged in bonding equals the minimum number of bonds needed to hold the atoms together, localized two-atom bond orbitals usually provide an adequate description (as shown in electron-dot representations or in conventional "valence" theory). However, when there are either electrons in half-filled orbitals, additional vacant orbitals, or too many or too few bonding electrons, the molecular orbital description is needed. Then, new concepts come into play: antibonding orbitals, nonbonding orbitals, multiple-bonds, triplet states, promotion, hybridization, and delocalization.

We will find that these generalizations give us a substantial basis for discussing the bonding in condensed phases, the subject of the next chapter.

Most of the substances around us are in the liquid or the solid state. Yet we have considered only gaseous molecules thus far. Liquids and solids are more difficult to deal with—a molecule of methane has only five atoms, but a small grain of sand may contain 10^{18} atoms. More difficult, yes, but fortunately, not in proportion to the number of atoms. In fact, the principles of bonding developed for gas molecules serve admirably as we classify the types of liquids and solids and as we explain their properties.

eight
solids and liquids

8-1 A general description of liquids and solids

The miracle of condensation is familiar. When night falls, vaporous water molecules join together to sparkle in a drop of dew. At a gust of winter weather, they join hands in the frigid array of a delicate snowflake. But then they scatter to the wind if caught by the warming sun.

Every substance has its own night, its own winter, and its own summer. Every element and every compound will condense to a liquid if cooled sufficiently. If cooled still further, every one of these liquids ultimately becomes a solid.* Conversely, every solid and every liquid can be vaporized if the temperature is raised sufficiently.

Liquids are readily recognized by their fluidity and solids by their rigidity. When a solid melts, its molar volume increases by a few percent (ice is a rare exception). More notable, however, is the fact that both liquids and solids are virtually incompressible compared with the responsiveness of a gas. Raising the pressure on a gas from one to two atmospheres halves its volume. Under the same pressure change, a typical liquid decreases in volume by about 0.01 percent and a solid by even less. The condensed phases are similarly resistant to temperature-induced volume change. Both liquids and solids behave as though their atoms and molecules are held together at spacings near their "natural size." Great effort is needed to crowd them closer together, and it takes work to pull them apart.

* Helium is a lone, willful exception. It does not become a solid at any temperature unless the pressure exceeds 25 atmospheres.

Within these generally uniform characteristics, liquids and solids display wide variety and individuality. Some solid substances are lustrous, malleable, and easily classed as *metals.* Others are crystalline, with regular crystal faces and sharp cleavage planes —some of these are classified as *salts* or *ionic crystals* and some as *covalent solids.* Still other solids are soft and retain many properties of the gaseous molecules from which they are condensed—these are *molecular crystals.* Table 8-1 shows the range of boiling points, melting points, and associated heat effects for these four types.

The molecular solids have melting and boiling points extending from almost absolute zero to above room temperature. In contrast, covalent solids, metals, and ionic solids generally melt at much higher temperatures. The energies holding the molecular solids and liquids together must be quite weak relative to those that cause covalent solids, metals, and ionic solids to condense. This conclusion is corroborated by the heats of fusion and vaporization.

The thermodynamic properties are quite informative. First, notice that *every ΔH entry is positive.* Every liquid has higher energy than its solid form at the melting point. Every gas has higher energy than its liquid form at equilibrium. Next, notice that *every ΔS entry is positive.* Every liquid is more random than its solid form at the melting point. Every gas is more random than its liquid form at equilibrium.

The magnitudes of the entropies are also meaningful. In the entire table, $\Delta H_{\text{fus.}}$ ranges from 5 to 11,000 calories, while $\Delta S_{\text{fus.}}$ ranges only from 2 to 17 cal/mole degK. The entropies of vaporization are even more nearly constant. Though $\Delta H_{\text{vap.}}$ ranges from 20 to 68,000 calories, all the values of $\Delta S_{\text{vap.}}$ lie between 16 and 27 cal/mole degK (if we omit helium).

These entropies paraphrase in quantitative thermodynamic language what everyone knows about solids and liquids. Solids involve atoms close together in regular lattice arrangements, whereas liquids have their atoms close together but packed in random fashion. The regularity of the solid makes it energetically more stable than the liquid and this energy effect accounts for the rigidity of the solid. Once the temperature is high enough for the randomness of the liquid to dominate the energy stability of the solid, melting occurs. Then the intrinsic disorder of the liquid permits it to flow, to take the shape of its container, and to act as a solvent for other molecules of quite different shape.

In turn, gases involve atoms far apart and even more disordered than in a liquid because of the positional randomness available to the gaseous particles. The liquid–gas entropy change

Table 8-1 Melting Points, Boiling Points, and Associated Heat Effects

Substance		Melting Point (°K)	Boiling Point (°K)	ΔH(kcal/mole)		ΔS(cal/mole deg K)	
Molecular	**Solids**			**Fusion**	**Vaporization**	**Fusion**	**Vaporization**
helium	He	1.4 (26 atm)	4.2	0.005	0.020	1.5	4.7
hydrogen	H_2	14	20	0.03	0.22	2.0	15.8
neon	Ne	24	27	0.080	0.431	3.26	17.5
nitrogen	N_2	63	77	0.17	1.33	2.7	17.2
argon	Ar	83	87	0.28	1.56	3.4	17.9
methane	CH_4	89	112	0.23	2.2	2.6	19.7
xenon	Xe	161	166	0.55	3.02	3.4	18.3
chlorine	Cl_2	172	239	1.53	4.88	8.9	20.4
n-nonane	C_9H_{20}	220	424	3.7	9.0	16.8	21.3
carbon tetrachloride	CCl_4	250	350	0.64	7.1	2.6	20.4
water	H_2O	273	373	1.4	11.3	5.3	30.1
benzene	C_6H_6	278	353	2.4	8.3	8.5	23.5
naphthalene	$C_{10}H_8$	353	491	4.6	9.7	12.9	19.7
Covalent Solids							
beryllium chloride	$BeCl_2$	678	760	3.0	25	4.5	30
germanium	Ge	1233	2973	8.3	—	6.7	—
silicon	Si	1683	2560	11	—	6.5	—
silica (quartz)	SiO_2	1883	—	2.0	—	1.1	—
carbon (graphite)	C(gr)	3273	5100	—	—	—	—
carbon (diamond)	C(dia)	dec. to C(gr)		—	—	—	—
Metals							
mercury	Hg	234	630	0.58	15.5	2.5	24.5
sodium	Na	371	1153	0.63	24.6	1.7	21.1
lithium	Li	353	1599	1	—	3	—
lead	Pb	600	2023	1.2	43.0	2.0	21.3
aluminum	Al	933	2600	2.6	67.6	2.8	26.0
silver	Ag	1234	2466	2.7	60.7	2.2	24.6
platinum	Pt	2042	4283	5.2	125	2.5	26.7
Ionic Solids							
potassium nitrate	KNO_3	610	673*	2.8	—	4.6	—
silver bromide	AgBr	703	1806	2.2	37.0	3.1	20.5
silver chloride	AgCl	728	1830	3.2	43.7	4.3	23.9
magnesium chloride	$MgCl_2$	987	1691	10.3	32.7	10.4	19.3
potassium chloride	KCl	1045	1680	6.1	38.8	5.8	23.1
sodium chloride	NaCl	1081	1738	6.8	40.8	6.3	23.5
barium chloride	$BaCl_2$	1235	1462	5.4	57	4.4	27

* Decomposes.

is dominated by the randomness of the gas, hence $\Delta S_{\text{vap.}}$ is approximately the same for all gases (near 21.6 cal/mole degK: this is Trouton's Rule) unless there is some special order in the liquid state. Water is the usual example cited—hydrogen bonding is the cause.

(b) THE ELEMENTS AND THE PERIODIC TABLE

An element has but one kind of atom, so there is no basis for charge separation or the formation of ionic solids. We are left with three classes: molecular solids, covalent solids, and metals. Figure 8-1 shows the distribution of these types across the periodic table. There are only about 15 elements that are obviously molecular solids—these are crowded into the upper right corner. In contrast, there are about 70 elements that are metallic—those off to the left in the periodic array. Between the metals and the molecular solids there is a zone of elements that includes the covalent solids as well as some solids not readily classified. Some elements, like arsenic and antimony, exhibit both molecular and metallic forms. Phosphorus displays both covalent and molecular solid forms. These borderline elements are particularly important because of their intermediate character, and we'll pay particular attention to them.

Compounds between two different nonmetallic elements always form molecular or covalent solids. A compound between a metallic and a nonmetallic element will usually form an ionic or a covalent solid. Two metals may form one or more metallic compounds or, more often, a range of metallic solutions of one element dissolved in the other.

With this general feeling for the geography of the solid state, we'll tackle the various bonding situations one at a time. The solids will be considered roughly in order of melting points, lowest first. This means we'll begin with the inert gases.

8-2 The inert gases

The elements in the last column of the periodic table are distinctively unreactive. Only a few weakly bonded compounds of xenon are known and only a single compound of krypton—no compounds of argon, neon, or helium have ever been prepared. These elements are well described by the family name "inert gases." The reason for this inertness is understandable in terms of the traditional bonding rules and orbital occupancy. All the valence orbitals are filled—there remains no capacity for normal chemical bond formation.

Yet these elements, too, liquify at a sufficiently low tempera-

Nonmetals: Covalent solids; Molecular solids

Metals

1 H_2																	2 He
3 Li	4 Be											5 B_{12}	6 C	7 N_2	8 O_2	9 F_2	10 Ne
11 Na	12 Mg											13 Al	14 Si	15 P_4	16 S_8	17 Cl_2	18 Ar
19 K	20 Ca	21 Sc	22 Ti	23 V	24 Cr	25 Mn	26 Fe	27 Co	28 Ni	29 Cu	30 Zn	31 Ga	32 Ge	33 As	34 Se	35 Br_2	36 Kr
37 Rb	38 Sr	39 Y	40 Zr	41 Nb	42 Mo	43 Tc	44 Ru	45 Rh	46 Pd	47 Ag	48 Cd	49 In	50 Sn	51 Sb	52 Te	53 I_2	54 Xe
55 Cs	56 Ba	57 La	72 Hf	73 Ta	74 W	75 Re	76 Os	77 Ir	78 Pt	79 Au	80 Hg	81 Tl	82 Pb	83 Bi	84 Po ?	85 At ?	86 Rn
87 Fr	88 Ra	89															

58 Ce	59 Pr	60 Nd	61 Pm	62 Sm	63 Eu	64 Gd	65 Tb	66 Dy	67 Ho	68 Er	69 Tm	70 Yb	71 Lu
90 Th	91 Pa	92 U	93 Np	94 Pu	95 Am	96 Cm	97 Bk	98 Cf	99 Es	100 Fm	101 Md	102 No	103 Lw

Figure 8-1 The nature of the elemental solids.

ture and solidify at still lower temperatures. What is the origin of the energy-lowering that causes this condensation? We reply without hesitation: the energy can be lowered relative to the gas phase if electrons of one atom have an opportunity to be attracted to other nuclei in the condensed state. The plot remains the same—simple electric interactions are at the heart of all chemistry.

Since the valence orbitals of each atom are all occupied, the only possible place for electron sharing is in the high-energy, empty orbitals of its neighbors. This assertion is readily verified with the one-electron, effective-charge model of the neon atom. In Table 2-1, the effective nuclear charge Z^* felt by the most easily removed neon electron was given as 2.52. The average radius of a valence 2p electron near a nucleus with $Z^* = 2.52$ is 1.05 Å. However, the 2p orbitals are all completely occupied; if another atom approaches, it must occupy the space defined by the outer orbitals. The 3s orbital is the best of those available—it has the lowest energy of any of the outer orbitals. How large is the 3s orbital in an atom with this same Z^*? That is easily calculated (using equation (2-4)) to be 2.83 Å. That radius should be approximately equal to the spacing of two (or more) neon atoms clustered together. This estimate is in reasonable agreement with the observed internuclear spacing in solid neon—3.18 Å for the nearest neighbors. Figure 8-2 portrays the situation. The valence electrons of the left atom occupy the vacant

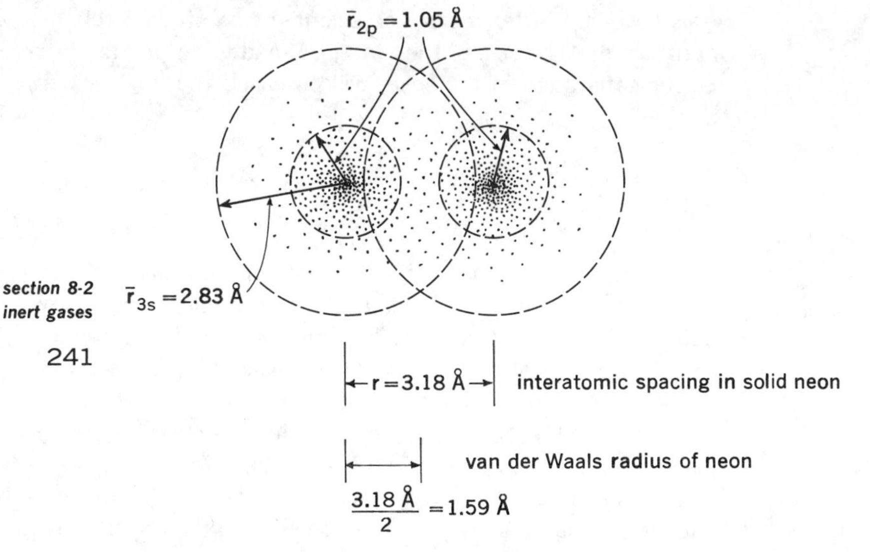

Figure 8-2 The spacing of neon atoms in the solid contrasted to average orbital radii.

3s orbitals of the right atom, and conversely. We see that it is intuitively natural to picture the two atoms as spheres in contact, each with a radius equal to half the internuclear distance.

Of course, attractive forces at such large separations can be expected to be quite weak. That is why the boiling points and the heats of vaporization of the inert gases are so low. These two factors, interaction distance and energy of interaction, are given in Table 8-2, together with comparative data for the

Table 8-2 Comparison of Covalent Bonds to van der Waals Bonds: Distance and Energy

Periodic Row	Halogen X_2	Inert Gas M	Distance (Å) r_{XX}	r_{MM}	Energy (kcal/mole) D_{X_2}	$\Delta H_{sub.}$(M)*
1	F_2	Ne	1.42	3.18	36	0.511
2	Cl_2	Ar	1.99	3.82	57	1.846
3	Br_2	Kr	2.28	4.02	46	2.557
4	I_2	Xe	2.66	4.40	36	3.578

* $\Delta H_{sub.}$(M) = heat of sublimation (solid → gas) for the crystalline inert gas.

adjacent halogen molecule. In every row, the availability of a half-filled valence orbital of a halogen atom permits close approach and a high bond energy. The completely filled valence orbitals of an inert gas permit only outer-orbital approach and very low bond energy. Because large energy differences are readily evident in many phenomena, they are distinguished by name. The valence-orbital interactions are, of course, called chemical bonds. The outer-orbital bonding is called "*van der Waals bonding,*" after the Dutch scientist who studied this type of interaction. The size of the atom, defined to be half the internuclear separation in the solid, will be called the *van der Waals radius.*

(a) HELIUM ATOMS: THE SIMPLEST VAN DER WAALS INTERACTION

We have indicated already that the bonding between inert gas atoms can be attributed to the sharing of electrons by two nuclei in orbitals outside the filled valence orbitals. Fortunately it is possible to analyze this type of interaction in more quantitative detail. We'll do so for helium atoms—the simplest case.

The probability distribution of a helium atom is spherically symmetrical, as shown in Figure 8-3(*a*). At any instant, however, the electron is located at some point in space. The figure represents the instantaneous electron positions at many different moments in time (or, better, in many different atoms at a given time). Figure 8-3(*b*) singles out one possible instantaneous configuration. This is a highly probable arrangement—the electrons are rather close to the nucleus and on opposite sides of it. The Figure 8-3(*c*) shows another possible configuration—less probable than Figure 8-2(*b*) because the electrons are on the same side of the nucleus. This tendency for electrons in atoms to stay apart is called *electron correlation.* It results from the repulsion between like charges, this repulsion making the high-energy configuration in 8-3(*c*) less probable than 8-3(*b*).

Of course, this same correlation of electron movement also occurs in aggregates of atoms. The results of this correlation can be seen through Figures 8-3(*d*) and 8-3(*e*). Both drawings show probable electron arrangements of the individual atoms

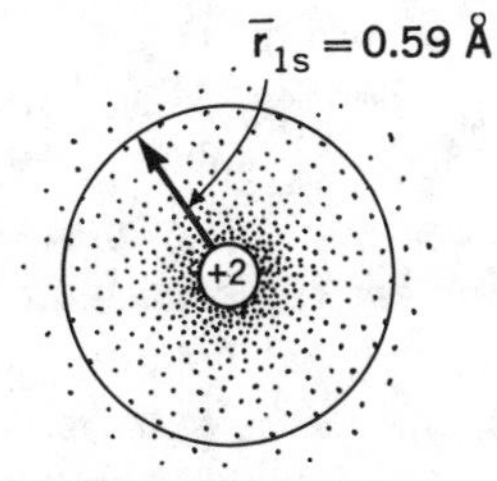

(*a*) Probability picture of helium atom

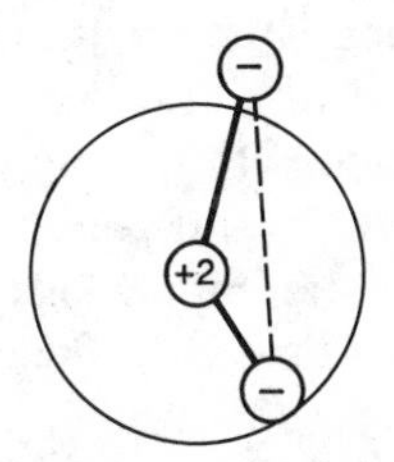

(*b*) Instantaneous view of a high probability arrangement of a helium atom

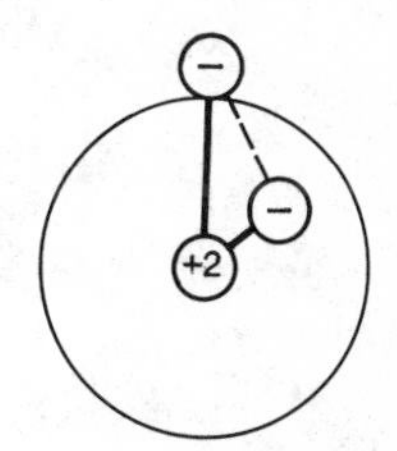

(*c*) Another instantaneous view of a helium atom — a low probability arrangement

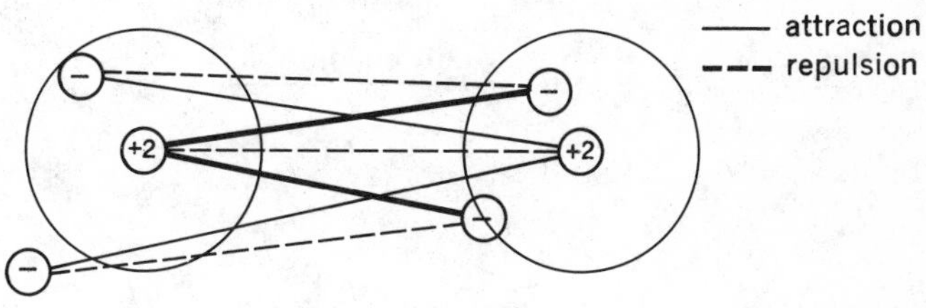

(*d*) An instantaneous view of the *new* forces that appear when two helium atoms approach each other: net attraction

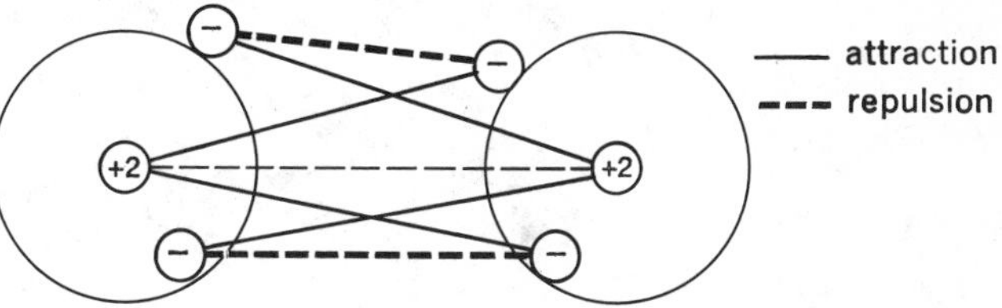

(*e*) An instantaneous view of the *new* forces that appear when two helium atoms approach: net repulsion

Figure 8-3 Van der Waals interactions between two helium atoms. Chemical bonds form if the potential energy is lowered when two atoms approach (Section 3-2). The weak bonding in liquid helium results from the attractive forces associated with an arrangement like that of (d). (Notice that in (d) and (e) only the new forces are shown.)

but with different orientations to each other. In (*d*) the new interactions are dominated by the attraction that occurs between the electrons in one atom and the nucleus of the other. This configuration lowers the energy. In (*e*) the new interactions are dominated by electron–electron repulsions. This configuration raises the energy. If configurations like (*d*) and (*e*) were equally probable, there would be no energy lowering as two helium atoms approach each other. But electrons are too fast on their feet to agree to that. The electrons on the left-hand nucleus correlate their motion with those on the right-hand nucleus, so that arrangements like (*d*) are more probable than arrangements like (*e*). Now, averaging over time, the energy is lowered as long as the nuclei are not brought so close together that nuclear–nuclear repulsions become overbearing.

All this was put into a usefully, quantitative, yet simple form by F. London. He recognized that the arrangements shown in Figures 8-3(*b*) and 8-3(*c*), when viewed from quite a distance, appear to be simple electric dipoles. Hence, at the relatively large internuclear distances found in liquid helium, the energies of the two configurations 8-3(*d*) and 8-3(*e*) can be reasonably approximated by the easily calculated dipole–dipole interaction energy. Figure 8-4 shows the nature of the simplification for the two configurations 8-3(*d*) and 8-3(*e*). All that now remains is to determine the probability of the favorable configuration (8-4(*a''*)) relative to that of the unfavorable configuration (8-4(*b''*)). This is

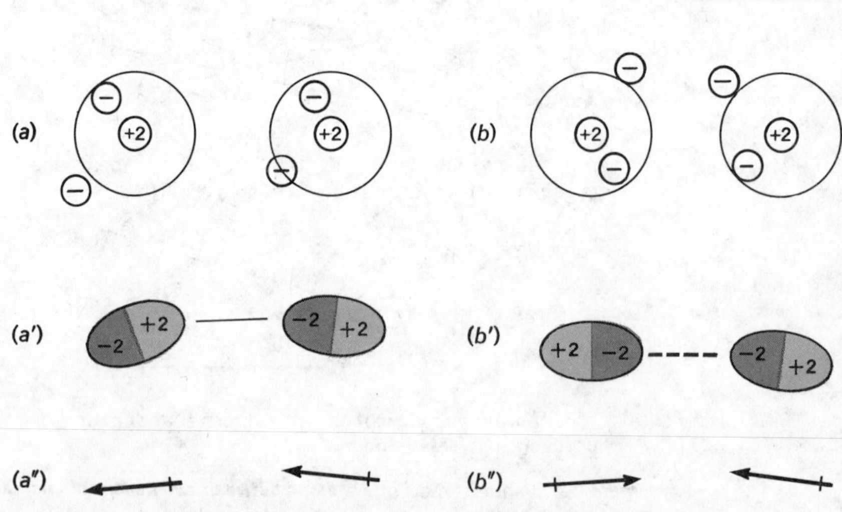

Figure 8-4 Dipole-dipole approximation in helium atom attractions. The instantaneous attraction (a) and repulsion (b) of two helium atoms (see Figs. 8-3(d) and 8-3(e)) are shown with representations of the instantaneous electric dipoles that result.

approximated by assuming that one of the atoms can be considered to be a mobile volume of charge that can be "polarized" into a dipole under the influence of the instantaneous dipole of its neighbor. Such an "induced dipole" will always be produced in a favorable energy relationship, so it produces an attractive configuration.

This model of the interaction between two helium atoms defines a useful approximation to the attractions that result from electron correlation at large internuclear separations. After its originator London, such attractions are named London Forces.

(b) THE OTHER INERT GASES

The London approximation readily predicts the way in which the energy should depend upon the distance of interaction.

$$V_{\text{London}} = -\frac{a}{r^6} \tag{8-1}$$

Of course (8-1) does not indicate the energy at short distances where repulsions are most important. An empirical approach is to add a positive term that causes energy to rise rapidly at small r. A convenient, hence frequently used form is

$$V = -\frac{a}{r^6} + \frac{b}{r^{12}} \tag{8-2}$$

The constants a and b can be selected to agree with appropriate measured quantities. Figure 8-5 shows the best available such energy–distance curves for the interaction of two identical inert gas atoms. Two features are immediately evident. First, the equilibrium distance (where the energy is a minimum) increases as we proceed down the periodic table to Xe. The atoms become larger as electrons are added and as the principal quantum number increases. Second, the strengths of the interactions increase with increasing atomic size. Thus, these curves agree with the trends in r_{MM} and $\Delta H_{\text{vap.}}(\text{M})$ shown in Table 8-2. They should—data like those in the table were used to determine the values of the constants a and b. Nevertheless, we can now examine and interpret their magnitudes. The data require that the polarizability increase rapidly with atomic size. Qualitatively this means that the larger the atom, the more volume available to its polarizable electrons, (the valence electrons) and the less tightly bound they are.

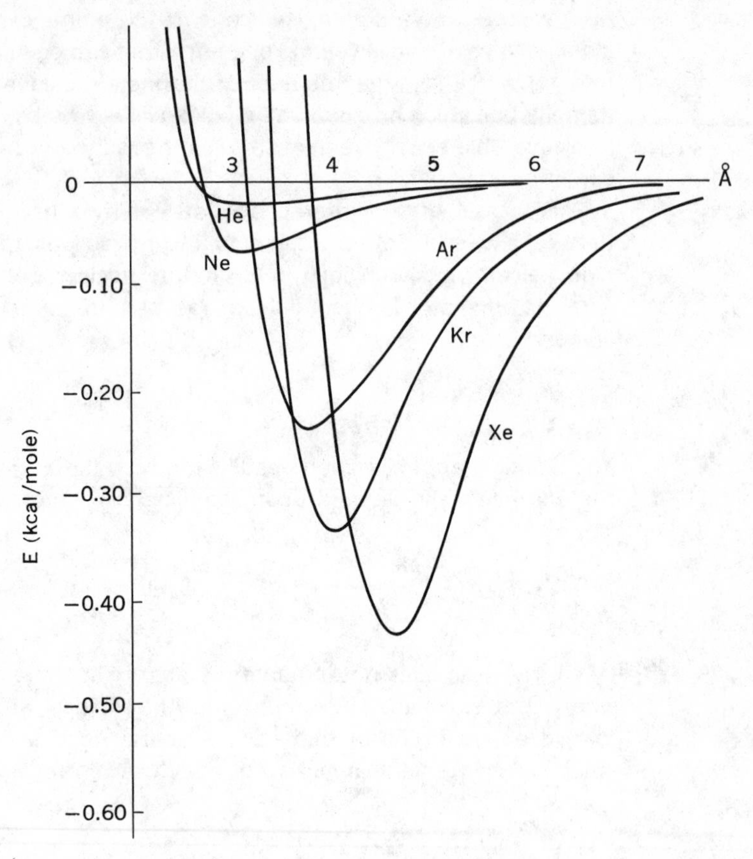

Figure 8-5 Interatomic energy curves for inert gases.

(c) INERT GAS CRYSTAL STRUCTURES

Figure 8-2 encourages us to picture inert gas atoms as spheres in contact. This proves to be a useful model, despite the quantum mechanical conclusion that atoms have no boundaries. It is particularly valuable in deducing the crystal structure.

Figure 8-6(*a*) shows how a third atom would nestle up close to two other like inert gas atoms if getting as close as possible is its aim. The vacant outer orbitals of each atom are so spacious and are occupied so incompletely by a single adjacent atom that there is plenty of room for a third neighbor. For simplicity we'll represent this situation with circles in contact, as in Figure 8-6(*b*). Now there is one sensible spot for a fourth atom. It can rest comfortably in the natural cradle atop the trio, thereby nudging up close to all three of its neighbors (8-6(*c*)).

Figure 8-6(*d*) makes a jump to ten atoms. The cluster of four atoms has been nestled into a larger group of atoms, so that the center atom now has six atoms around it and three cradled above. The shaded atom finds itself well surrounded on the sides and top—and to everyone's advantage, since so many nearest neighbors find themselves at the "in-contact" distance. In fact, there is only one other place left near the shaded atom—three more atoms could be placed below the first plane, matching the three above (see 8-6(*e*)). The trio above and the trio below have been moved away to clarify the arrangements. Now the shaded atom has twelve neighbors at the "in-contact" distance. More important, a continuation of this type of growth will ultimately give every atom the same 12 neighbor environment that the shaded atom possesses.

The packing finally developed is seen every day in the grocery store, wherever neatness prevails and oranges are on sale. It is one of the most effective ways of packing spheres. It is called

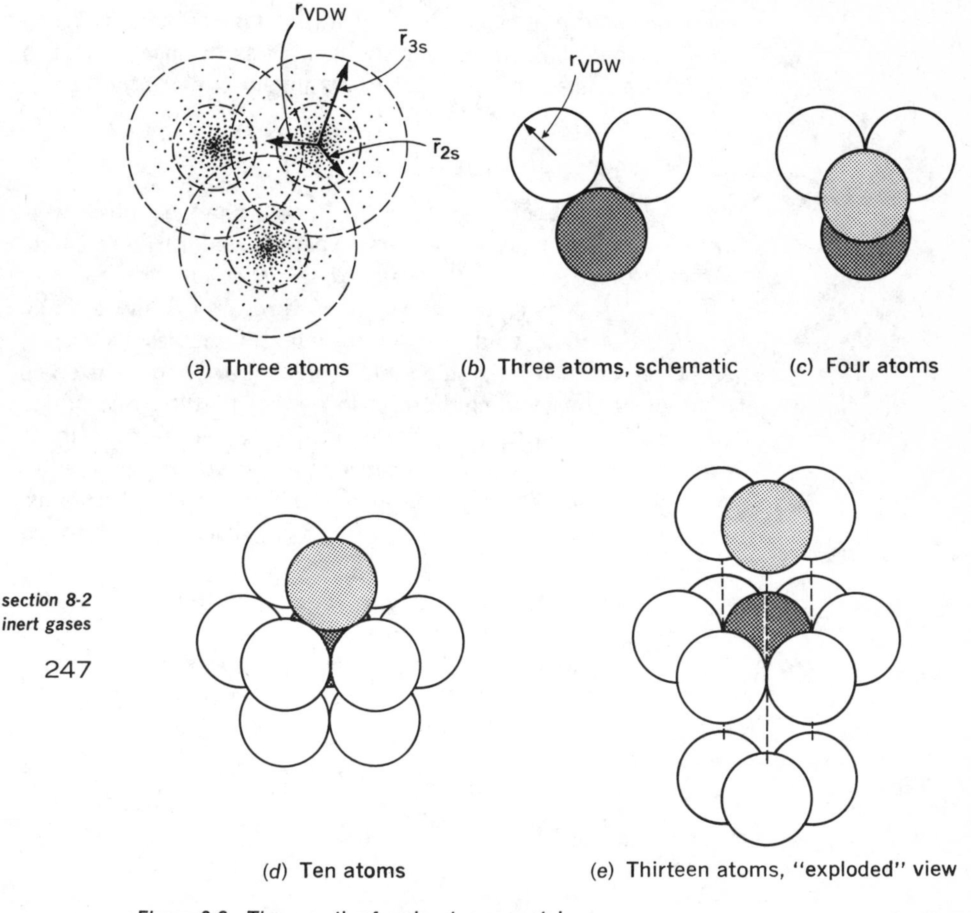

(*a*) Three atoms (*b*) Three atoms, schematic (*c*) Four atoms

(*d*) Ten atoms (*e*) Thirteen atoms, "exploded" view

Figure 8-6 The growth of an inert gas crystal.

"cubic closest packing," and X-ray studies show that all the inert gas atoms crystallize in this arrangement. The r_{MM} internuclear distances listed in Table 8-2 were measured in these crystal studies. The "in-contact" radius to be assigned to each atom is $\frac{1}{2}r_{MM}$. As we noted earlier, this is a convenient indication of atomic size for an inert gas.

There is much to learn from this closest-packing-of-spheres model of inert gas crystals. We explained the existence of bonding by the desire of nuclei to place themselves near the electrons of other atoms under the circumstances that only vacant outer orbitals are available. The vacant outer orbitals then exert no directive influence on the packing of neighbors. This is contrary to the principles developed in Chapter Six for covalent bond formation. Half-filled valence orbitals form bond angles connected either to the directional properties of the parent orbitals or to the repulsion of electron pairs. The situation for the inert gases is reminiscent of the clustered structures found in the orbital-excess, electron-deficient boron compounds, B_5H_9 and $B_{10}H_{14}$ (see Section 7-5b). Whenever there are lots of vacant orbitals ineffectively filled, closest-packed arrangements are found.

(d) INERT GAS LIQUIDS

The melting phenomenon is pictured as a change from a well-ordered, closely packed solid array, as pictured in Figure 8-6, to a highly disordered, though still closely spaced, arrangement. Figure 8-7 shows, in two dimensions, how a solid and a liquid might differ if we could see them with submicroscopic vision. Any given atom in the liquid state tends to have about the same number of nearest neighbors as in the solid lattice, but irregularly spaced and with voids here and there. The extent of these voids can be seen in the difference in liquid and solid densities or, more meaningfully, in the molar volumes of the two states. The molar volumes are given in Table 8-3. The liquids expand, on

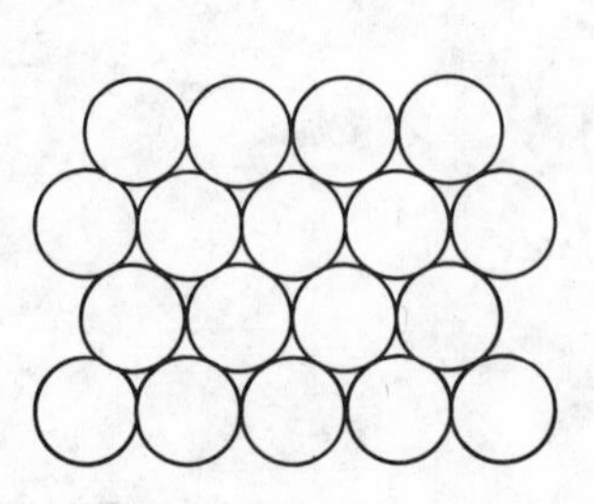

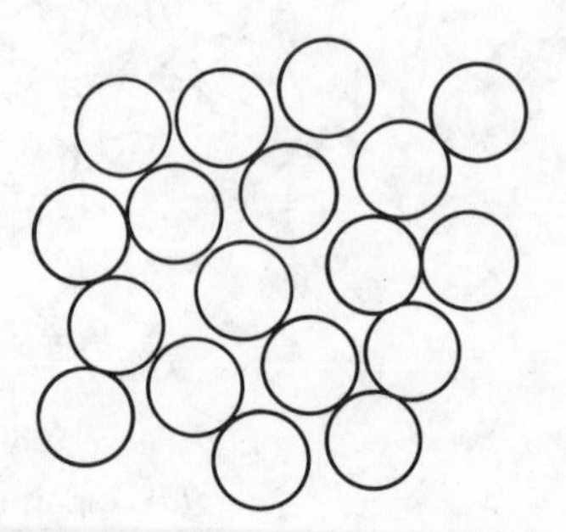

Figure 8-7 Two-dimensional representations of a solid and a liquid.

Table 8-3 Molar Volumes of Solid and Liquid Inert Gases

	Solid V_s(cc/mole)	Liquid V_l(cc/mole)	$\frac{V_l}{V_s}$
He	—	31.7	—
Ne	13.9	16.8	1.21
Ar	22.6	24.2	1.07
Kr	27.7	32.2	1.16
Xe	36.7	48.6	1.32
Rn	50.4	55.5	1.10

the average, about 15 percent for these elements. Of course there are not really voids in the liquid since none of the atoms has a boundary. Rather, the atoms are disarrayed and moved away from the optimum, energy-minimum distance by a few percent. Note that it only takes about a 5 percent increase in the average radius assigned to each atom to increase the molar volume by 15 percent.

One other datum in Table 8-3 catches attention. The molar volume of helium is quite out of line—far bigger than suggested by extrapolation from the other inert gases. This special feature is attributed to the extra mobility of this element due, in turn, to its low mass and weak interatomic forces. These same aspects of helium account for its failure to crystallize except under pressure, for the existence of two "kinds" of helium, and for some unique viscosity properties of one of these liquid states. None of these special phenomena can be interpreted without a quantum mechanical description of the liquid state.

8-3 Molecular solids and liquids

When two chlorine atoms come together, they have the opportunity to form a strong covalent bond through the sharing of two electrons in half-filled valence orbitals. When two chlorine molecules come together, they find themselves in about the same fix as when two argon atoms approach each other. Through the sharing of electrons, each atom in a Cl_2 molecule has already filled all its valence orbitals. The only way it can interact and share electrons with another chlorine molecule is through its outer orbitals—those outside the valence orbitals. Just as for the inert gases, we can expect such interactions to take place at rather large distances and with very weak bonds. Thus, the filled-orbital occupancy achieved through chemical bonding causes most molecules to condense as loosely bound liquids and solids in which the integrity of the gaseous molecule is retained. These are called molecular liquids and solids and, like the inert gas condensed phases, they are held together by weak attractions called van der Waals forces.

(a) THE HALOGENS

Although the forces between chlorine molecules are similar to those between argon atoms, the chlorine crystal must take account of the non-spherical molecular shape of Cl_2. Consequently none of the halogen crystals have the simple grocery store, orange-counter arrangement found in the inert gas crystals. Molecular shape affects the crystal packing in all molecular crystals, so only a few have the simple cubic, closest-packed structure.

The chlorine, bromine, and iodine crystal structures are all known and are quite alike. Figure 8-8 shows the packing in solid chlorine, which consists of interconnected parallel layers of Cl_2 molecules. One such layer is shown in Figure 8-8 as the shaded atoms. There is a layer below (the unshaded atoms) and a layer above (not shown), each displaced halfway between atoms.

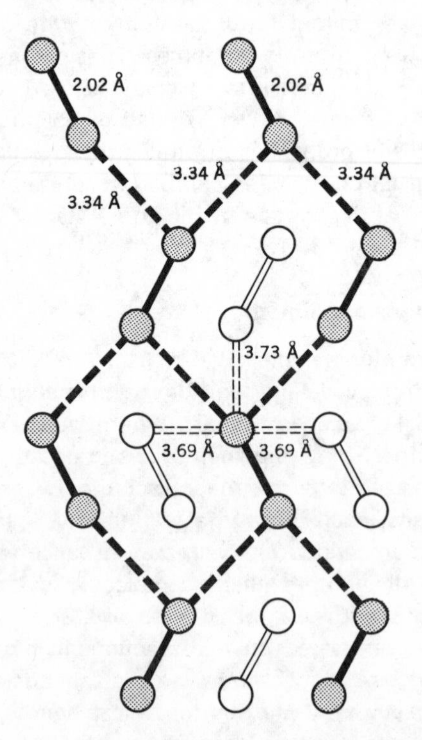

Figure 8-8 Crystalline chlorine: a molecular crystal. The shaded molecules are in one layer and the white molecules are in a lower layer.

Within the plane of a single layer, there are two nearest-neighbor distances. Each chlorine atom has one neighbor at a distance of 2.02 Å, very close to the gas phase Cl_2 internuclear distance 2.00 Å. In addition, each atom has two neighbors at 3.34 Å, one almost parallel to the Cl—Cl bond direction and one almost perpendicular. Then, each chlorine atom has neighbors in the layer above and below it at distances of 3.69, 3.73 Å, and longer. These distances are most informative.

The shortest distance, 2.02 Å, shows that molecules of chlorine are still present in the solid, with bond order virtually unchanged from the gas phase. The Cl—Cl distances between layers, 3.69 and 3.73 Å, are quite close to the internuclear distance in solid argon, 3.82 Å. The intermediate distance of 3.34 Å is longer by 60 percent than the Cl_2 distance, but shorter by 13 percent than the argon—argon distance. It seems that within the layers there is a significant component of polyhalogen bonding of the type described in Section 7-4b. Table 8-4 provides

Table 8-4 Comparisons Between Solid Halogens and Their Adjacent Inert Gases

$\frac{X_2}{M}$	$R_{XX'}$,* Å	$\frac{R_{XX'}}{R_{MM}}$	$\Delta H_{sub.}(X_2)$ (kcal/mole)	$\frac{\Delta H_{sub.}(X_2)}{\Delta H_{sub.}(M)}$
F_2/Ne	—	—	1.9	3.7
Cl_2/Ar	3.34	0.87	6.4	3.5
	3.69	0.97		
	3.73	0.98		
Br_2/Kr	3.30	0.82	9.9	3.9
I_2/Xe	3.56	0.81	14.9	4.2
	4.35	0.99		
	4.40	1.00		

* Nearest nonbonded neighbor distances.

further evidence of this extra bonding. If the two chlorine atoms merely looked like two argon atoms to other chlorine molecules, the heat of sublimation ought to be about double that of solid argon. Instead, it is almost four times as large, as are all the other halogen–inert gas sublimation heat ratios. Furthermore, the bromine and iodine nearest-neighbor distances are also quite short compared with the adjacent inert gas distances. The distances between layers for I_2 are almost exactly equal to the xenon—xenon crystal spacing. We can conclude that within layers there is halogen—halogen bonding like that encountered in the polyhalogens, while between the layers the interaction is due to London forces like those that bind the inert gas crystals together.

(b) VAN DER WAALS RADII

Crystal structures like that of the halogens give us a basis for assigning size to an atom, though we know it has no edges. Operationally, the nonbonded nearest-neighbor distances indicate how close two atoms will approach each other. For example, the 3.69 Å Cl—Cl distance in solid chlorine seems to define the size of chlorine atoms when they do not bond (the 3.34 Å distances probably involve some bonding). If we divide the 3.69 Å between the two atoms, each would be assigned a radius of about 1.8 Å. The size, so determined, is called the *van der Waals radius.*

The value in this determination of the nonbonded size of a chlorine atom is that it proves to be reasonably applicable to other crystals that also involve nonbonded Cl—Cl contacts (as, for example, in solid CCl_4). Furthermore, nonbonded contact distances between chlorine and other atoms can be estimated as the sum of the van der Waals radius of chlorine plus the van der Waals radius of the other atom, similarly determined.

Figure 8-9 shows commonly accepted van der Waals radii for the elements that form molecular crystals. These radii represent averages of the nonbonded distances found in different solids. They help us understand molecular crystal arrangements, but, even more important, they guide us as we study the molecules themselves. If a molecule is prepared in which two nonbonded atoms are held closer than the sum of their van der Waals radii,

1 H 1.2			2 He
7 N 1.5	8 O 1.4	9 F 1.4	10 Ne 1.6
15 P 1.9	16 S 1.9	17 Cl 1.8	18 Ar 1.9
33 As 2.0	34 Se 2.0	35 Br 2.0	36 Kr 2.0
51 Sb 2.2	52 Te 2.2	53 I 2.2	54 Xe 2.2

Figure 8-9 Van der Waals radii of some nonmetals (given in angstroms).

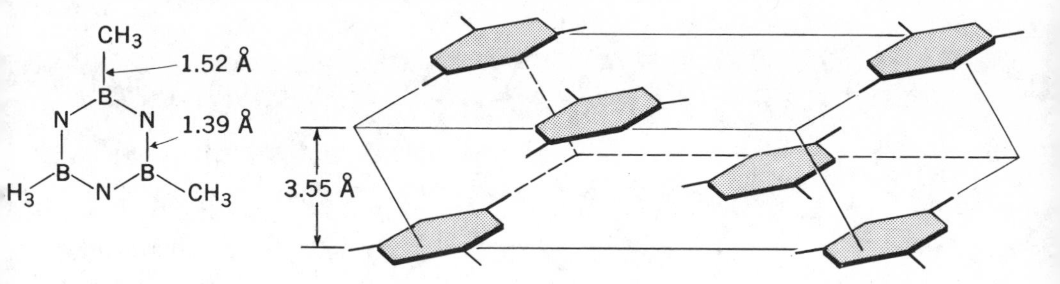

(a) Molecular structure (b) Crystal structure

Figure 8-10 The crystal and molecular structure of B-trimethylborazole. The B—N bond lengths are remarkably similar to those found in benzene, with which it is isoelectronic. The interlayer spacing in the crystal is slightly greater than that in the graphite modification of boron nitride (see Fig. 8-15), probably as a result of the methyl groups.

that molecule will tend to be unstable. Also, the flexibility of a molecule will be constrained to avoid bent structures that bring nonbonded atoms too close together. These ideas are so useful that most chemists use molecular models embodying these van der Waals radii—they are called "space-filling" models, to distinguish them from ball-and-stick models that show only bonded distances.

(c) SOME MOLECULAR CRYSTALS

Figures 8-10 to 8-13 show examples of more complicated molecular crystals. The first example, B-trimethylborazole, consists of planar, hexagonal molecules stacked, one above another, in a head-to-tail arrangement. The hexagonal ring has alternating boron and nitrogen atoms, and the six B—N bond lengths are identical, within experimental uncertainty, at 1.39 Å. This is the same as the C—C bond length observed in benzene. The nonbonded distances shown extend from the methyl group carbons to a nitrogen atom of an adjacent molecule, 3.28 and 3.35 Å,

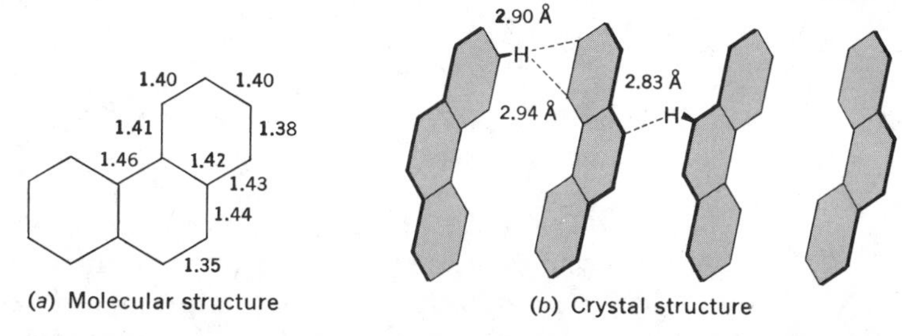

(a) Molecular structure (b) Crystal structure

Figure 8-11 The crystal and molecular structures of phenanthrene, $C_{14}H_{10}$. Hydrogen atoms have been omitted from the drawings.

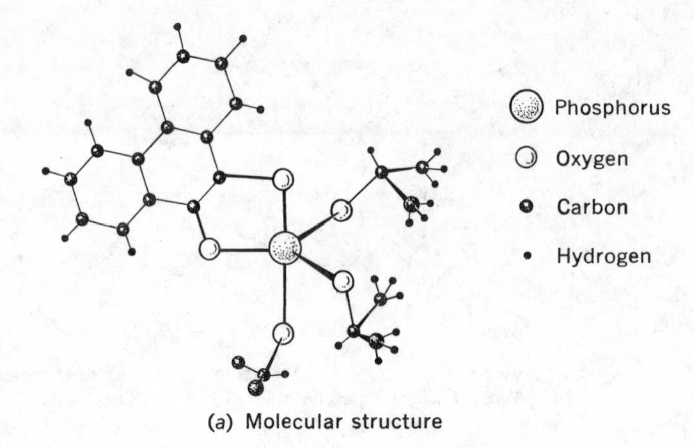

(*a*) Molecular structure

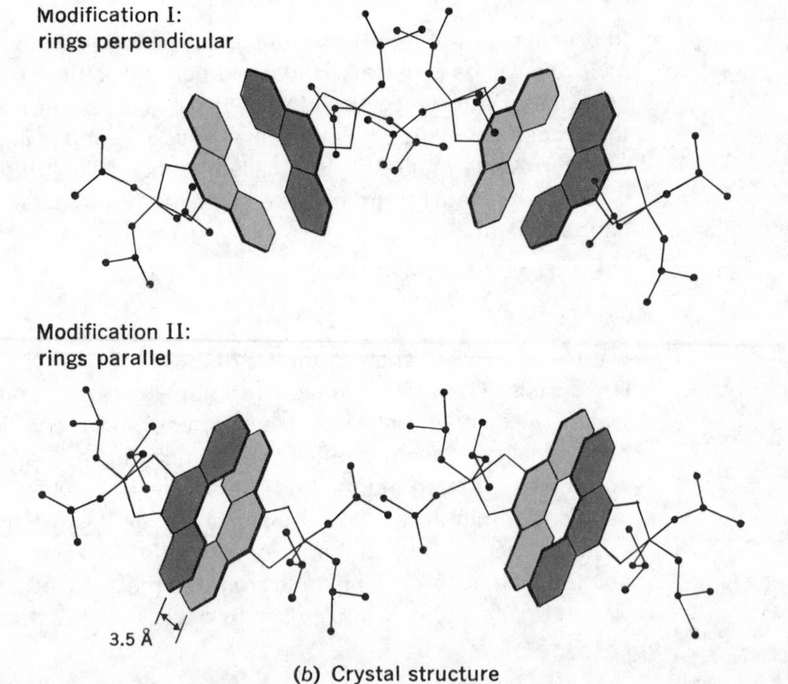

(*b*) Crystal structure

Figure 8-12 The crystal and molecular structure of 2,2,2-triisopropoxy-4,5-(2′,2″-biphenyleno)-1,3,2-dioxaphospholene.

respectively. These distances cannot immediately be interpreted in terms of van der Waals radii since the hydrogen atoms on the carbon are not located. However, a useful estimate of the van der Waals radius of a methyl group, hydrogens included, is 2.0 Å. Adding this to the van der Waals radius of 1.5 Å for nitrogen, from Figure 8-9, leads us to expect contact distances of 3.5 Å, only about 0.2 Å larger than actually observed. The vertical distance between two molecules is 3.55 Å.

Figure 8-11 shows the packing of phenanthrene, $C_{14}H_{10}$, in its crystal lattice. In a particular molecule, there are nine different C—C bond lengths determined, ranging from 1.347 to 1.455 Å. These lengths can be rationalized with resonance and with molecular orbital arguments. The molecules are not stacked in parallel planes, but are tilted such that the C—H bonds of one molecule point towards the plane of a neighbor at almost normal incidence. The nearest contacts are hydrogen—carbon distances of 2.90 and 2.83 Å. This awkward, tilted packing of flat molecules is found in other aromatic molecular crystals: e.g., benzene, naphthalene and anthracene. Substitutions, however, can cause the rings to become coplanar (e.g., in trinitrobenzene).

Figure 8-12 shows a complex molecule (58 atoms) that incorporates a phenanthrene ring. Figure 8-12(*a*) shows the molecule and 8-12(*b*) shows its two crystal forms. In I, the rings are oriented and spaced as in phenanthrene itself. In II, adjacent rings are parallel, with a separation of 3.5 Å. This is reminiscent of the ring spacings in trimethyl borazole and graphite crystals. In both I and II, the $CH(CH_3)_3$ groups interlock neatly.

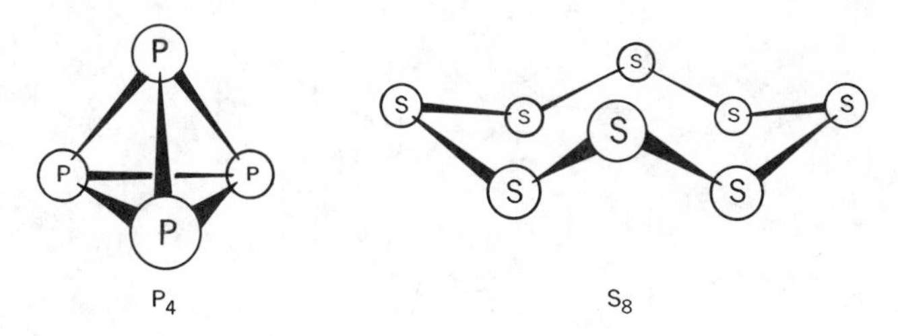

Figure 8-13 Molecular structures of elemental phosphorus and sulfur found in both solids and in the gas phase.

Figure 8-13 shows the molecular units found in solid phosphorus ("white" phosphorus) and solid sulfur. The P_4 and S_8 molecules that make up these solids are found both in solution and in the gas phase.

8-4 Covalent solids

Some solids seem to be held together in exactly the way molecules are bonded, except in an endless network. The conventional bonding principles suffice to explain the observed structures and their properties.

(a) DIAMOND AND GRAPHITE: JEKYLL AND HYDE

Carbon is an element particularly suited to bonding in the solid state. Each carbon atom, as we have seen, is capable of forming four equivalent bonds directed to the corners of a tetrahedron. Methane, CH_4, uses up this bonding capacity by forming four carbon—hydrogen bonds. A more complicated molecule is neopentane, $C(CH_3)_4$, in which the central carbon is now bonded to four other carbons. If we imagine larger and larger molecules built up in this way, we will eventually end up with a three-dimensional network of carbon atoms, each tetrahedrally bonded to four other carbon atoms. This is the form of elemental carbon known as diamond. Strong covalent bonds are formed between each pair of atoms—in fact diamond might be thought of as one giant molecule C_n. The bond lengths are 1.544 Å, exactly the same as in ethane. Hence the individual bond energies must be similar to that in ethane, about 83 kcal/mole.

A second form of elemental carbon, graphite, has quite a different structure. Carbon atoms are bound to one another to form two-dimensional sheets. The C—C distances are all equal, 1.42 Å, so each carbon is bonded to three others by sp^2 hybrid bonds as in benzene. Figure 8-14 shows the structure of

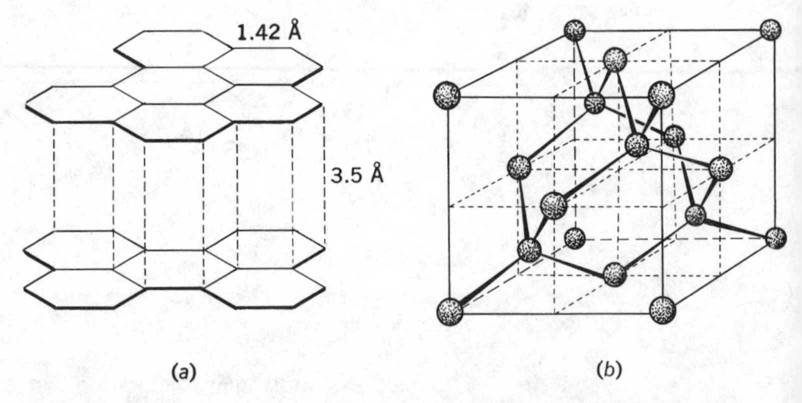

Figure 8-14 Carbon structure: (a) graphite, showing fragments of two adjacent layers; (b) diamond.

graphite. Notice the six-membered ring that is the basic repeating unit. Fragments of two layers are shown. The spacing between the layers, 3.5 Å, is much greater than the bonded C—C distance. In fact, it is the same as the distance between the phenanthrene rings shown in Figure 8-11—it is a van der Waals distance. So graphite is a van der Waals crystal in one dimension and a covalent crystal in the other two.

With this understanding of the bonding in diamond and graphite, it is interesting to contrast their properties. Their familiar appearances strike a chord that pervades the comparison. Diamond is clear and its brilliant sparkle casts its role as a precious gem; graphite is a pedestrian grey. Diamond is hard and is used as an abrasive; graphite rubs off the lead of a pencil to make marks on a paper and it is used as a dry lubricant. Diamond is an electrical insulator (though a bit expensive for the purpose); graphite is used as an electrical contact (a "wiper") to the commutators of motors and generators.

Let's begin with stability. In both the sp^3 bonding of diamond and the sp^2-π bonding of graphite, all the valence orbitals are fully engaged in bonding. The crystals should have comparable stability. They do. Though diamond is unstable with respect to graphite, the energy difference is very small, only about half a kilocalorie.

$$C(\text{diamond}) \rightarrow C(\text{graphite}) \qquad \Delta H = -0.4532 \text{ kcal/mole} \tag{8-4}$$

On the other hand, diamond has a higher density than graphite, 3.51 gm/cc compared with 2.25 gm/cc. This is because of the loose packing between the chickenwire layers in graphite (see Fig. 8-14). This implies that as the pressure is raised, the stability of diamond will improve relative to that of graphite. This is the secret of synthetic diamonds—at pressures above several thousand atmospheres, diamond becomes more stable, and if the temperature is raised to hasten the process, graphite changes into diamond.

Diamond is hard because its strong covalent bonds link each atom to every other in a three-dimensional network. Graphite, on the other hand, has hexagons of carbon atoms tightly bonded in planar sheets, but the sheets are quite loosely held together. Graphite breaks (or cleaves) along these planes quite easily and these sheets slide over each other and over other molecular surfaces so readily that graphite has a slippery character and acts as a lubricant.

Diamond's covalent bonds keep all of the bonding electrons localized between the atoms they are binding together. This localization implies that diamond is a poor electrical conductor. Graphite is quite different. It places a quarter of its electrons into a π molecular orbital that extends along the plane of the hexagons from one end of the crystal to the other. In this two-dimensional M.O., the π electrons are completely delocalized. With their high mobility they are quite responsive to external electric fields, joyfully rolling downhill to the positive end of the field. This mobile charge makes graphite a good electrical conductor in the two dimensions parallel to the bonded sheets. These mobile electrons also account for both the absorption of light and the lustre of graphite. Diamond has such strong bonds that photons of light in the visible spectral region do not have enough energy to reach an antibonding orbital and permit light absorption. When a diamond is colored, it is because there are impurities present.

(b) SILICON, GERMANIUM, AND TIN

These elements, Si, Ge, and Sn, are below carbon in the periodic table. They have identical electronic structures and have similar bonding in the solid. All three exist in the diamond structure, although tin also has two other, more complex crystal structures that are metallic in character. None of the three, however, forms a sheet structure like graphite. This peculiarity of carbon is undoubtedly due to the special strength of π bonds in the first row of the periodic table. Figure 7-4 showed the unique bond energy per bond order for C—C and N—N multiple bonds that probably accounts for the stability of graphite.

Like diamond, extremely pure silicon and germanium solids are excellent insulators. However, with only minute concentrations of suitable impurities, the electrical properties change dramatically. An element like phosphorus, for example, at a concentration of only one part in 10^7 in germanium, gives the crystal electrical conductivity because the phosphorus atom with its five valence electrons, must take the lattice position of a silicon atom, which has only four electrons. This impurity conductivity characterizes the impure solid as a *semiconductor*—like neither a metal nor a covalent solid—and its special behavior makes transistors possible. To discuss semiconductors further, we must await a discussion of the bonding in metals.

(c) BORON NITRIDE—ARTIFICIAL CARBON

Network solids are not limited to elements. Boron nitride, BN, provides a fine example. Figure 8-15 shows the two structures that have been observed, both with the empirical formula BN. One is just like graphite and the other like diamond, except that boron and nitrogen atoms alternate in the lattice positions.

Let's look at the atomic structures of these atoms to see how the bonding can be explained. For the graphite structure, both boron and nitrogen can easily form the necessary sp^2 hybrid orbitals by each promoting one s electron into a p state. Now each atom can bond to three neighbors in the planar hexagonal

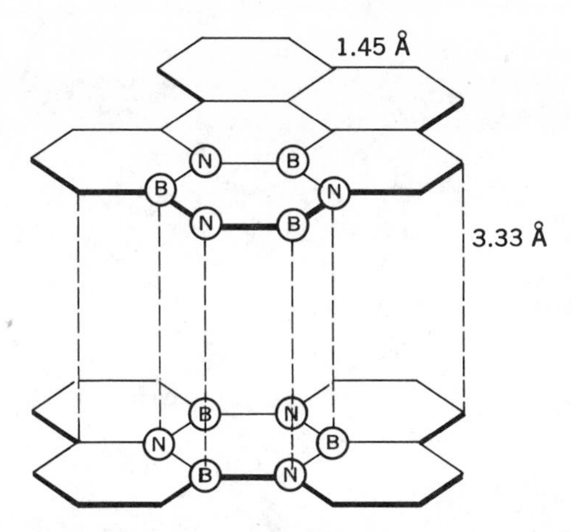

Figure 8-15 The graphite modification of boron nitride. The layers are arranged so that unlike atoms are above each other. Notice the remarkable similarity of in-plane and interplane distances between BN and graphite (Fig. 8-14). The diamond modification of boron nitride can be prepared from the graphite form at high temperature and pressure.

network for which sp^2 hybrid orbitals are perfectly oriented. Further, the boron $\pi(p_z)$ orbital is vacant; it is an electron acceptor. But it is surrounded by three nitrogen $\pi(p_z)$ orbitals ready to act as electron donors. This situation is just like that in B-trimethylborazole shown in Figure 8-10 in which the bond lengths duplicate those of benzene. With such units interlinked as in Figure 8-15, there are exactly enough electrons in the π molecular orbitals to duplicate the graphite bonding.

Needless to say, boron nitride in the graphite structure has properties similar to those of graphite. When BN is prepared in the diamond structure, it is even harder than diamond itself. Thus the chemist uses his knowledge of bonding to extend his horizons and, sometimes, to improve upon Nature.

(d) QUARTZ—A SILICON-OXYGEN NETWORK SOLID

A great variety of network solids are formed by silicon and oxygen. They are called *silicates*, and they make up 87 percent of the Earth's crust. We will discuss at this time only one of these—the prototype, which is called silica, SiO_2. Silica occurs in three different crystal forms, quartz, tridymite, and cristobalite. The most familiar of these, quartz, is found in mineral deposits as crystals or as a crystalline constituent of many rocks, such as granite. In the quartz crystal (shown in Fig. 8-16), each silicon atom is surrounded by a tetrahedron of covalently bonded oxygen atoms. Each oxygen atom is bonded to two silicons. The structure is linked in three dimensions by an

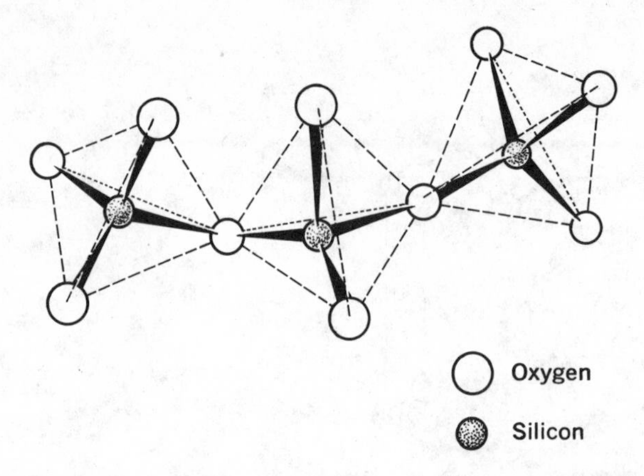

Figure 8-16 The crystal structure of quartz, SiO_2. Each silicon atom is surrounded by a tetrahedron of oxygen atoms (r(Si—O) = 1.61 Å). Each oxygen atom serves to join two tetrahedra together. In the low-temperature form of quartz, these chains of tetrahedra are helically arranged.

Si—O—Si network of bonds, so a hard, high-melting crystal results.

(e) SILICONE POLYMERS: ONE-DIMENSIONAL CRYSTALS

Again man's ingenuity plus Nature's guidance pay off in partnership. The silicon—oxygen linkages that account for the strength of quartz can be built into a one-dimensional, thread-like crystal. These compounds, called *silicones,* have the general structure

$$-\underset{\displaystyle R}{\overset{\displaystyle R}{\underset{|}{\overset{|}{Si}}}}-O-(\underset{\displaystyle R}{\overset{\displaystyle R}{\underset{|}{\overset{|}{Si}}}}-O)_n-\underset{\displaystyle R}{\overset{\displaystyle R}{\underset{|}{\overset{|}{Si}}}}-O- \qquad (8\text{-}5)$$

in which R is any organic group and *n* is as large as possible. These compounds are properly treated in this chapter because the value of *n* may be so large that the molecular weight may be several million. The chain skeleton of this molecule has the strong silicon—oxygen bonds, but its one-dimensional structure gives the molecule flexibility and fiber-like properties. With R = methyl, this polymer can be made into a rubber-like gum by occasional oxidative cross-linking, as pictured in Figure 8-17. This silicone rubber is, in some respects, superior to natural rubber. It can withstand much higher temperatures than rubber and so can be used on the heat shield of a space capsule. It also withstands much lower temperatures without cracking,

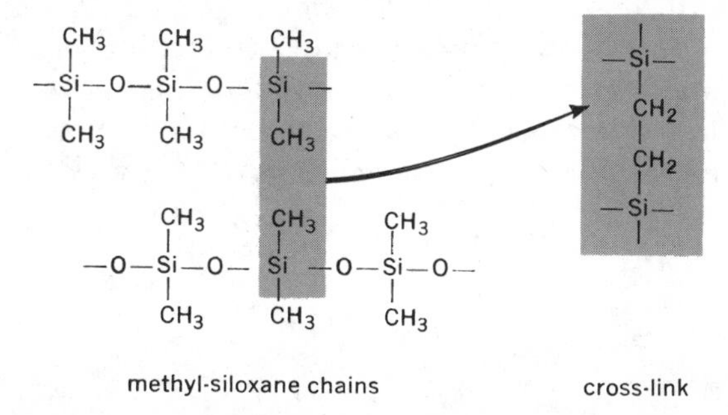

Figure 8-17 Cross-linking in methyl-siloxane to form silicone rubber. Hydrogen is removed by oxidation.

so it can be used in cold climates and in high-flying aircraft to seal doors and windows.

When the cross-linking is achieved through silicon—oxygen bonds, the polymer is called a *siloxane*. These compounds are even more resistant to heat, oxidation, and chemical attack. They are used in heat-resistant enamels, coatings, and in housepaints.

8-5 Hydrogen bonded solids

In Section 7-4g it was noted that when hydrogen bonding can occur as a molecule condenses, the crystal *always* has a structure that takes advantage of that possibility. That is a relatively minor triumph for hydrogen bonds compared with their crucial role in biologically important molecules. It is no exaggeration to claim that life on our planet would have assumed radically different forms—if any at all—were hydrogen bonding not present in water and in the proteins and nuclei acids that compose living cells and that transmit hereditary traits. Thus one of the weakest bonds we know, the hydrogen bond, occupies a preeminent position in the biological scheme of things.

(a) HYDROGEN BONDING IN ICE IS NICE

Water, our most abundant chemical, is ideally suited for the formation of hydrogen bonds. The oxygen atom in gaseous H_2O has two bonds to hydrogen at an angle of 104.5°, and two unshared electron pairs. This geometry permits the formation of four hydrogen bonds at tetrahedral angles. The ability to form tetrahedral bonds accounts for the special stability of the diamond lattice. Water forms a similar solid with, of course, bonds considerably weaker than the covalent C—C bonds. Nevertheless, ice is much more strongly held together than such near relatives as F_2O and Cl_2O, both of which form weak van der Waals solids.

The environment of one oxygen atom in ice is illustrated in Figure 8-18, and a larger fragment of the crystal is shown in

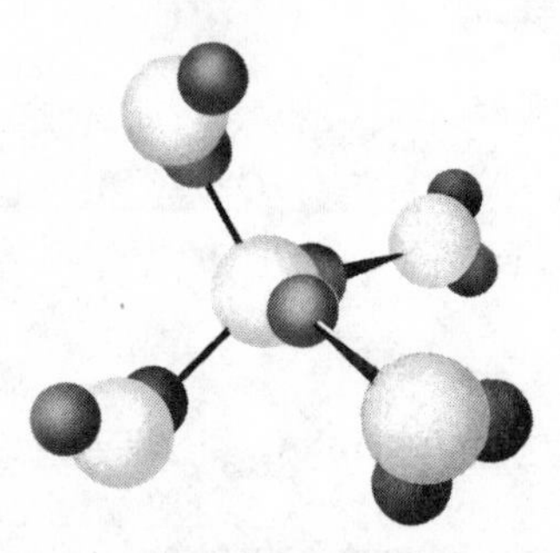

Figure 8-18 The coordination of one water molecule in ice.

Figure 8-19. The latter drawing emphasizes that fact that we can't pin down a given pair of hydrogens to one specific oxygen. The energy of the crystal is not affected by an irregular arrangement of the bonds as long as the number remains four per oxygen atom. This irregularity results in a residual entropy or disorder in ice, even at very low temperatures. An interesting aspect of the ice structure is the presence of open "channels" down through the centers of the puckered, six-membered rings (see Fig. 8-19). This explains the density of ice, which is particularly low due to this open structure. Liquid water lacks the regularity of ice (even though considerable hydrogen bonding is still present), so this liquid is more dense than its solid. This very unusual property causes ice to float, accounting for, among other things, the fact that in cold climates rivers and lakes don't freeze from the bottom, which would kill off all the water life and wreck the ice-skating.

A rough idea of the energy of a hydrogen bond in ice may be obtained from the heat of sublimation, the energy required to convert one mole of molecules from the solid to the gas phase. Table 8-5 compares the heats of sublimation of the hydrides of groups IV and VI. The ratios of these heats for H_2Se/GeH_4 and H_2Te/SnH_4 are the same, 1.32; none of these

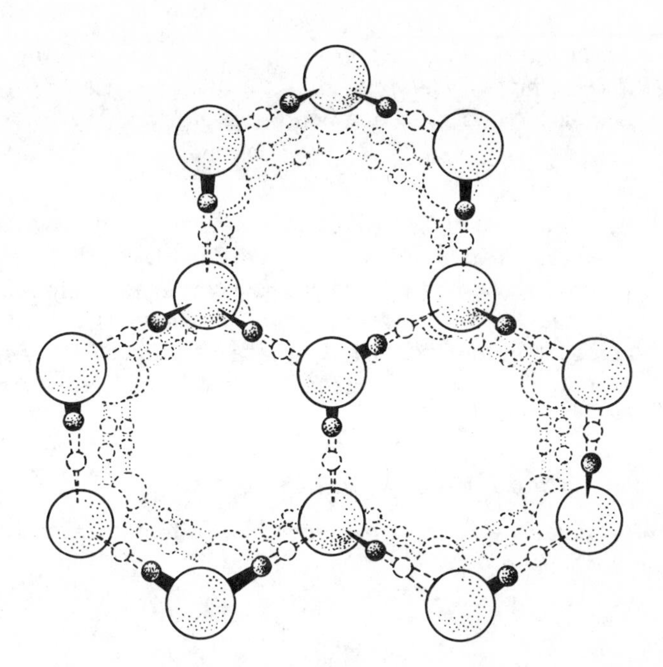

Figure 8-19 Crystal structure of ice. Each hydrogen lies between two oxygen atoms at one of the two positions shown. The arrangement is random and hydrogen atoms change positions, but always maintain around each oxygen atom four tetrahedrally placed hydrogen neighbors, two close (1.0 Å) and two far (1.8 Å). Notice the open channels, characteristic of the ice structure, that account for the low density of ice.

molecules is considered to form hydrogen bonds. If no hydrogen bonding were present in H_2O, its heat of sublimation would be expected to be 1.32 times that of CH_4.

$$\Delta H_{\text{sub.}}(H_2O) = 1.32 \cdot \Delta H_{\text{sub.}}(CH_4) = 2.6 \text{ kcal/mole}$$

The discrepancy between this predicted value and the experimental result of 12.20 kcal/mole (12.20 − 2.6 = 9.6 kcal/mole) might account for the extra binding energy of the hydrogen bonds. Since there are two bonds per molecule, each bond would have an energy of $\frac{1}{2} \cdot 9.6$ or *4.8 kcal.* This can be taken

Table 8-5 Heats of Sublimation of Some Isoelectronic Hydrides

Group IV	$\Delta H_{\text{sub.}}$ (kcal/mole)	Ratio VI/IV	$\Delta H_{\text{sub.}}$ (kcal/mole)	Group VI
CH_4	2.0	—	12.20	H_2O
GeH_4	4.0	1.32	5.3	H_2Se
SnH_4	5.0	1.32	6.6	H_2Te

as a lower limit, since there is another natural assumption that gives a higher value. Since *all* the bonds in ice are hydrogen bonds, the energy of one bond is given simply by

$$\tfrac{1}{2}\,\Delta H_{\text{sub.}} = \tfrac{1}{2} \cdot 12.20 = 6.10 \text{ kcal/mole}$$

In this approximation we assume that no van der Waals forces are present in ice. This gives us an upper limit. Hence the hydrogen bond energy in ice is somewhere between 5 and 6 kcal/mole. This simple approach places the H_2O—H_2O hydrogen bond in the middle of the range obtained by many different experiments and calculations of widely different complexity (4–7 kcal/mole).

(b) HYDROGEN BONDING IN SOLID ACIDS AND AMIDES

An organic acid is so named because it contains the carboxyl functional group —C═O.
|
OH

Some simple carboxylic acids are

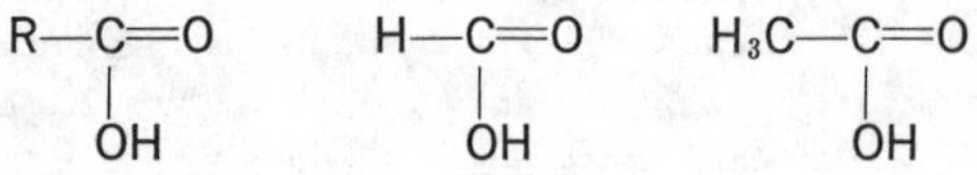

R—C═O	H—C═O	H_3C—C═O
|	|	|
OH	OH	OH
general acid	formic acid	acetic acid

In aqueous solution each of these molecules can release a proton:

$$\text{RCOOH} = \text{H}^+(\text{aq}) + \text{RCOO}^-(\text{aq}) \qquad (8\text{-}6)$$

$$\text{CH}_3\text{COOH} = \text{H}^+(\text{aq}) + \text{CH}_3\text{COO}^-(\text{aq}) \qquad (8\text{-}7)$$

The carboxyl group is capable of acting both as a proton donor and a proton acceptor in hydrogen bond formation. In the gas phase, the formation of dimers results:

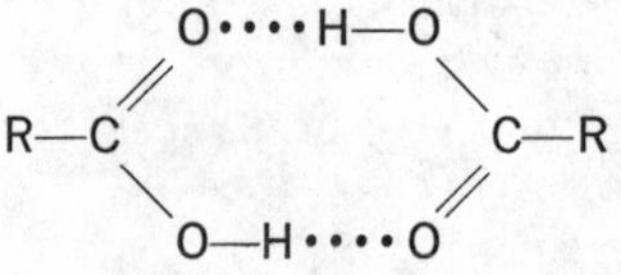

```
          O····H—O
        //        \
R—C                C—R        (8-8)
        \         //
          O—H····O
```

In the solid, hydrogen bonded chains are common. Figure 8-20(*a*) shows a fragment of such a chain in solid formic acid. The zigzag chains are held together by weak van der Waals forces.

Another type of compound can be obtained if the —OH part of the carboxyl group is replaced by an amine group —NH_2.

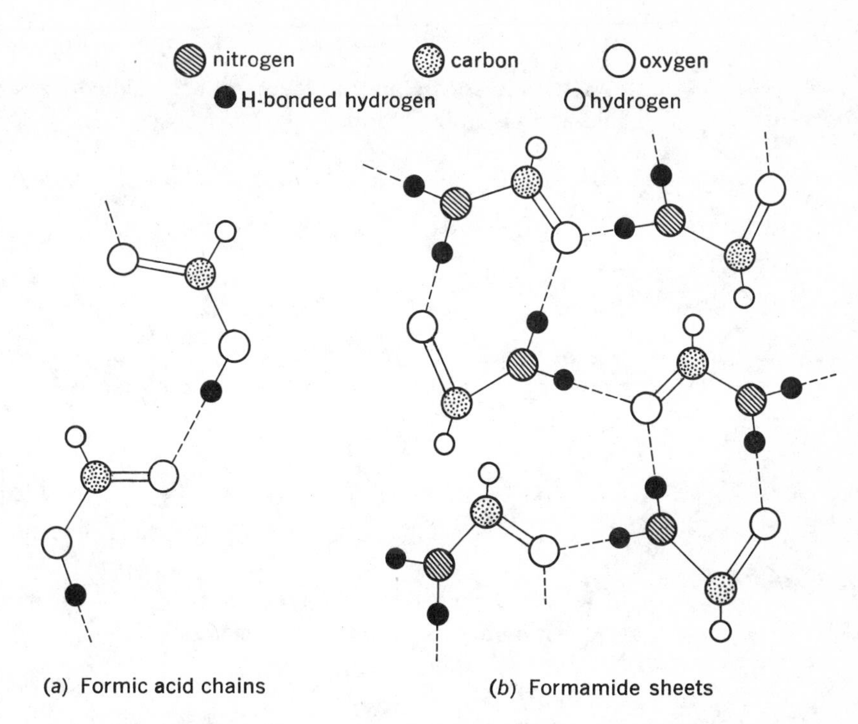

Figure 8-20 Structure of formic acid, HCOOH, and formamide, $HCO(NH_2)$, in the solid. Hydrogen bonds are shown by dotted lines.

Such compounds are called *amides*. The following are examples:

$$\underset{\text{general amide}}{R{-}\overset{\overset{\displaystyle O}{\|}}{C}{-}NH_2} \qquad \underset{\text{form}\textit{amide}}{H{-}\overset{\overset{\displaystyle O}{\|}}{C}{-}NH_2} \qquad \underset{\text{acet}\textit{amide}}{H_3C{-}\overset{\overset{\displaystyle O}{\|}}{C}{-}NH_2}$$

The amine group, with two protons, is extremely active in hydrogen bonding circles. Like acids, it can form zigzag chains

$$\cdots O{=}C(H){-}N(H){-}H\cdots O{=}C(H){-}N(H){-}H\cdots \tag{8-9}$$

which, in turn, may be joined to one another through the second hydrogen atom attached to each nitrogen atom. This results in the formation of *sheets*, loosely held together by van der Waals bonds. Figure 8-20(*b*) shows one of these cross-linked sheets.

Amino acids are organic molecules that contain both a carboxylic acid and an amine group.

$$\begin{array}{c} \text{H} \quad \text{O} \\ | \quad \| \\ \text{—C—C—OH} \\ | \quad\quad \\ \text{NH}_2 \quad\quad \end{array}$$

These interesting molecules can either donate a carboxyl proton or the amine group can accept a proton

$$\underset{\text{acid function}}{\overset{\text{O}}{\overset{\|}{\text{—C}}}\text{—OH} \rightarrow \overset{\text{O}}{\overset{\|}{\text{—C}}}\text{—O}^- + \text{H}^+} \qquad \underset{\text{base function}}{\text{—NH}_2 + \text{H}^+ = \text{—NH}_3^+}$$

In addition, opportunities for hydrogen bonding abound. The simplest members of the amino acids are

$$\underset{\text{general amino acid}}{\text{R—}\overset{\text{H}}{\overset{|}{\underset{\underset{\text{NH}_2}{|}}{\text{C}}}}\text{—}\overset{\text{O}}{\overset{\|}{\text{C}}}\text{—OH}} \qquad \underset{\text{glycine}}{\text{H—}\overset{\text{H}}{\overset{|}{\underset{\underset{\text{NH}_2}{|}}{\text{C}}}}\text{—}\overset{\text{O}}{\overset{\|}{\text{C}}}\text{—OH}} \qquad \underset{\text{alanine}}{\text{H}_3\text{C—}\overset{\text{H}}{\overset{|}{\underset{\underset{\text{NH}_2}{|}}{\text{C}}}}\text{—}\overset{\text{O}}{\overset{\|}{\text{C}}}\text{—OH}}$$

Amino acids are the building blocks of which proteins are made. Proteins are the building blocks of which you are made. They can be called "macromolecules"—the smallest protein has a molecular weight of about 6000, the largest, about 7,000,000. Table 8-6 lists several important types of proteins

Table 8-6 Biological Functions of Some Proteins

General Type	Biochemical Role
Enzymes	Catalysis: hydrolysis, oxidation, synthesis.
Structural	Hair, wool, feathers, muscle, silk, what have you.
Respiratory	Oxygen transport and storage (hemoglobin).
Antibodies	Defend organism against foreign agents: attacks by bacteria and viruses.
Hormones	Regulation of metabolism.
Nucleoproteins	Control of hereditary transmission, protein synthesis (chromosomes).

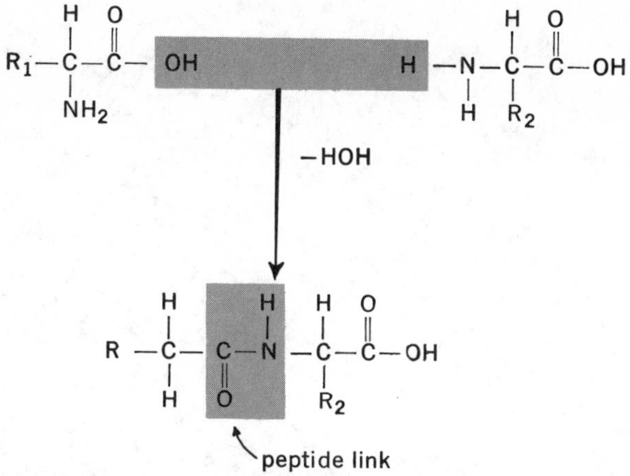

$$R_1-\underset{NH_2}{\overset{H}{C}}-\overset{O}{\overset{\|}{C}}-OH \;+\; H-\underset{H}{N}-\underset{R_2}{\overset{H}{C}}-\overset{O}{\overset{\|}{C}}-OH \xrightarrow{-HOH} R-\underset{H}{\overset{H}{C}}-\underset{O}{\underset{\|}{C}}-\overset{H}{N}-\underset{R_2}{\overset{H}{C}}-\overset{O}{\overset{\|}{C}}-OH$$

peptide link

Figure 8-21 Formation of a dipeptide.

and summarizes a few of their functions. This list gives an idea of their great importance.

Amino acids are linked together to form proteins by the elimination of one molecule of water from the OH group of one acid and the NH_2 group of the other (Fig. 8-21). This coupling is known as a *peptide link*. It has the general form shown here.

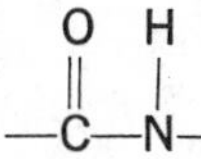

$$-\overset{O}{\overset{\|}{C}}-\overset{H}{\overset{|}{N}}-$$

If only two amino acids are joined by a peptide link we have a dipeptide. Proteins, containing thousands of amino acid groups are *polypeptides.*

These long polypeptide chains coil themselves into a helix, a structure that may be visualized by imagining the chain being

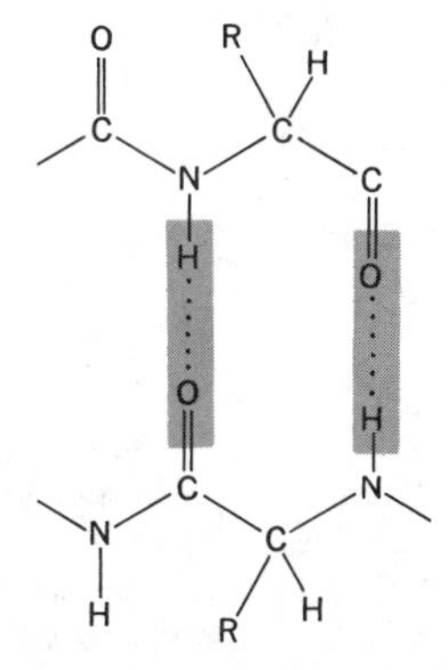

Figure 8-22 Hydrogen bonding between peptide linkages.

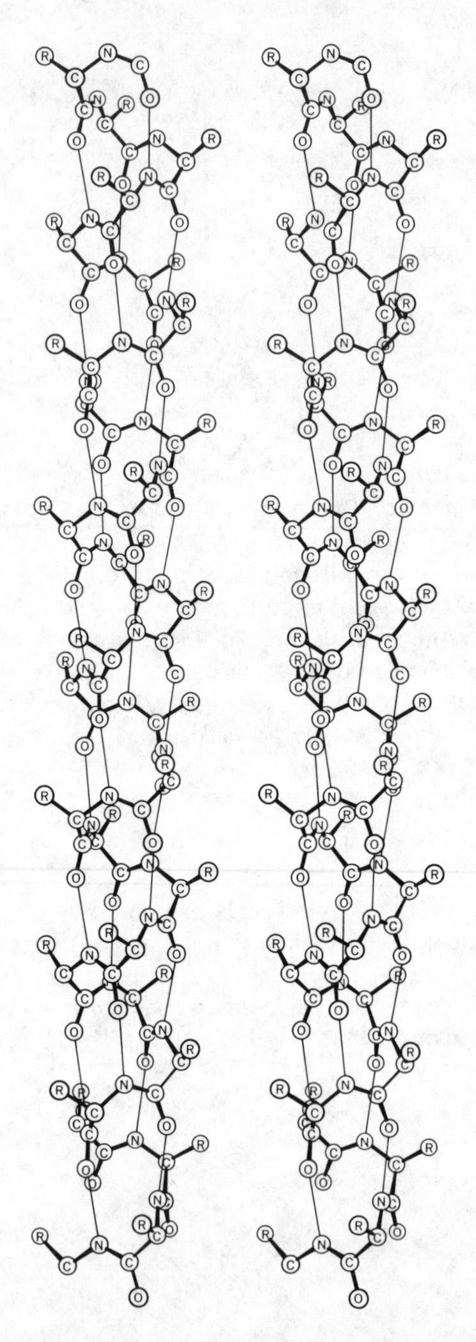

Figure 8-23 Alpha helix. To view this computer-drawn stereo representation, place a business size envelope between the two views, put your nose on the edge of the envelope, and allow your eyes to bring the two views into one. For details of the structure, see text. The authors are greatly indebted to Dr. C. K. Johnson of Oak Ridge National Laboratory for providing us with this figure.

wound in a spiral manner down the length of a small rod. The stability of this structure is due to hydrogen bonds between peptide links, as illustrated in Figure 8-22. These bonds form between a $-\overset{\displaystyle O}{\overset{\|}{C}}-$ group that interacts with an $-\overset{\displaystyle H}{\overset{|}{N}}-$ group about three amide linkages up the chain, in the next turn of the spiral. Figure 8-23 is a computer-drawn stereo picture of a general helix. It can be viewed in stereo by placing a piece of cardboard between the two views and letting your eyes bring the two images together. The stabilizing hydrogen bonds are shown as light lines, and the conventional bonds as dark lines. The hydrogen atoms themselves are omitted to keep the diagram relatively simple. This general structure can accommodate any side chains R to represent any protein. With 3.6 amide groups per turn of the helix, the structure is called an α helix.

(d) TWO HELICES ARE BETTER THAN ONE, TRALA

Genetic information is stored by the genes, a particular form of a macromolecule commonly known as DNA (deoxyribonucleic acid). The information (e.g., green hair, six toes) carried by the chromosomal DNA molecules is transferred to the appropriate protein-manufacturing cells by another large molecule RNA (ribonucleic acid) which is closely related to DNA. These two types of molecules are at the heart of the reproduction systems from amoeba on up (or on down). Because of their enormous molecular size, however, their structures are only gradually coming to light. Once again it is found that hydrogen bonding has a central role in the structure of these molecules. The investigation of the functions and structures of DNA and RNA molecules represents one of the most exciting aspects of contemporary biochemistry—or as it is beginning to be called, *molecular biology.*

The backbones of these species are sugar molecules linked together by phosphate groups, as illustrated schematically in Figure 8-24 and in more detail in Figure 8-25. The wide variety of possible molecules results from the different side chains that can be tacked on. These side chains are organic bases—so called because the nitrogen atoms contained in them are good electron donors. Two general types of bases are found, pyrimidine bases and purine bases. The parent compounds are illustrated in Figure 8-26. The replacement of one or more of the hydrogens (marked with arrows) by $—OH$, $—NH_2$ or $—CH_3$ groups accounts for the many different members of these families found in biological systems. Only four bases, however, pre-

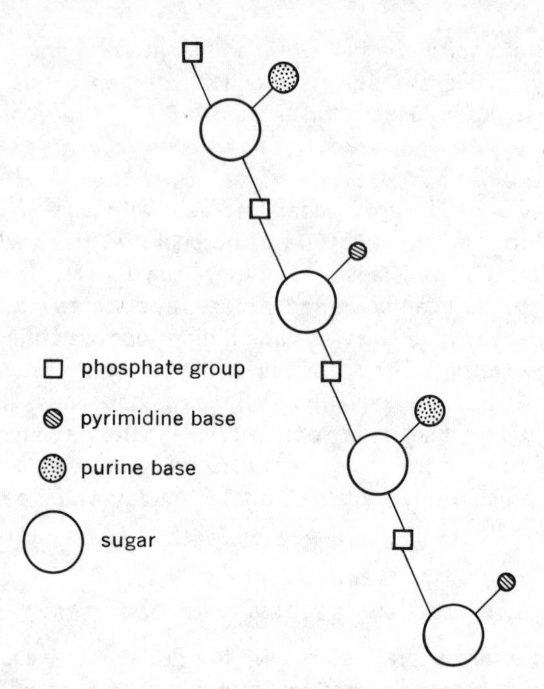

Figure 8-24 General arrangement of nucleic acid.

dominate in DNA and RNA, two purines and two pyrimidines, although innumerable other bases form a small percentage of the total. The number of base units may be as high as 10^8 in chromosomal DNA and as low as 80 in some RNA molecules. The information-carrying capacity of these species is related to

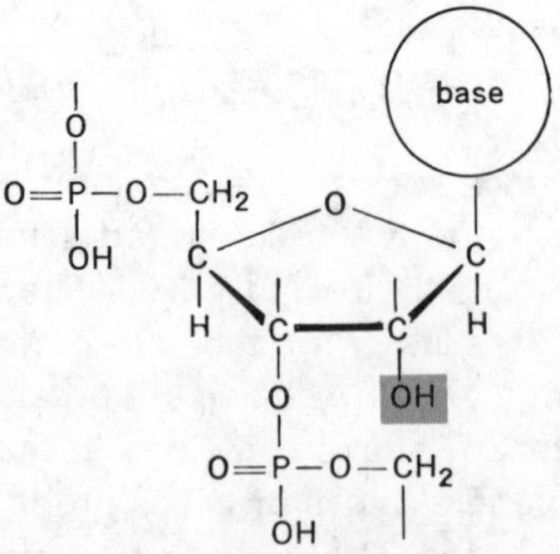

base
O
O=P—O—CH$_2$ O
OH C C
H C C H
O OH
O=P—O—CH$_2$
OH

Figure 8-25 Backbone of ribonucleic acid: five-membered sugar rings are hitched together with phosphate groups. The replacement of the OH group, shown shaded, with a hydrogen atom transforms this to the backbone of DNA, deoxyribonucleic acid. This simple change has a profound effect on the biological functions of this macromolecule.

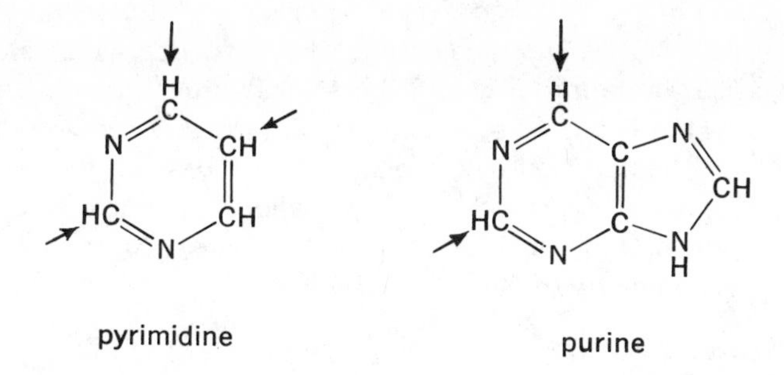

Figure 8-26 Parent compounds of bases found in DNA and RNA. Different compounds result from the substitution of —NH_2, —OH and —CH_3 groups for the hydrogens marked by arrows.

the *order* in which the bases are attached. It only takes a few different bases distributed among one hundred million sites to give the molecule an enormous vocabulary.

The structure of a nucleic acid is a hydrogen bond delight. It consists of a pair of intertwined helices held together by hydrogen bonds between pairs of bases on separate chains. The requirements of the geometry are such that just any old pair of bases won't do. The space between helices is properly bridged by a hydrogen bond only when it is from a pyrimidine base to a purine base. Two pyrimidine bases are too small and

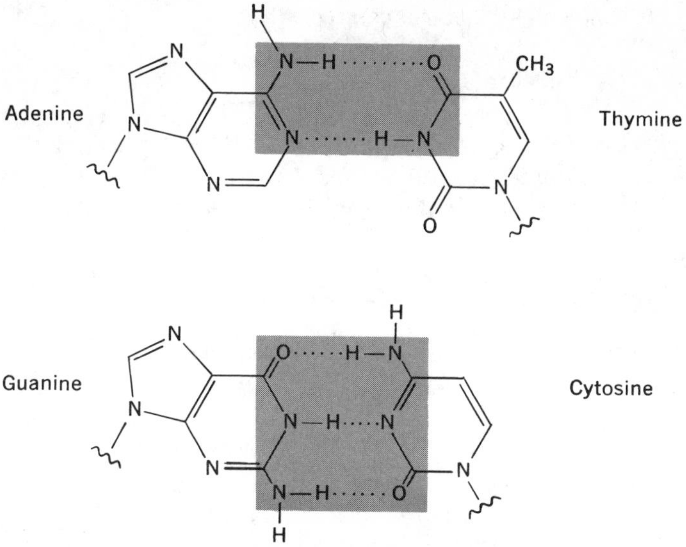

Adenine Thymine

Guanine Cytosine

Figure 8-27 Base pairing by hydrogen bond formation. These four bases predominate in DNA and RNA and the illustrated combinations are the most frequent.

two purines too large. Figure 8-27 illustrates the pairing arrangement that is most common. This high degree of choosiness about the appropriate partners for hydrogen bonding requires that the two chains be *complementary:* every base on one chain must be matched by a specific partner base on the other helix. This matching, or lock-and-key relationship, provides the mechanism for molecular reproduction, which is the cornerstone of the chemistry of life.

8-6 Metals

Most elements are metals. Referring back to Figure 8-1, we see that, roughly speaking, any element to the left of carbon is metallic. Carbon has its 2s and 2p valence orbitals half-filled. The elements to the left of carbon have fewer electrons—they have electron-deficient, orbital-excess occupancies. This orbital-excess situation is the salient property of metals. But before we discuss why a metal is a metal, we should examine the characteristics of the metallic state.

(a) CHARACTERISTIC PROPERTIES OF METALS

The physical attributes of metals are so familiar that classification is easy. The characteristics that identify a metal are

—a bright, lustrous appearance
—high electrical conductivity
—high thermal conductivity
—malleability

The electrical and thermal conductivities are most readily assessed quantitatively. Table 8-7 lists the electrical conductivities of representative solids. Except for graphite, which is quite

Table 8-7 Electrical Conductivities of Various Types of Solids

Substance	Solid Type	Conductivity $(\text{ohm cm})^{-1}$
Silver	metal	$6 \cdot 10^5$
Zinc	metal	$2 \cdot 10^5$
Graphite	covalent	$5 \cdot 10^4$
Sodium chloride	ionic	10^{-7}
Diamond	covalent	10^{-14}
Quartz	covalent	10^{-14}
Sulfur	molecular	10^{-17}
Paraffin	molecular	$2 \cdot 10^{-19}$

unusual, there is an enormous difference between the electrical conductivity of metals and that of any other type of crystal.

Another distinctive property of a metal is its high thermal conductivity, as anyone knows who has stirred a hot cup of coffee with a silver spoon. Table 8-8 lists the thermal conductivities for the 17 elements with the highest electrical conductivities.

Table 8-8 Electrical and Thermal Conductivities of the Elements with Highest Electrical Conductivities

Element	Electrical Conductivity $(\text{microohm cm})^{-1}$	Thermal Conductivity (cal/cm sec degK)
Ag	0.63	1.01
Cu	0.60	0.99
Au	0.46	0.70
Al	0.38	0.50
Ti	0.31	
Ca	0.29	0.3
Na	0.24	0.32
Mg	0.22	0.38
Rh	0.22	0.21
Mn	0.20	
Mo	0.19	0.35
Ir	0.19	0.14
W	0.18	0.4
Be	0.17	0.38
Zn	0.17	0.27
K	0.16	0.23
Co	0.16	0.16

The most obvious feature of the table is that the four elements at the top of the list, those with the highest electrical conductivities, are also the four with the highest thermal conductivities. An immediate implication is that electrical and thermal conductivity in metals are related, at least for the best conductors.

As dramatically revealed in Table 8-7, the electrical conductivity of a metal is its most distinctive property. An explanation is found in our understanding of the chemical bonding in metals, as exemplified by lithium.

(b) BONDING IN LITHIUM METAL

Lithium has a single valence electron and four valence orbitals. We have already discussed the bonding in Li_2 (Section 3-4a) in terms of the molecular orbitals formed from two 2s orbitals. However, each lithium atom has additional bond-forming capacity because of its vacant orbitals. This electron-deficient situation is what causes boron to settle into clustered structures (as in B_5H_9 and $B_{10}H_{14}$, see Section 7-5b). By clustering, each atom gets to share its few electrons with several neighbors—and to share theirs. This clustering characterizes the structures of metals.

It is possible to explain the bonding and properties of lithium through a molecular orbital treatment that considers only the 2s orbitals. It should be kept in mind, though, that the vacancy of the 2p orbitals is an essential requirement (otherwise, fluorine and hydrogen would be predicted to be metallic). Consideration of all of the valence orbitals is necessary for all metals other than the alkali metals. Nevertheless, the essential ideas that underlie our understanding of the metallic state are well expressed in this simple 2s molecular orbital argument.

Figure 8-28 shows a sequential development of the molecular orbitals that can be formed from two, three, and four 2s orbitals arranged in a linear array. In 8-28(*a*) we see the Li_2 problem—one bonding and one antibonding orbital are formed, and only the bonding orbital is occupied. In 8-28(*b*) a linear Li_3 arrangement is considered. A third M.O. now divides the energy spacing, but it is a nonbonding orbital. With three electrons, the orbital occupancy does not require the use of the antibonding orbital. The linear Li_4, shown in 8-28(c) produces another pair of bonding and antibonding M.O.'s, and the spacing of the energy levels is decreased further. Again we find that half the orbitals are bonding orbitals and, with one electron per atom, only these need be occupied.

Finally, as the linear array is made longer and longer, the M.O.'s come closer and closer together in energy until they give a band of levels that is virtually a continuum. However, the number of levels is always just equal to the number of component orbitals. With one valence electron per atom, only half the M.O.'s in the band of levels need be occupied—the bonding M.O.'s.

There are two characteristics of this band of M.O.'s that are crucially connected to the properties of the metallic state. First, each M.O. extends over the entire array of lithium atoms (in three dimensions, over the entire crystal). This implies electron mobility on a macroscopic scale. Second, there is no energy gap between occupied and vacant M.O.'s in the band. This means that very small perturbations of energy levels can cause occupancy changes that will show up in the physical properties. For example, an electric field can change the relative energies of levels that involve electron motion with or against the field. Figure 8-29 portrays this in a schematic way. If we apply an electric field that favors momentum states that move the elec-

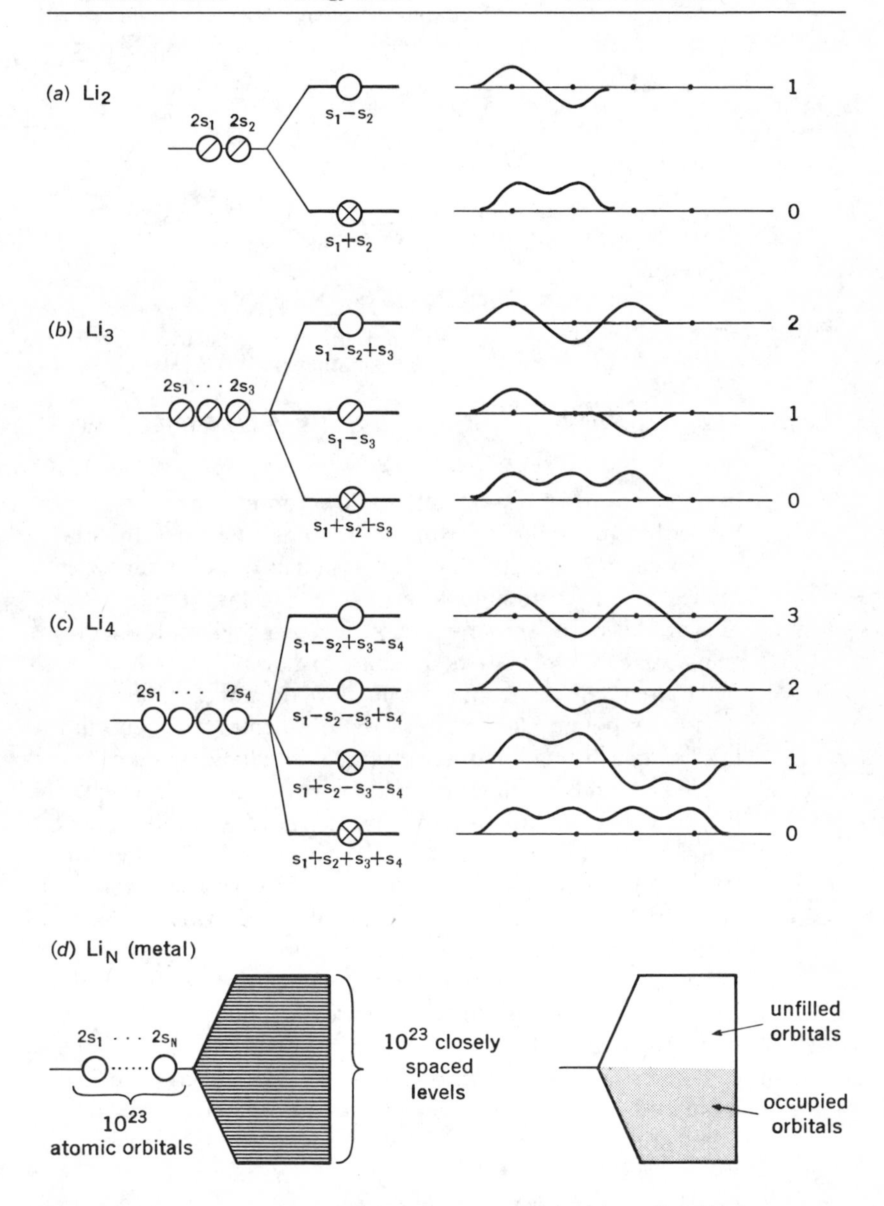

Figure 8-28 Molecular orbital development of the band theory of metals.

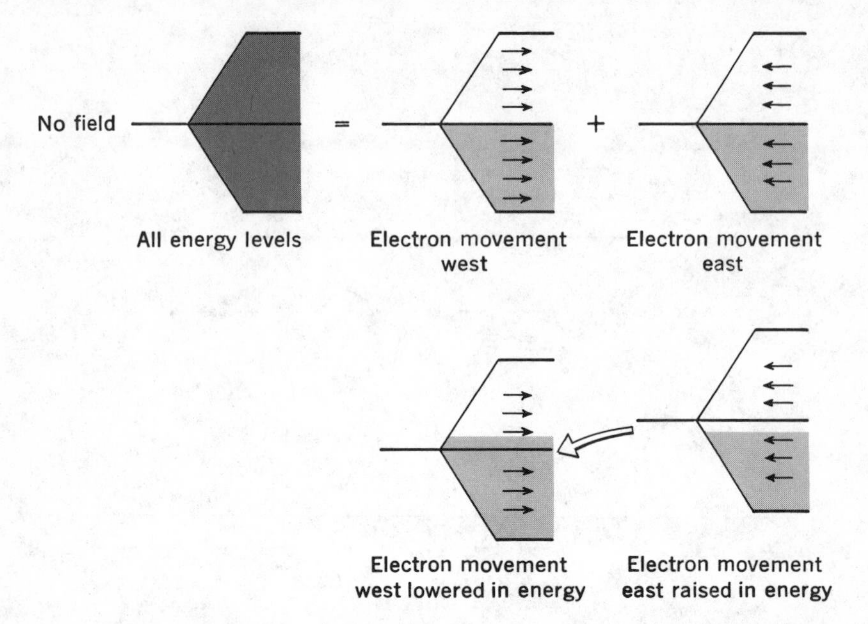

Figure 8-29 Effect of an electric field on electron movement in a metal.

trons westward, those states are lowered in energy relative to those that move the electrons to the east. Because the states are so closely spaced, the smallest such displacement brings many vacant westbound levels below filled eastbound levels. The orbital occupancy readjusts, and we find more electrons going west than east—electric current flows.

Whether current continues to flow depends upon whether these westbound electrons are removed from the west end of the crystal (through another electrical conductor) and replaced at the east end (from, for example, an electrochemical cell). The high electrical conductivity of metals depends, then, upon the mobility of electrons in M.O.'s that extend over the whole crystal and upon the absence of an energy gap between occupied and unoccupied orbitals. The absence of an energy gap occurs because the band is only partially filled.

The spatial extent of the orbitals explains the mobility of the electrons as reflected in the electrical conductivity. This same mobility explains the thermal conductivity of metals. If one end of a metal is heated, the electrons at that end acquire higher momentum. This extra momentum is quickly conveyed to the other end as the conducting electrons race back and forth. If the electrical conductivity is very high, the thermal conductivity will also be high. If the electrical conductivity is quite low, then momentum will have to be conducted through the lattice vibrations, which are less effective.

(c) BONDING IN BERYLLIUM METAL

Our discussion of lithium required that the M.O. band be only partially filled in order to obtain metallic conductivity. This is achieved in lithium: the N_0 2s orbitals give N_0 M.O.'s, but only $\frac{1}{2}N_0$ of the orbitals need be occupied because there is one valence electron per atom. Beryllium, with two electrons per atom, has just enough electrons to fill completely the N_0 M.O.'s in the 2s band. Yet, beryllium is a metal.

This example demonstrates the active role of the other vacant valence orbitals. If we had added, in Figure 8-28, the 2p orbitals, they would be placed some 40 kcal higher for the separated atoms. Figure 8-30 shows the more complete M.O. diagram for lithium as well as that for beryllium. In each case, the lowest bonding M.O.'s in the 2p band intermingle with the uppermost antibonding M.O.'s in the 2s band. This is unimportant for lithium because the lowest band is only half-filled. For beryllium,

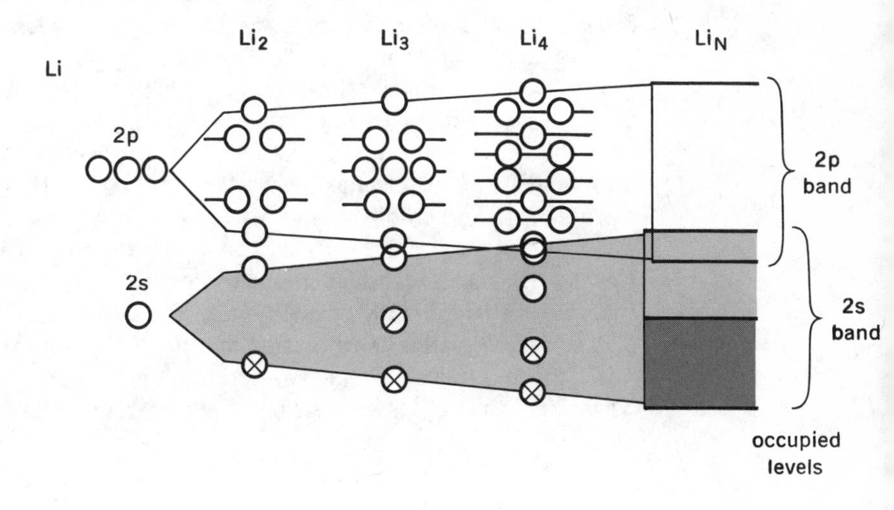

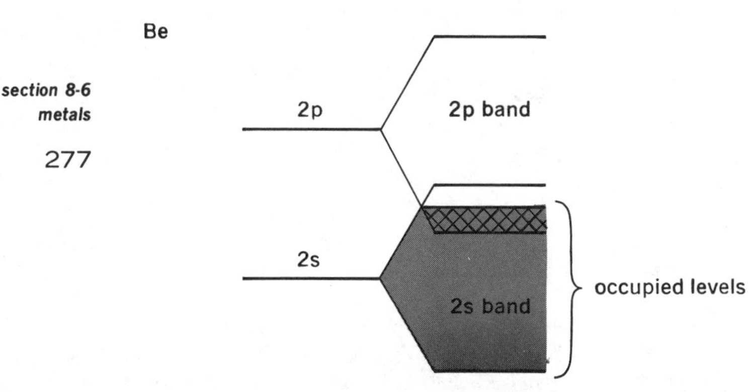

Figure 8-30 Metallic M.O.'s for lithium and beryllium, including p-orbitals.

however, the number of electrons would just fill the 2s M.O.'s if they did not intermingle in the band generated from the 2p orbitals. Because of the crossover, however, some of the 2p M.O.'s are occupied, as shown in Figure 8-30, so neither the 2s band nor the 2p band is fully occupied. Metallic properties result.

(d) CRYSTAL STRUCTURES OF METALS

With the electron-deficient, orbital-excess electron configurations that characterize metals, they can be expected to form condensed structures in which each atom seeks as many neighbors as possible. Just as for the inert gas solids, the packing of spheres indicates how atoms can pack to maximize the number of nearest neighbors. Figure 8-6 shows the build-up of one such arrangement. Examination of this array from another angle shows that it contains atoms arranged at the corners of a cube with other atoms centered in each face of the cube (see Fig. 8-31). Consequently this crystal type is called face-centered-cubic closest packing (f.c.c.).

There are other ways, however, to stack oranges and pack atoms close together. One of them involves only a subtle change from the f.c.c. arrangement. The f.c.c. packing was obtained by arranging six atoms around the first, all in a plane. Then three more atoms were cradled above and three more below. No fuss was made about it, but Figure 8-32 shows that there are actually two different ways in which the second trio could be added. It could be cradled directly below the upper trio or it could be cradled equally well after being rotated through 60°. In each arrangement the central atom (i.e., *every* atom) has twelve nearest neighbors. They differ, however, in their next-nearest-neighbor

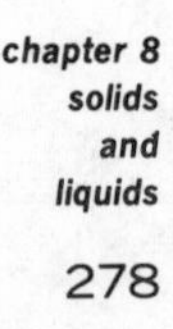

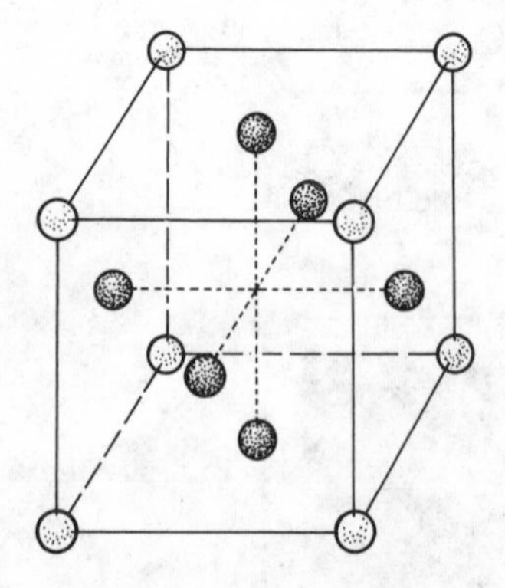

Figure 8-31 Face-centered-cubic closest packing, "f.c.c." For clarity, atoms in the face centers are shaded.

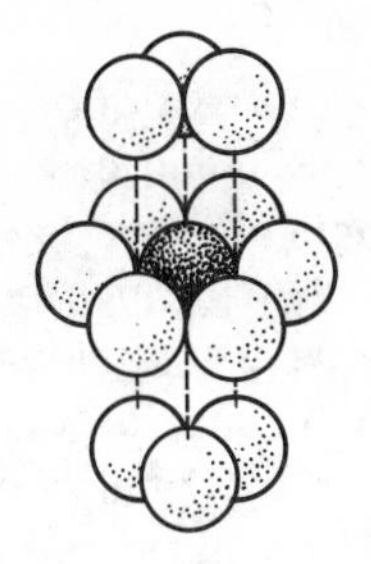

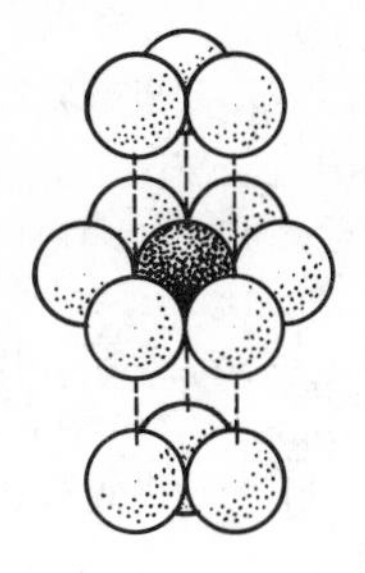

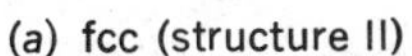

(b) hcp (structure III)

Figure 8-32 Two closest packing crystal structures: (a) face-centered-cubic closest packed and (b) hexagonal closest packed. Notice that the top and the bottom layers are identical in the h.c.p. structure, but are rotated 60° relative to each other in the f.c.c. structure.

relationships. The f.c.c. arrangement is the one in which the third layer is rotated relative to the first layer (see Fig. 8-32(*a*)). The other arrangement, in which the third layer is placed directly under the first, is called *hexagonal closest packing* (h.c.p.).

There is one more structure that is somewhat less effective. Instead of packing twelve neighbors next to each atom, eight can be spaced around it. They occupy, then, the corners of a cube with the central atom at its center, as shown in Figure 8-33. This arrangement is called the *body-centered-cubic* structure (b.c.c.).

(e) ORBITAL OCCUPANCY AND METALLIC CRYSTAL STRUCTURES

Sodium, magnesium and aluminum are good examples to study. The first of these, sodium, has one valence electron with which

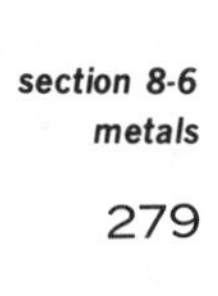

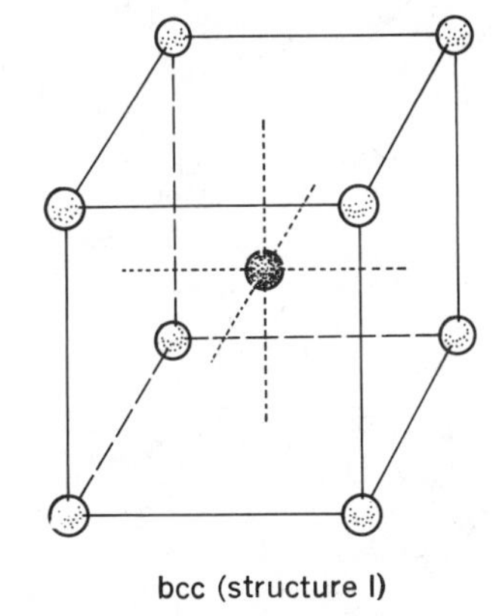

bcc (structure I)

Figure 8-33 Body-centered-cubic packing, "b.c.c." The body center position is shaded for clarity.

to form bonds to other atoms, and the rest of its valence electrons are vacant. This orbital-excess situation calls for a closest-packing structure but, with only one s electron to be shared so widely, a sodium atom settles for only eight nearest-neighbors, the body-centered-cubic structure. We'll call this structure I.

Magnesium has a ground state orbital occupancy of $1s^2$, $2s^2 2p^6$, $3s^2$. However by paying a promotional energy price of 80 kcal/mole, both of the 3s electrons become available for bonding (see Fig. 8-34). With a 2s and a 2p electron available for sharing in the many vacant orbitals of its neighbors, magnesium selects one of the twelve-nearest-neighbor arrangements, the hexagonal closest packing. We'll call this packing structure II.

Aluminum has one more electron and orbital occupancy $1s^2$, $2s^2 2p^6$, $3s^2 3p$ in its ground state. At a cost of 83 kcal three electrons can be made available for bonding (again, see Fig. 8-34). With three valence electrons to share in its four 3s and 3p orbitals, aluminum also can afford twelve nearest-neighbors, but it picks the face-centered-cubic alternative. This, then, becomes structure III.

Astonishingly, the crystal structures of most of the metals correlate with these three examples. Figure 8-35 extends the

Element	Sodium	Magnesium	Aluminum
	3s 3p	3s 3p	3s 3p
Ground state	...∅ ○○○	...⊗ ○○○	...⊗ ∅○○
Bonding state	...∅ ○○○	...∅ ∅○○	...∅ ∅∅○
Promotion energy (kcal/mole)	0	80	83
Number of s and p electrons for bonding	1	2	3
Crystal structure	I bcc	II hcp	III fcc

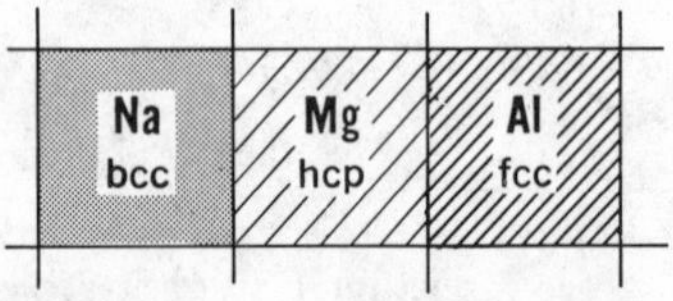

Na	Mg	Al
bcc	hcp	fcc

Figure 8-34 The bonding states and crystal structures of sodium, magnesium and aluminum.

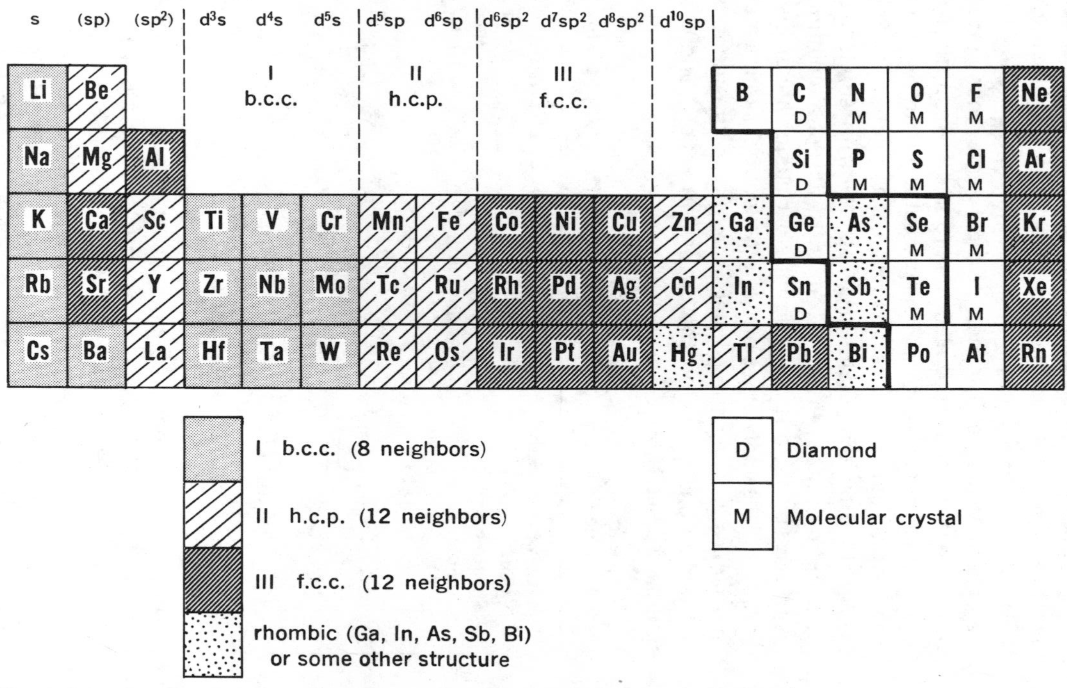

Figure 8-35 Distribution of crystal types in the periodic table (at room temperature).

small piece of the periodic table included in Figure 8-34 to show the crystal structures I, II, and III as they are found throughout the periodic table. Examine, first, the elements from titanium to zinc in the first long row. Each of the elements, titanium, vanadium, and chromium, has a bonding state in which there are some d electrons (3, 4, or 5) and a single s electron. With a single s electron, the structure is the same as that of sodium, structure I. We shall see that the d electrons help to strengthen the crystal, but apparently they are not influential in determining the crystal structure. The next two elements, manganese and iron, have a number of d electrons (5 and 6, respectively) and one s and one p electron. Their crystal structure is II, that of magnesium, our prototype sp crystal. Then cobalt, nickel, and copper have bonding states $d^{6,7,8}\ sp^2$. With three electrons in the s and p orbitals, the structure is III, that of aluminum. Zinc reverts to structure II because the special stability of the filled d orbitals leaves only one s and one p electron, like magnesium. These rules relating the crystal structure to the number of s and p electrons are called the Engel Correlation* and they are summarized in Table 8-9.

Table 8-9 The Engel Correlation for Predicting Metallic Crystal Structures

Number of Unpaired Valence s and p Electrons in the Bonding State	Structure	Number of Nearest Neighbors
1	I (b.c.c.) body-centered-cubic	8
2	II (h.c.p.) hexagonal closest packing	12
3	III (f.c.c.) face-centered-cubic	12

The most glaring discrepancies in this simple set of rules are found in the alkaline earth elements (Ca, Sr, and Ba) and three members of the aluminum family (Sc, Y, and La). For the latter, careful consideration of the promotional energies shows, however, that the bonding states are actually compatible with the Engel Correlation. Table 8-10 shows the promotional energies to three possible bonding states for the aluminum metals, each giving three bonding electrons. For aluminum, the sp^2 state gives three bonding electrons at a promotional energy cost of 83 kcal. Both the d^2s state (structure I) and the dsp state (struc-

* These rules were formulated by N. Engel, a Danish chemist.

Table 8-10 Promotional Energies for the Aluminum Metals (in kilocalories above the ground state)

Element	d^2s(I)	dsp(II)	sp^2(III)	Engel Prediction	Observed
Al	very high	190	83	III	III
Sc	33	45	high	I or II	II (20° C) I (high temp.)
Y	31	43	high	I or II	II (20° C) I (high temp.)
La	6	40	high	I or II	II (20° C) I (high temp.) III (high temp.)

ture II) are far too expensive, 190 kcal or more. For the next three elements, the d^2s and dsp promotional energies are quite moderate. These electronic arrangements dictate structures I and II. In fact, both are observed for all three elements. Lanthanum, trying its best, also displays a high-temperature structure III phase, but it is clear that the promotional energies to the sp^2 bonding state are prohibitively high at room temperature.

A similar, more detailed study of calcium, strontium, and barium reveals that structures I are predicted, which correspond to the structure shown for barium in Figure 8-35 and to known high-temperature structures of calcium and strontium. Strontium also shows structure II as a high-temperature form, again in accord with the Engel Correlation. The non-existence of structure II for calcium and the surprising stability of its structure III remain, now, a lone discrepancy.

On the right-hand side of the periodic table, the Engel Correlation runs out of gas beyond zinc as the orbital excess begins to dwindle. Now the elements begin to wonder whether they want to be metals or nonmetals.

(f) ORBITAL OCCUPANCIES AND HEATS OF SUBLIMATION

The heat of sublimation of a metallic crystal is determined by the number of valence electrons participating in the bonding. This is readily apparent in Figure 8-36 which gives the sublimation heats for the first three columns of the periodic table. This table also shows that d orbitals contribute to the bonding, despite the fact that they do not exert directive influence on the crystal arrangement.

The trend in heats of sublimation across the periodic table can be examined in a number of ways. Figure 8-37 shows two of these. In the upper plot, the heats are plotted against a row in the periodic table for part of the way across. In the lower

Li (I)[a] 2s **32.2**[b]	Be (II) 2s2p **53.5** (2×26.8)[c]	B (*) $2s2p^2$ **129** (3×43)
Na (I) 3s **23.1**	Mg (II) 3s3p **31.5** (2×15.8)	Al (III) $3s3p^2$ **67.9** (3×22.6)
K (I) 4s **18.9**	Ca (III, I) 3d4s **36.6** (2×18.3)	Sc (II, I) 3d4s4p **73** (3×24.3)
Rb (I) 5s **18.1**	Sr (III, I, II) 4d5s **33.6** (2×16.8)	Y (II, I) 4d5s5p **94** (3×31)
Cs (I) 6s **16.3**	Ba (I) 5d6s **35.7** (2×17.8)	La (II, I, III) 5d6s6p **96** (3×32)

* Twelve-sided, clustered unit with covalent bonds.
[a] Observed crystal structures: I = bcc; II = hcp; III = fcc.
[b] Heat of sublimination, kcal/mole.
[c] Heat of sublimination, kcal/mole per bonding electron.

Figure 8-36 Heats of sublimation and heats of sublimation per bonding electron for the metals of the first three columns of the periodic table.

plot, the sublimation energy per bonding electron is shown. Yet another way of examining the data would be to add to the values of $\Delta H_{\text{sub.}}$ the promotional energy needed to obtain the bonding state. All these approaches show a rising stability for metallic crystals up to the fifth column, and then a decline. Nevertheless, the data show that metallic stability is enhanced as the number of bonding electrons increases, and that the energy per bonding electron is 20–40 kcal per electron, or 40–80 kcal per electron pair. These energies are normal chemical bond energies, but, because the electron pairs must bind many nearest neighbors, individual atom–atom distances are characteristic of quite weak bonds. For example, the bond energy in the diatomic molecule Na_2 is 17 kcal/mole and its bond length is 3.08 Å. The metallic sodium bond energy is 23.1 kcal per electron, or 46.2 kcal per electron pair. This bond energy is distributed over bonds to eight nearest neighbors or as 5.8 kcal per interaction. The bond length should be much longer in the metal—as it is, 3.72 Å. All the alkali metal elements have metallic bonds about 15–20 percent longer than their diatomic bond lengths.

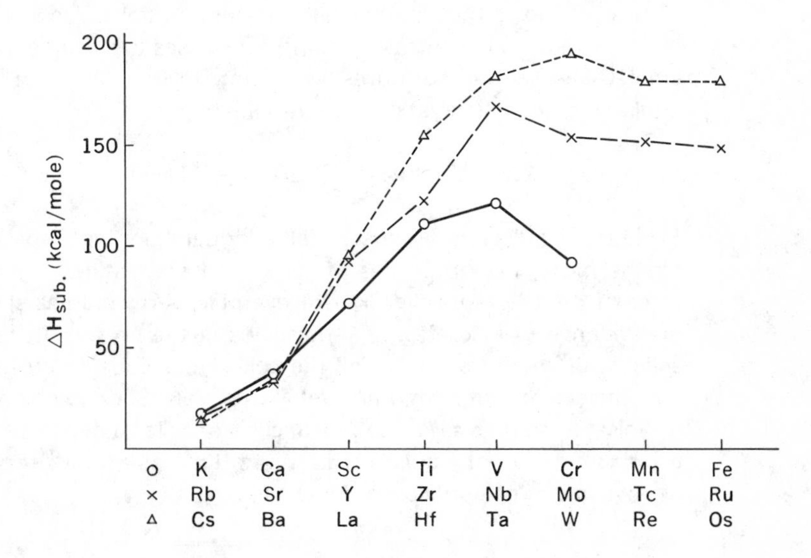

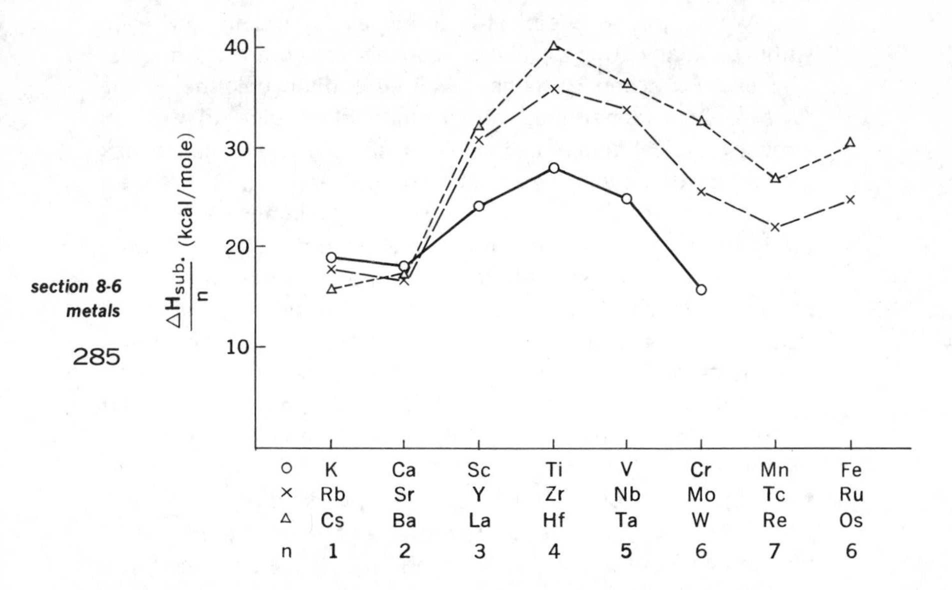

Figure 8-37 Variations of $\Delta H_{sub.}$ and $\Delta H_{sub.}$ per electron among metals.

8-7 Ionic solids

The gaseous, diatomic molecules $Cl_2(g)$ and NaCl(g) both form chemical bonds by electron-pair sharing in half-occupied valence orbitals. The bond energies are 57 and 98 kcal/mole, respectively. Chlorine condenses to a colored, molecular crystal, but only if the temperature is lowered to the liquid-air range. Figure 8-8 shows that the crystal contains Cl_2 molecules weakly bound together. In contrast, NaCl(g) condenses to a transparent crystal even at temperatures exceeding 1000°K. The heat of sublimation of this crystalline material is high.

$$\text{NaCl(solid)} \rightarrow \text{NaCl(gas)} \qquad \Delta H = +48 \text{ kcal} \qquad (8\text{-}10)$$

Evidently NaCl(g) has substantial residual bonding capacity. Furthermore, the properties of solid NaCl are distinctively different from those of solid Cl_2—for example, X-ray studies show no evidence of discrete NaCl molecules in the crystal. In fact. solid NaCl is different from all the solid types considered thus far—molecular, covalent, and metallic crystals. Sodium chloride is typical of an *ionic solid,* the fourth class of solids listed in Table 8-1. Examination of NaCl, first as a gas, then as a solid, serves as a valuable introduction to ionic solids.

(a) SALT—A TYPICAL SALT

The substance sodium chloride has the household name "salt." This very common chemical is found all around us and even within us. Many dried-up lake beds furnish commercial amounts. The endless ocean contains dissolved sodium chloride to the tune of $\frac{1}{2}$ mole per liter, and human blood evolved with an almost identical concentration. This ubiquitous substance has properties that typify most solid compounds formed between an element far to the left in the periodic table and an element far to the right. Because NaCl is familiar and typical, its name was gradually transferred to this general class—compounds such as KBr, KCl, $MgCl_2$, CaF_2, and so on, have all been called "salts" since the time of the alchemists. We shall, however, use the name *ionic solids*, a term that more meaningfully represents the bonding that explains their properties.

The gaseous NaCl molecule has a high bond energy that is attributed to ionic bond character. This ionic character gives the gaseous NaCl molecule a high electric dipole moment, 8.97 D. This corresponds to the movement of eight tenths of an electron charge through the bond length 2.36 Å, from the sodium to the chlorine atom. Thus the molecule is reasonably well characterized as a positive sodium ion embedded in the outer electron distribution of a negative chloride ion. As in the case of LiF, discussed in section 3-4e, this skewed electron distribution actually holds the bonding electrons closer, on the average, to sodium than the average radius of a neutral sodium atom.

From an orbital occupancy point of view, the sodium atom in NaCl(g) has considerable residual bonding capacity. It has, after all, three vacant valence orbitals and, to the extent that the electron distribution is skewed away, perhaps four. We can expect that the sodium atom should bond readily to a good electron donor—like, for example, a chlorine atom in another NaCl(g) molecule. Sodium chloride should readily form molecular dimers in the gas phase. In fact it does, and the bonds are strong: 44.6 kcal is the enthalpy required to break apart the dimer.

$$Na_2Cl_2(g) \rightarrow 2NaCl(g) \qquad \Delta H = +44.6 \text{ kcal} \qquad (8\text{-}11)$$

The Na_2Cl_2 structure is not known, but chemists are confident that it has a condensed structure like one of those shown in Figure 8-38. In (*a*), the dimer is pictured as two rather weakly bound NaCl molecules, each with bond length the same as that of the monomer, 2.36 Å. The representation (*b*) envisages the same square arrangement, but with four equal bonds, each a bit longer than the 2.36 diatomic bond length. The third version (*c*) is like (*b*)—the bond lengths are equal, but a planar, trapezoidal geometry is presumed. Of these, the equal-bond-length models (*b*) and (*c*) are generally preferred, and one of them is likely to be correct. (For example, approximate calculations of the bond energy based upon the trapezoidal configuration gave the correct energy with an internal Cl—Na—Cl angle of 105° and uniform bond lengths of 2.59 Å.) If this is so, already in the dimer we no longer discern individual NaCl molecules. Each sodium atom is equally bonded to two chlorine atoms.

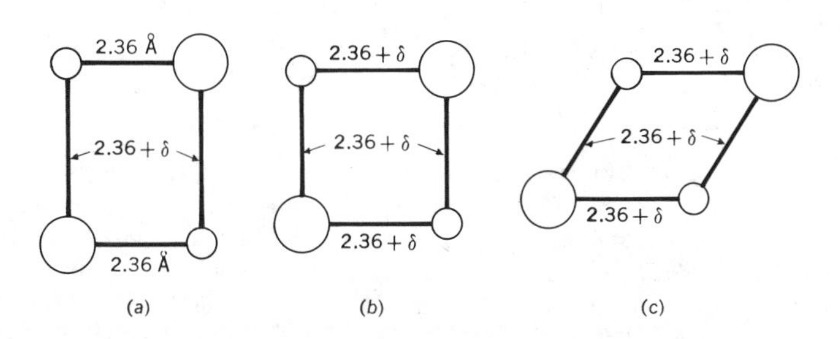

Figure 8-38 Possible structures of NaCl dimer.

Since the dimerization energy is so large and since each sodium still has a lot of unfilled valence orbital space, we can expect further aggregation. In fact, the situation reminds us of the inert gases with their completely vacant extra-valence orbitals, and of the metals with their almost-vacant valence orbitals. Both cases display close-packing arrangements without specific bonding. It is almost obvious that two dimers like (*b*) could cluster, one above the other, to give a stable tetramer. And certainly the upper square group would be rotated 90° so that each sodium had a negatively charged chlorine above it, rather than a similar, positively charged sodium atom (see Fig. 8-39(*a*)). This clustering will continue until each sodium atom is

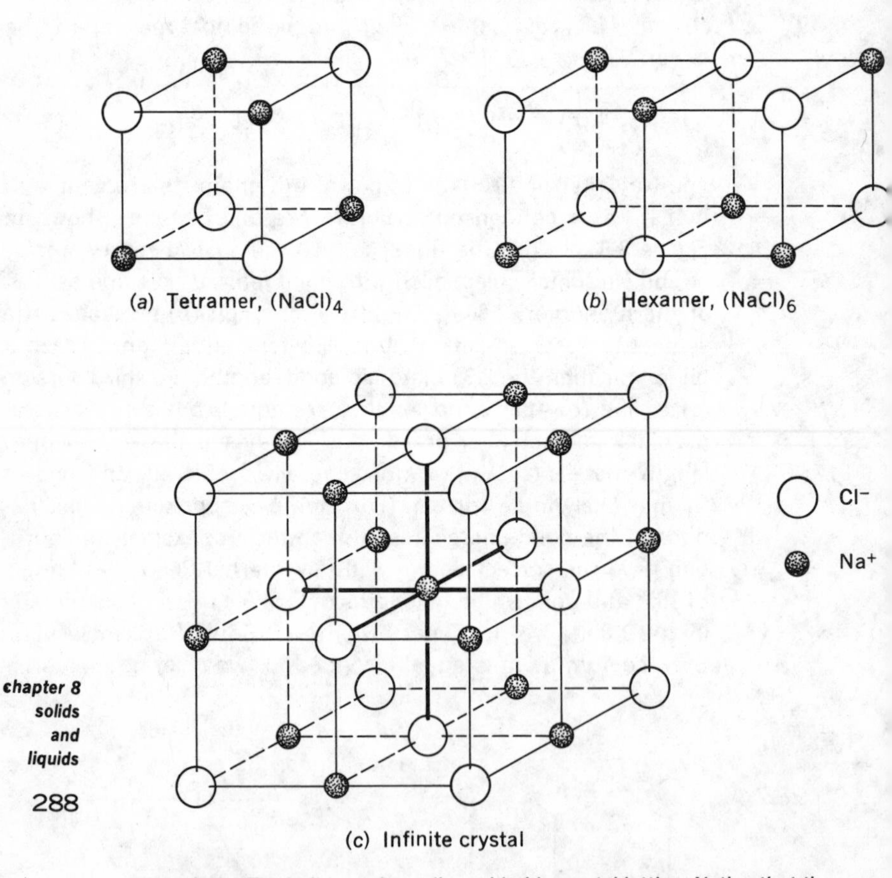

Figure 8-39 The build-up of a sodium chloride crystal lattice. Notice that the chloride ions define a face-centered cube. The sodium ion in the center of the cube is surrounded by six equidistant chloride ions (heavy lines). This position in the f.c.c. lattice is known as an octahedral hole. If the lattice were extended, it would be clear that the sodium ions themselves have a f.c.c. arrangement with chloride ions in octahedral holes. The NaCl structure is, then, two interleaved f.c.c. arrangements of ions.

surrounded by six chlorine atoms and each chlorine atom by six sodium atoms (see Fig. 8-39(c)). With the skewed charge distribution appropriate to the different electronegativities, this arrangement permits each Na^+ to bask in the comfortable negative environment provided by six nearest Cl^- neighbors. The Na^+–Cl^- distances are longer than in the diatomic molecule (2.814 Å is observed in the crystal, 2.36 Å in the gas), indicating a weaker bond. This is more than compensated, however, by the fact that each sodium atom bonds equally to six neighbors. Thus we are led to the most characteristic aspect of ionic crystals. Their structures are best described without mentioning molecules (despite the stability of the parent molecules such as NaCl(g)). Instead, the crystal is aptly pictured as a closest-packed array of positive and negative ions, alternating in the lattice to give maximum electrostatic attraction between nearest neighbors. Such a lattice is called an *ionic* solid.

(b) OTHER IONIC CRYSTALS

Without being too explicit, Figures 8-38 and 8-39 suggest that Cl^- is much larger than Na^+, the building blocks of the NaCl lattice. This is consistent with our estimates of size based upon the one-electron, effective charge approximation. Since the ionization energy of Cl^- is quite low (the electron affinity of Cl is 83 kcal/mole) and that of Na^+ is very high (the *second* ionization energy of Na is 1091 kcal/mole), Cl^- must be much larger than Na^+. That means our packing problem is more complicated than it was with inert gas or metal atoms all of uniform size. Figure 8-40 shows how the packing in a single layer of the NaCl crystal arrangement is affected by the ion sizes if they differ. The sizes pictured in (*a*) are labelled "just right." The X^- ions are crowded right up against M^+, but are not repelling each other excessively. In contrast, (*b*) is a poor fit. If the X^- ions move in close to M^+, they crowd together, repelling each other and raising the energy. If they hold each other apart, then they are not very close to the oppositely charged M^+. Finally, (*c*) doesn't allow the X^- ions close enough to feel any X^-–X^- repulsion at all.

Figure 8-40 suggests that if M^+ is too large (as in (*c*)) or if M is too small (as in (*b*)), a different MX crystal packing might be obtained. In fact when M^+ is quite large, as in CsCl, another arrangement is found in which Cs^+ has *eight* nearest neighbors of Cl^- (instead of six as in NaCl). The CsCl lattice is shown in Figure 8-41(*a*). On the other hand, if M^+ is too small, only four negative ions pack around the positive ions. Zinc sulfide provides two common examples of this situation. The arrangement

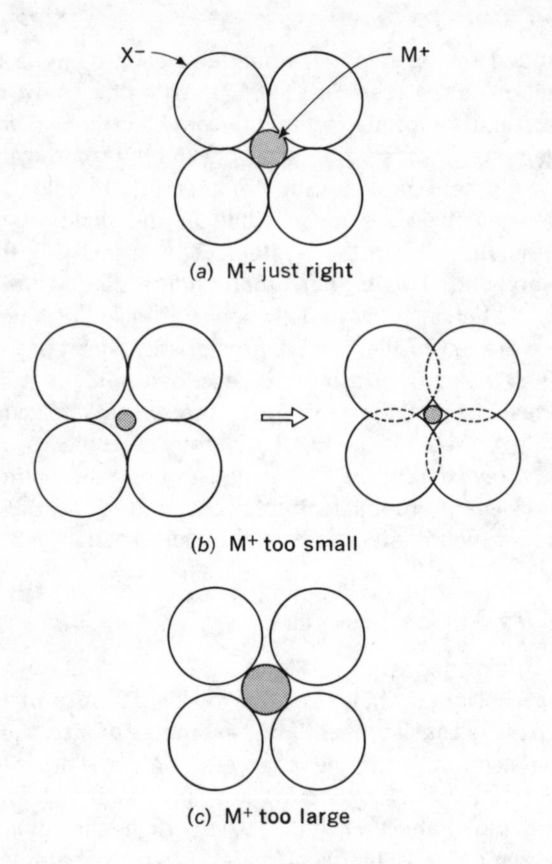

Figure 8-40 Effect of relative ion size on the NaCl lattice.

shown in Figure 8-41(*b*) is called *zinc blende*, or cubic ZnS. This lattice is just like diamond except that the atoms of Zn and S alternate. Figure 8-41(*c*) is the *wurtzite* form of zinc sulfide, or hexagonal ZnS. In each of these lattices, every atom has four nearest neighbors of opposite kind—the result of the large size of sulfur relative to zinc.

Figures 8-39 and 8-41 show ways of packing positive and negative ions so that each has 4, 6, or 8 nearest neighbors of opposite charge. The number of nearest neighbors is called the *coordination number.* The observed coordination is determined by the relative sizes of the ions. If the positive ion is much smaller than the negative ion, the coordination numbers of 4 are observed. If the positive ion is quite large, then 8 will probably be observed. For most MX crystals, the NaCl lattice is found, coordination number 6. These rules apply for alkali halides, alkaline earth oxides, alkaline earth sulfides, and a number of other substances, as shown in Table 8-11.

It goes without saying that more complicated crystal struc-

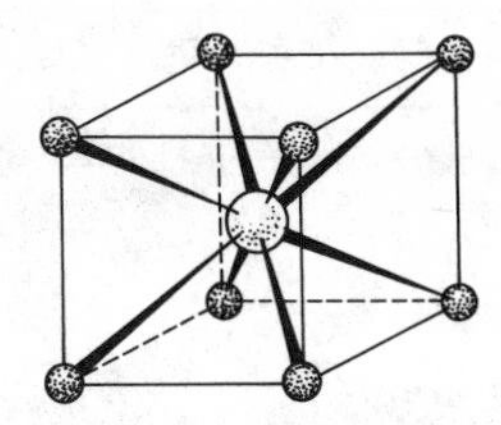

(a) CsCl lattice (coordination number 8)

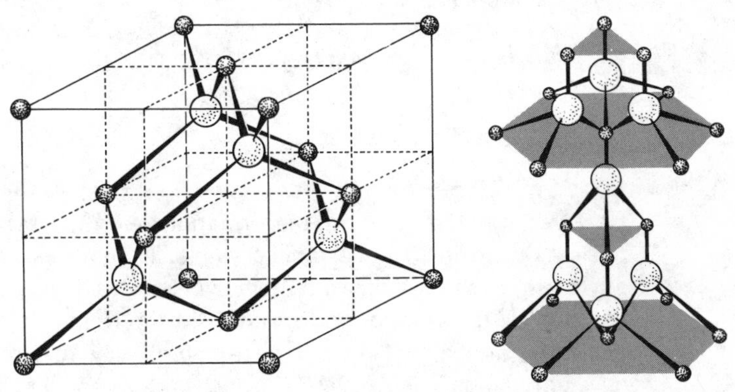

(b) Zinc blende lattice (coordination number 4)

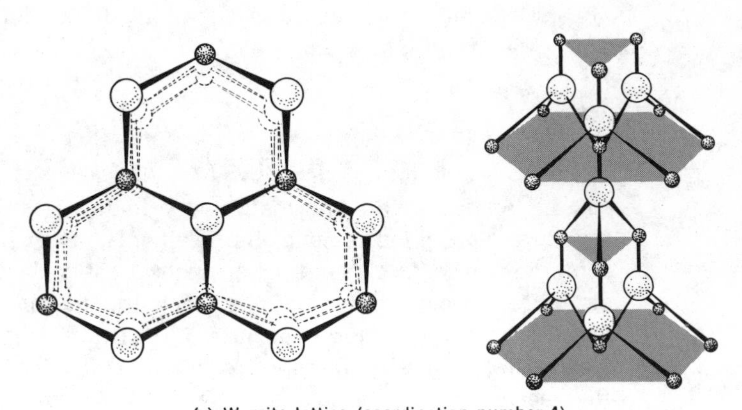

(c) Wurzite lattice (coordination number 4)

Figure 8-41 Crystal lattices for MX crystals with coordination numbers eight and four. (a) CsCl lattice: a b.c.c. arrangement (see Fig. 8-33) in which the negative ions (anions) occupy body center positions in the positive ion (cation) lattice. Alternatively, it may be described as two interleaved simple cubic lattices. This lattice is not closest packed, but represents a slightly less efficient packing arrangement. (b) Zinc blende lattice: a f.c.c. lattice of cations with negative ions in alternate tetrahedral sites. The cubic nature of the structure is shown in the first part of the figure and a different view similar to Figure 8-32 is shown in the second. If the atoms in the lattice are identical, it becomes the diamond structure (see Fig. 8-14(b)). (c) Wurzite lattice: a h.c.p. arrangement (see Fig. 8-32) with alternate tetrahedral sites occupied. This becomes the ice structure (see Fig. 8-19) when the atoms are identical.

Table 8-11 Crystal Lattices for Ionic MX Crystals

Coordination number 8—CsCl lattice
6—NaCl lattice
4Z—zinc blende lattice
4W—wurtzite lattice

LiF (6)	LiCl (6)	LiBr (6)	LiI (6)	BeO (4W)	BeS (4Z)
NaF (6)	NaCl (6)	NaBr (6)	NaI (6)	MgO (6)	MgS (6)
KF (6)	KCl (6)	KBr (6)	KI (6)	CaO (6)	CaS (6)
RbF (6)	RbCl (6)	RbBr (6)	RbI (6)	SrO (6)	SrS (6)
CsF (6)	CsCl (8)	CsBr (8)	CsI (8)	BaO (6)	BaS (6)
AgF (6)	AgCl (6)	AgBr (6)	AgI (4Z)	ZnO (4Z, W)	ZnS (4Z, W)
				CdO (6)	CdS (4Z, W)
					HgS (4Z)

tures are needed for MX_2 compounds, such as $CaCl_2$ or Na_2O. However, most of them are understandable on the same simple basis as used for the MX ionic crystals. The ions pack so as to surround positive ions with negative ions and vice versa, as dictated by the compound formula and the relative sizes of the two ions, and as influenced by the tendency toward ionic or covalent bonds. Figure 8-42 shows two common crystal structures, fluorite (as shown by CaF_2, SrF_2, BaF_2, $SrCl_2$, $BaCl_2$, ZrO_2, ThO_2, UO_2, etc.) and rutile (as shown by TiO_2, MnO_2, GeO_2, SnO_2, MgF_2, MnF_2, NiF_2, ZnF_2, etc.).

(c) IONIC LATTICES AND CLOSEST PACKING

The various ionic lattices described in Figures 8-39, 8-41, and 8-42 can be related in an informative way to the closest-packed structures observed among the elements. Both the face-centered-cubic (f.c.c.) and hexagonal-closest-packed (h.c.p.) structures involve trios of atoms, each trio nesting a fourth atom. These four atoms enclose a space—a "hole"—with tetrahedral symmetry. Close examination of Figure 8-32 reveals that each atom participates in the formation of eight such "tetrahedral holes." These voids in the closest-packed lattices suffice to define three of the ionic lattices we have described.

Since there are eight tetrahedral holes around each atom and each such hole is shared by four atoms, there are twice as many tetrahedral holes as there are atoms. If a closest-packed array of doubly charged positive ions were prepared in the f.c.c. arrangement, negative ions could be placed in these tetrahedral holes to give an electrically neutral crystal. For example, suppose the positive ions in the f.c.c. lattice were Ca^{+2} ions and the negative ions were F^- ions. Then we would have to fill every tetrahedral hole to achieve electrical neutrality. The lattice thus

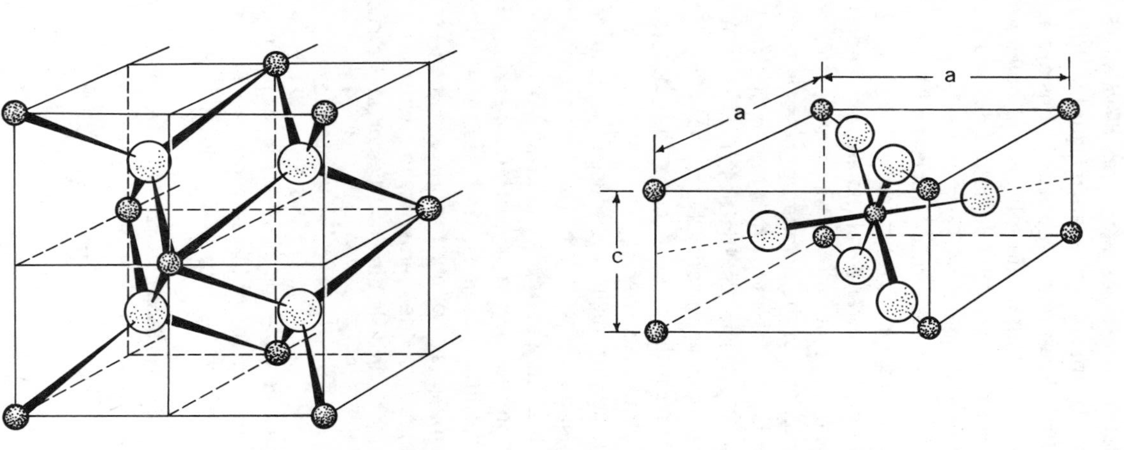

(*a*) Fluorite (CaF_2) (coordination number 4)   (*b*) Rutile (TiO_2) (coordination number 6(3))

Figure 8-42 Crystal lattices for MX_2 crystals. (a) Fluorite lattice: the anions occupy every tetrahedral site in a f.c.c. array of cations. For clarity, only half of the face-centered cube is drawn. (b) Rutile lattice: the cations are at the corners and body center positions of a rectangular cell with two equal sides, a, and one side, c, of a different length. Each cation is coordinated by six anions in an octahedral arrangement, while each anion is coordinated by three cations (hence the designation 6(3)).

prepared proves to be the fluorite lattice (CaF_2) pictured in Figure 8-42(*a*). On the other hand, suppose the positive ions were Zn^{+2} ions and the negative ions were S^{-2} ions. Then we would need to fill only half of the available four-sided vacancies, to give a one-to-one lattice. With Zn^{+2} ions in the f.c.c. packing and with half the tetrahedral holes alternately filled with S^{-2} ions, one of the zinc sulfide lattices portrayed in Figure 8-41(*b*) results, the zinc blende lattice. In a similar way, if Zn^{+2} ions are arranged in the h.c.p. packing and alternate tetrahedral holes are filled with S^{-2} ions, the other zinc sulfide lattice is obtained, the wurtzite lattice shown in Figure 8-41(*c*).

These relationships between ionic lattices and the elemental lattices are quite pervasive. For example, the f.c.c. packing contains, in addition to tetrahedral holes, octahedral holes formed by clusters of six atoms. If Na^+ ions are arranged in an f.c.c. lattice and each one of the octahedral holes is filled with a Cl^- ion, the lattice is neutral and it has the NaCl structure, as portrayed in Figure 8-39. The CsCl lattice (Fig. 8-41(*a*)) can be obtained from the body-centered-cubic lattice (Fig. 8-33) by placing a Cl^- at the center of each cube (the body-centered position) and Cs^+ at every corner.

Since the five ionic lattices just discussed are so common (CaF_2, ZnS, zinc blende and wurtzite, NaCl, and CsCl), we see that very many ionic lattices involve positive ions in closest-packed or body-centered-cubic arrays. These packing arrangements provide interstitial positions, or holes, that are highly effective in clustering ions of opposite charge as nearest neighbors. Just as observed for the elements, this clustering seems to be a dominant factor in fixing ionic lattice crystal structures.

(d) MOLTEN SALTS

The comparison between Cl_2 and NaCl is again informative as we consider the liquid state obtained when an ionic solid is melted. Table 8-12 shows the heat effects that accompany

Table 8-12 Heats of Fusion and Sublimation of Cl_2 and NaCl

	$\Delta H^\circ_{fus.}$	$\Delta H^\circ_{sub.}$	$\frac{\Delta H^\circ_{fus.}}{\Delta H^\circ_{sub.}}$
Cl_2	1.5	6.4	0.23
NaCl	6.8	47.6	0.14
	$\Delta S^\circ_{fus.}$	$\Delta S^\circ_{sub.}$	$\frac{\Delta S^\circ_{fus.}}{\Delta S^\circ_{sub.}}$
Cl_2	8.9	29.3	0.30
NaCl	6.3	29.8	0.21

fusion and sublimation (the sum of $\Delta H_{\text{fus.}} + \Delta H_{\text{vap.}}$) of these two substances.

The heat of sublimation of solid NaCl is 7.5 times that of solid Cl_2, showing that the forces holding the ionic crystal together are much larger than those binding the van der Waals molecular crystal. This is explained in terms of chemical bonding of the ionic type in the NaCl crystal. Yet the heat of fusion of solid NaCl is a smaller fraction of its heat of sublimation (14 percent) than the corresponding fraction for Cl_2 (23 percent). This suggests that the forces holding the ionic crystal together are fairly intact in the liquid.

The entropies are even more surprising. The entropies of sublimation of solid Cl_2 and NaCl are about the same, because this quantity is dominated by the randomness of the gaseous state for each compound. Since the vapor in each case is made up of diatomic molecules (Cl_2 or NaCl), their entropies are similar. The entropies of fusion show, however, that melting solid sodium chloride, with its regular packing of positive and negative ions, increases randomness even less than melting solid chlorine. This relatively small entropy of fusion shows that the structure of the molten salt must be rather similar to that of the crystal. This implies that the ions are still present and that the sodium ions are still fairly well surrounded by chloride ions (and vice versa). To be sure, the packing is less precise and the ions have the mobility of the liquid state, but they still maintain an electrostatically favorable arrangement.

The presence of these ions in the molten salt is more positively and dramatically shown by electrical conductivity measurements. Whereas solid NaCl is an excellent insulator, molten NaCl is an excellent conductor (not as good as a metal, but very much better than most liquids and comparable to or higher than aqueous salt solutions). We conclude that molten salts contain ions, regularly but not rigidly distributed, and through their mobility, these ions are able to conduct an electrical current.

(e) THE BONDING IN IONIC CRYSTALS

It is frequently said, and many chemists believe, that ionic bonding is the simplest and best understood bonding type. Yet some disturbing aspects are revealed if one penetrates the attractive veneer. Attractive because of its simplicity—that is what makes the ionic description so popular. Consider, for instance, the bonding in the ionic NaCl molecule. This molecule is crudely represented as a positive and a negative charge placed at the measured bond distance. The charge needed to give the measured experimental dipole moment is 0.79 electronic charges. It is child's play to calculate the energy needed to separate such charges from that bond length to infinity.

$$E = \frac{q_1 q_2}{r} = \frac{[(0.79)(4.8 \cdot 10^{-10})]^2}{2.36 \cdot 10^{-8}}$$
$$= 6.1 \cdot 10^{-12} \text{ ergs/molecule}$$
$$= 88 \text{ kcal/mole} \qquad (8\text{-}12)$$

One is immediately struck by the fact that this very crude calculation is only 10 kcal away from the measured bond energy for NaCl, 98 kcal/mole. That this close agreement is fortuitous is readily proved. First, the calculation gives the energy needed to break NaCl into two *ions*, whereas the bond dissociation energy refers to the formation of two *neutral* atoms. The energy needed to form ions is larger:

$$\begin{array}{rlrl} NaCl(g) & \rightarrow Na(g) + Cl(g) & \Delta H & = 98 \text{ kcal} \\ Na(g) & \rightarrow Na^+(g) + e^-(g) & \Delta H & = 118.5 \text{ kcal} \\ Cl^-(g) + e^- & \rightarrow Cl^-(g) & \Delta H & = -83.4 \text{ kcal} \\ \hline NaCl(g) & \rightarrow Na^+(g) + Cl^-(g) & \Delta H & = 133 \text{ kcal} \end{array} \qquad (8\text{-}13)$$

Now the agreement is not so good—88 kcal is only $\frac{2}{3}$ of the experimental energy, (8-13). But wait—what were we calculating as the two ions were separated? We calculated a potential energy—that associated with the placement of two opposite charges a fixed distance apart. Where is the virial theorem? It tells us that a potential energy change $\Delta\bar{V}$ is always accompanied by a kinetic energy change $\Delta\bar{T}$. In Section 3-2, we learned that $\Delta\bar{V}$ and $\Delta\bar{T}$ are simply related to each other, and to the total energy change $\Delta\bar{E}$ (3-7):

$$\Delta\bar{E} = +\tfrac{1}{2}\Delta\bar{V} = -\Delta\bar{T}$$

According to this argument, the 88 kcal change in potential energy would be accompanied by a -44 kcal change in electronic kinetic energy. The total energy change $\Delta\bar{E}$ would then be the sum, only 44 kcal. Now the discrepancy with (8-13) is quite alarming.

But wait—we aren't finished yet. If NaCl is presented as a pair of charges set at a separation of 2.36 Å, it is still fair to ask, why don't they fall together? Experimentally, of course, they do not, so we are obliged to admit that there must be some repulsion between them in addition to the attractive part represented by (8-12). Such a repulsive term will reduce still further the 44 kcal bond-energy estimate. Now the agreement is so poor that we can no longer enthuse about a pure electrostatic model for the bonding in NaCl.

However, the devotees to this approach do not yet abandon the electrostatic model. "Polarizability" is added—that is, the spherically symmetric charge distributions appropriate to a pure ionic model are allowed to distort. Such a distortion of Cl^-, in the presence of a positive charge Na^+, will lower the energy still more—perhaps enough to account for the remaining discrepancy. This distortion causes the electron distribution to be less skewed toward chlorine, that is, it changes toward a more covalent bond description. We see that the ionic description acts only as a convenient starting point and, as more refined considerations are taken into account, the model approaches the covalent description (with ionic contribution) but from the other side.

All this is relevant to ionic crystals because quantitative treatments of crystal energies begin with the same simplicity as in the NaCl diatomic problem. As one looks more closely, though, the same caveats and adjustments are needed. Calculations of the "crystal lattice energy" are reasonably successful only for the alkali halide crystals. The famed Born–Haber cycle provides, for these cases, an experimental estimate of the lattice energy U.

$$\begin{array}{ccc} M^+(g) + X^-(g) & \xrightarrow{U} & MX(\text{solid}) \\ \downarrow \; E_I \; \downarrow & & \uparrow E_{III} \\ M(g) + X(g) & \underset{E_{II}}{\rightarrow} & M(\text{solid}) + \tfrac{1}{2}X_2(g) \end{array} \qquad (8\text{-}14)$$

Since energy is a state function,

$$U = E_I + E_{II} + E_{III} \qquad (8\text{-}15)$$

and every term on the right-hand side of (8-15) is experimentally accessible. Such values of U can then be compared to calculations of the type described for NaCl, with similar correction terms and adjustments. The parametric freedom in the treatment is considerable (for example, in the repulsion term), and sometimes special considerations are invoked: extra repulsion in lithium salts, specially large covalent contribution in silver salts, and so on. A rather startling quirk occurs in the application of the Born–Haber cycle to oxides and sulfides. To carry through the calculation (to derive U for comparison to calculations) it is necessary to assign an electron affinity to gaseous

O^{-2} and S^{-2}. In each case, the number needed is negative; that is, neither ion is stable in the gas phase. This means that the gaseous ion cannot exist, so the integrity of the Born–Haber determination of U is undermined. This does not, however, disturb the proponents of the ionic description of salt crystals—an interesting commentary on scientists and the progress of science. Let us hope that the descriptions emphasized here are more sturdy but, in any event, let us beware.

8-8 Chemical bonds and the future

Quantum mechanics has given structure and coherence to chemistry by providing a basis for understanding the chemical bond. This central concept can be developed logically and convincingly, beginning with the simpler molecules that are fully clarified by the theory. Then the various bond types can be seen to be firmly connected as each leads outward in its own direction toward more complex molecules only recently placed on the chemical shelf.

Despite this optimistic state of our knowledge, we have much to learn and there remains a strong dependence upon empirical tie-points. New experimental advances keep the subject exciting. It is alive with developments, extending from *ab initio* computations on increasingly complex molecules to discoveries of entirely new and unpredicted bond types. Thus the wheel of chemistry turns ever faster, but it rides smoothly on its sturdy hub, the concept of the chemical bond.

appendices

Ionization Energies of the Elements (kcal/mole)

$$A^{+(n-1)} \rightarrow A^{+n} + e^- \quad \Delta H = E_n$$

At. No.	Elem.	E_0^*	E_1	E_2	E_3	E_4	E_5	E_6	E_7	E_8
1	H	17.4	313.6	—	—	—	—	—	—	—
2	He	—	566.8	1254	—	—	—	—	—	—
3	Li	14	124.3	1744	2823	—	—	—	—	—
4	Be	—	214.9	419.9	3548	5020	—	—	—	—
5	B	(7)	191.3	580.0	874.5	5980	7843	—	—	—
6	C	28	259.6	562.2	1104	1487	9034	11300	—	—
7	N	—	335.1	682.8	1094	1786	2257	12727	15377	—
8	O	34	314.0	810.6	1267	1785	2624	3184	17044	20088
9	F	79	401.8	806.7	1445	2012	2634	3623	4268	21990
10	Ne	—	497.2	947.2	1500	2241	2913	3641	—	—
11	Na	(12.5)	118.5	1091	1652	2280	3192	3969	4806	6093
12	Mg	—	176.3	346.6	1848	2521	3256	4301	5186	6134
13	Al	(11)	138.0	434.1	655.9	2767	3593	4391	5567	6562
14	Si	(32)	187.9	376.8	771.7	1041	3844	4730	5682	6990
15	P	(18)	241.8	453.2	695.5	1184	1499	5083	6072	7132
16	S	48	238.9	540	807	1091	1672	2030	6711	7582
17	Cl	83.3	300.0	548.9	920.2	1230	1564	2230	2635	8032
18	Ar	—	363.4	637.0	943.3	1379	1730	2105	2860	3308
19	K	(10), (20)	100.1	733.6	1100	1405	1905	2299	2721	3574
20	Ca	—	140.9	273.8	1181	1550	1946	2514	2952	3305
21	Sc	—	151.3	297.3	570.8	1700	2120	2560	3205	3667
22	Ti	(9)	158	314.3	649.0	997.2	2301	2767	3252	3966
23	V	(22)	155	328	685	1100	1499	2975	3482	4013

Ionization Energies of the Elements (kcal/mole) (Continued)

$$A^{+(n-1)} \rightarrow A^{+n} + e^- \quad \Delta H = E_n$$

At. No.	Elem.	E_0^*	E_1	E_2	E_3	E_4	E_5	E_6	E_7	E_8
24	Cr	(23)	156.0	380.3	713.8	1140	1683	2099	3713	4266
25	Mn	—	171.4	360.7	777.0	1200	1753	2306	2744	4520
26	Fe	(13)	182	373.2	706.7	1310	—	2375	2998	3482
27	Co	(22)	181	393.2	772.4	—	—	1916	3067	3759
28	Ni	(29)	176.0	418.6	810.9	—	—	—	—	3874
29	Cu	(41.5)	178.1	467.9	849.4	—	—	—	—	—
30	Zn	—	216.6	414.2	915.6	—	—	—	—	—
31	Ga	—	138	473.0	708.0	1480	—	—	—	—
32	Ge	—	182	367.4	789.0	1050	2153	—	—	—
33	As	—	226	466	653	1160	1444	2940	—	—
34	Se	—	225	496	738	989	1568	1891	3574	—
35	Br	77.6	273.0	498	828	—	1377	2043	2375	4451
36	Kr	—	322.8	566.4	851	—	1492	1810	2560	2906
37	Rb	>5	96.31	634	920	1213	1637	1946	2288	3136
38	Sr	—	131.3	254.3	1005	1300	1651	2094	2444	2820
39	Y	—	147	282.1	473	1425	1776	2145	2675	2975
40	Zr	—	158	302.8	530.0	791.8	1898	2280	2675	3205
41	Nb	—	158.7	330.3	579.8	883	1153	2375	2883	3252
42	Mo	23	164	372.5	625.7	1070	1411	1570	2906	3528
43	Tc	—	168	351.9	681.2	—	—	—	—	—
44	Ru	—	169.8	386.5	656.4	—	—	—	—	—
45	Rh	—	172	416.7	716.1	—	—	—	—	—
46	Pd	—	192	447.9	759.2	—	—	—	—	—
47	Ag	—	174.7	495.4	803.1	—	—	—	—	—

48	Cd	—	207.4	389.9	864.2	—	—	—	—	—
49	In	—	133.4	435.0	646.5	1250	—	—	—	—
50	Sn	—	169.3	337.4	703.2	939.1	1829	—	—	—
51	Sb	—	199.2	380	583	1020	1291	2491	2744	—
52	Te	—	208	429	720	880	1383	1683	3159	—
53	I	70.9	241	440.3	—	761	1637	1914	2398	3920
54	Xe	—	279.7	489	740	1015	1384	1914	2352	2906
55	Cs	>4	89.8	579	807	—	—	—	2491	2813
56	Ba	—	120.2	230.7	818.7	—	—	—	—	2929
57	La	—	129	263.6	442.1	—	—	—	—	—
58	Ce	—	149.9	284	461	768	—	—	—	—
59	Pr	—	131	—	235	—	—	—	—	—
60	Nd	—	131	—	—	—	—	—	—	—
61	Pm	—	—	—	—	—	—	—	—	—
62	Sm	—	130	258	—	—	—	—	—	—
63	Eu	—	131	259	—	—	—	—	—	—
64	Gd	—	142	277	—	—	—	—	—	—
65	Tb	—	155	—	—	—	—	—	—	—
66	Dy	—	157	—	—	—	—	—	—	—
67	Ho	—	—	—	—	—	—	—	—	—
68	Er	—	—	—	—	—	—	—	—	—
69	Tm	—	—	—	—	—	—	—	—	—
70	Yb	—	144	279	—	—	—	—	—	—
71	Lu	—	141	339	—	—	—	—	—	—
72	Hf	—	161	344	—	—	—	—	—	—
73	Ta	—	182	374	—	—	—	—	—	—
74	W	12	184	408	—	—	—	—	—	—
75	Re	4	181	383	—	—	—	—	—	—
76	Os	—	201	392	—	—	—	—	—	—
77	Ir	—	207	392	—	—	—	—	—	—

Ionization Energies of the Elements (kcal/mole) (Continued)

$A^{+(n-1)} \rightarrow A^{+n} + e^- \quad \Delta H = E_n$

At. No.	Elem.	E_0^*	E_1	E_2	E_3	E_4	E_5	E_6	E_7	E_8
78	Pt	—	207	428	—	—	—	—	—	—
79	Au	—	213	473	—	—	—	—	—	—
80	Hg	—	240.5	432.4	788	1660	1891	—	—	—
81	Tl	—	140.8	470.9	687	1169	—	—	—	—
82	Pb	—	170.9	346.6	736	976	1587	—	—	—
83	Bi	—	168	384.7	589.4	1045	1291	2036	—	—
84	Po	—	194	447	629.6	—	—	—	—	—
85	At	—	219	464	675.7	—	—	—	—	—
86	Rn	—	247.8	493	678	—	—	—	—	—
87	Fr	—	88.3	519	773	—	—	—	—	—
88	Ra	—	121.7	233.9	—	—	—	—	—	—
89	Ac	—	159	279	461	—	—	—	—	—
90	Th	—	—	265	461	677.5	—	—	—	—
91	Pa	—	—	—	—	—	—	—	—	—
92	U	—	92	—	—	—	—	—	—	—
93	Np	—	—	—	—	—	—	—	—	—
94	Pu	—	118	—	—	—	—	—	—	—
95	Am	—	138	—	—	—	—	—	—	—

* E = electron affinity; parenthetical values are not experimental, calculated only.

Values from W. Finkelnburg and W. Humbach, *Naturwiss.*, **42,** 35 (1955); C. E. Moore, Circular of the National Bureau of Standards 467, *Atomic Energy Levels,* Vol. III, 1958; R. W. Kiser, *Tables of Ionization Potentials,* U.S. Atomic Energy Commission TID 6142 (1960).

Ionization Energies of Molecules (kcal/mole)*

Diatomics

H_2	355.8	CO	323.1	F_2	362	HF	363.7
N_2	359.3	CN	348.9	Cl_2	264.8	HCl	293.8
O_2	278.5	NO	213.3	Br_2	243.3	HBr	268.0
OH	304.0	CH	256	I_2	214.0	HI	239.4

Triatomics

CO_2	318.0	N_2O	297.5	H_2O	290.4
OCS	257.6	NO_2	225.5	H_2S	241.2
CS_2	232.5	O_3	295.2	H_2Se	227.8
SO_2	284.6	HCN	320.8	H_2Te	210.7

Polyatomics

methyl chloride	CH_3Cl	260.1	methane	CH_4	299.3
methyl bromide	CH_3Br	242.8	ethane	C_2H_6	268.6
methyl iodide	CH_3I	220.0	propane	C_3H_8	255.3
methanol	CH_3OH	250.2	n-butane	C_4H_{10}	245.1
methanthiol	CH_3SH	217.7	i-butane	C_4H_{10}	243.8
methyl amine	CH_3NH_2	206.9	n-pentane	C_5H_{12}	238.7
nitromethane	CH_3NO_2	255.5	cyclopentane	C_5H_{10}	242.8
acetonitrile	CH_3CN	281.8	ethylene	C_2H_4	242.6
formaldehyde	H_2CO	250.6	propylene	C_3H_6	224.4
acetaldehyde	CH_3CHO	235.5	1-butene	C_4H_8	220.9
formic acid	HCOOH	254.8	*trans*-2-butene	C_4H_8	210.6
acetic acid	CH_3COOH	239.2	*cis*-2-butene	C_4H_8	210.6
methyl formate	$HCOOCH_3$	249.4	1-pentene	C_5H_{10}	219.0
methyl formamide	$HCOONH_2$	236.4	cyclopentene	C_5H_8	207.8
			acetylene	C_2H_2	263.1
toluene	$C_6H_5CH_3$	203.5	benzene	C_6H_6	213.3
naphthalene	$C_{10}H_8$	187.3			

* K. Watanabe, T. Nakayama, and J. Mottl, *J. Quant. Spectroscopy and Energy Transfer*, **2,** 369–82 (1962).

Electron Affinities of Molecules

$$M^- \rightarrow M + e^- \qquad \Delta H = E_0 \text{ (kcal/mole)}$$

Diatomics

O_2	3.5, 10	CH	37	SH	53
CN	88.1	OH	42	SiH	(34)*
NO	21	NH	5.1	PH	(21)*
C_2	71.5				

Triatomics	
C_3	42
N_3	72–81
SCN	50
NH_2	33.3

Polyatomics	
SF_5	84.4
SF_6	34.4
C_6H_5	50.7
$C_6H_5CH_2$	20.8

* Parenthetical values are calculated only; no experimental data.

appendix c: electron affinities of molecules

Bond Energies, Bond Lengths, and Force Constants for Diatomic Molecules

D-1 First Row—First-Row Molecules

Each cell lists: molecule, D_0 (kcal/mole), r_0 (Å), k_0 (mdyne/Å)

	H	Li	Be	B	C	N	O	F
H	H_2 108 0.76 5.1	LiH 58 1.61 0.95	BeH 53 1.35 2.1	BH 79 1.24 2.8	CH 81 1.12 4.1	NH 86 1.05 6.0	OH 102 0.98 7.1	HF 134 0.92 8.8
Li		Li_2 25 2.68 0.25	LiBe	LiB	LiC	LiN	LiO	LiF 137 1.56 2.5
Be			Be_2 17	BeB	BeC	BeN	BeO 100 1.33 7.3	BeF 124 1.36 5.6
B				B_2 69 1.59 3.5	BC	BN 152 1.28 8.1	BO 173 1.21 13.3	BF 186 1.27 7.8
C					C_2 144 1.24 11.8	CN 188 1.18 15.8	CO 256 1.13 18.7	CF 127 1.27 7.2
N						N_2 225 1.10 22.4	NO 162 1.15 15.5	NF 62.6 1.32 5.6
O							O_2 118 1.21 11.4	OF 5.4
F								F_2 36 1.44 4.5

D-2 Second Row—Second-Row Molecules

Each cell lists: molecule, D_0 (kcal/mole), r_0 (Å), k_0 (mdyne/Å)

	Na	Mg	Al	Si	P	S	C
Na	Na_2 18 3.08 0.17						
Mg	NaMg	Mg_2					
Al	NaAl	MgAl	Al_2 40				
Si	NaSi	MgSi	AlSi	Si_2 76 2.25			
P	NaP	MgP	AlP	SiP	P_2 11.6 1.90 5.5		
S	NaS	MgS	AlS 2.03	SiS 151 1.93 4.9	PS 70	S_2 101 1.89 4.9	
Cl	NaCl 98 2.36 1.2	MgCl 63 1.8	AlCl 118 2.14 2.0	SiCl 77 2.6	PCl	SCl	Cl_2 57.1 2.00 3.2

D-3 First Row—Second-Row Molecules

Each cell: molecule; D_0 (kcal/mole); r_0 (Å)

	Na	Mg	Al	Si	P	S	Cl
H	NaH 48 1.89	MgH 47	AlH 68 1.65	SiH 75 1.52	PH 1.42	SH 82 1.35	HCl 103.1 1.27
Li	LiNa	LiMg	LiAl	LiSi	LiP	LiS	LiCl 111.9
Be	NaBe	BeMg	BeAl	BeSi	BeP	BeS	BeCl 137
B	NaB	MgB	BAl	BSi	BP	BS 119 1.61	BCl 128 1.72
C	NaC	MgC	AlC	CSi 104	CP 140 1.56	CS 175 1.53	CCl
N	NaN	MgN	AlN	SiN 105 1.57	NP 139 1.49	NS 116 1.50	NCl
O	NaO	MgO 94	AlO 120 1.62	SiO 192 1.51	PO 144 1.47	SO 124 1.49	OCl 64 1.55
F	NaF 114	MgF 107	AlF 159 1.65	SiF 130 1.60	PF	SF	FCl 61.4 1.63

D-4 First- and Second-Row Bromides and Iodides

M	D_{MBr} (kcal/mole)	r_{MBr} (Å)	D_{MI} (kcal/mole)	r_{MI} (Å)
H	87.4	1.42	71.4	1.62
Li	100.2		84.6	
Be				
B	128	1.88	104	
C	95.6			
N	67			
O	56.1		43	
F	57			
Na	86.7	2.50	72.7	2.71
Mg	59			
Al	106.6	2.30	88	
Si	70			
P				
S				
Cl	52.1		49.6	2.38
Br	46.1	2.29	42.8	
I	42.8		35.6	2.67

D-5 First- and Second-Row Ions

Ion	Bond Energy (kcal/mole)	Length (Å)
H_2^+	61	1.06
He_2^+	71	1.08
HBe^+		1.31
HB^+		1.21
HC^+	83	1.13
HN^+		1.08
HO^-		0.98
HAl^+		1.60
HP^+		1.43
HCl^+	104	1.32
CN^+		1.17
CN^-		1.07
CO^+	228	1.12
N_2^+	146 or 201	1.12
NO^+		1.06
NS^+		1.25
O_2^+	149	1.12
O_2^-		1.28
O_2^{-2}		1.49
P_2^+		2.25
Cl_2^+	101	1.89

Molecular Dipole Moments and Bond Lengths (Gas Phase)

(D = Debye Units = 10^{-18} esu-cm)

Alkali Halides			Other Diatomics			Polyatomics				Polyatomics	
	μ (D)	r_e (Å)		μ (D)	r (Å)		μ (D)	r (Å)	θ		μ (D)
LiF	6.28	1.56	LiH	5.88	1.61	H_2O	1.85	0.96	104.5°	CH_3F	1.82
LiCl	7.09	2.02	OH	1.66	0.98	H_2S	0.95	1.33	93.3°	CH_3Cl	1.94
LiBr	7.23	2.17				H_2Se	0.4	1.46	91.0°	CH_3Br	1.79
LiI	7.39	2.39	HF	1.91	0.92	H_2Te	<0.2	1.7	89.5°	CH_3I	1.64
			HCl	1.07	1.27	F_2O	0.2	1.42	103.2°		
NaF	8.12	1.93	HBr	0.79	1.42					CH_2F_2	1.96
NaCl	8.97	2.36	HI	0.38	1.62	NH_3	1.47	1.01	107.3°	CH_2Cl_2	1.60
NaBr	9.09	2.50				NF_3	0.23	1.37	102.1°	CH_2Br_2	1.5
NaI	9.21	2.71	ClF	0.88	1.63	NCl_3	0.6	~1.76		CH_2I_2	1.11
			BrF	1.29	1.76	PH_3	0.55	1.42	93.1°		
KF	8.55	2.17	IF	1.6		PF_3	1.0	1.54	104°	CHF_3	1.60
KCl	10.24	2.67				PCl_3	0.8	2.04	100.5°	$CHCl_3$	1.00
KBr	10.60	2.82	BrCl	0.57		$SbCl_3$	3.9	2.32		$CHBr_3$	1.00
KI	10.86	3.05	ICl	0.6	2.32					CHI_3	0.8
						O_3	0.53	1.28	116.8°		
RbF	8.51	2.27	IBr	1.21		NO_2	0.3	1.19	134.1°	H_2CO	2.30
RbCl	10.48	2.79				N_2O	0.18	1.13/1.19	180°	Cl_2CO	1.19
RbBr	10.78	2.94	TlF	4.2	2.08	SO_2	1.61	1.43	119.5°	Cl_2CS	0.28
RbI	11.26	3.18	TlCl	4.4	2.48	ClO_2	0.78	1.49	119°		
										HCOOH	1.52
CsF	7.85	2.35	NO	0.15	1.15	ClF_3	0.6	1.60/1.70	87.5°/180°	CH_3COOH	1.75
CsCl	10.36	2.91	CO	0.13	1.13	BrF_3	1.0	1.72/1.81	86.2°/180°	CH_3NO_2	3.46
CsBr	10.85	3.07	ClO	1.70		BrF_5	1.51	1.68/1.80	85°/90°	HCN	2.95
CsI	11.50	3.32									
			SrO	8.90	1.92						
			BaO	7.93	1.94						

Bond Lengths (Å) and Average Bond Energies (kcal/mole) for Hydrides and Halides of Formula XY_n

	Bond Length	Average Bond Energy		Bond Length	Average Bond Energy		Bond Length	Average Bond Energy		Bond Length	Average Bond Energy
Group III											
BH_3	—	88.9	BF_3	1.30	160.6	BCl_3	1.75	115.8	BBr_3	1.87	87.9
AlH_3	—	—	AlF_3	—	147.1	$AlCl_3$	—	111.3	$AlBr_3$	—	86.6
GaH_3	—	—	GaF_3	1.88	—	$GaCl_3$	—	67.2	$GaBr_3$	—	72.1
InH_3	—	—	InF_3	—	—	$InCl_3$	—	88.0	$InBr_3$	—	68.6
Group IV											
CH_4		99.3	CF_4		117	CCl_4		78.0	CBr_4		64.7
SiH_4	1.48	77.3	SiF_4	1.55	140	$SiCl_4$	2.01	95.6	$SiBr_4$	2.15	78.8
GeH_4	1.53	69.2	GeF_4	1.67	—	$GeCl_4$	2.08	81.4	$GeBr_4$	2.29	67.1
SnH_4	—	60.4	SnF_4	—	—	$SnCl_4$	2.30	75.3	$SnBr_4$	2.44	64.6
			SiF_2	—	147.4	$SnCl_2$	2.42	—	$SnBr_2$	2.55	—
Group V											
NH_3	1.01	93.4	NF_3	1.37	66.5	NCl_3	—	—	NBr_3	—	—
PH_3	1.44	76.7	PF_3	1.54	117	PCl_3	2.04	76.0	PBr_3	2.18	62.9
AsH_3	1.52	70.9	AsF_3	1.71	116.3	$AsCl_3$	2.16	73.8	$AsBr_3$	2.31	61.2
SbH_3	1.71	61.4	SbF_3	2.03	—	$SbCl_3$	2.36	75.0	$SbBr_3$	2.52	63.1
			PF_5	1.57(a)* 1.53(e)*	91.3	PCl_5	2.19(a)* 2.04(e)*	62.1	PBr_5		54.7
						SbCl	2.34(a)* 2.29(e)*	60.5			

Bond Lengths (Å) and Average Bond Energies (kcal/mole) for Hydrides and Halides of Formula XY_n (Continued)

	Bond Length	Average Bond Energy		Bond Length	Average Bond Energy		Bond Length	Average Bond Energy		Bond Length	Average Bond Energy
Group VI											
OH_2	0.957	110.8	OF_2	1.42	51.3	OCl_2	1.70	49.3	OBr_2	—	—
SH_2	1.33	87.9	SF_2	—	—	SCl_2	1.99	64.8	SBr_2	—	—
SeH_2	1.46	75.7	SeF_2	—	—	$SeCl_2$	—	60.1	$SeBr_2$	—	56.4
TeH_2	1.7	63.7	TeF_2	—	—	$TeCl_2$	2.36	—	$TeBr_2$	2.51	—
			SF_4	1.65(a)* 1.55(e)†							
			SF_6	1.58							
			SeF_6	1.70							
			TeF_6	1.84							

* (a) and (e) refer to apical and equatorial bonds of a trigonal bipyramid.
† Trigonal bipyramid with one equatorial position empty.

Bond Lengths (Å), Bond Dissociation Energies (kcal/mole), and Dipole Moments (μ (Debye)) in Selected Polyatomic Molecules

G-1 X-NO Molecules

Molecule	μ	D_{X-N}	r_{X-N}	D_{N-O}	r_{N-O}
NO	0.153	—	—	162	1.15
HNO		49	1.06		1.21
NNO	0.167		1.13		1.19
ONO	0.316	—	—	73	1.19
FNO	1.81	55	1.52		1.13
ClNO	1.9		1.97		1.14
BrNO	1.8		2.14		1.15
ONNO		4			

G-2 X-NO$_2$ Molecules

Molecule	μ	D_{X-N}	r_{X-N}	D_{N-O}	r_{N-O}
$(ONO_2)^-$	0	—	—	—	1.24
HNO_2			0.98		1.46
FNO_2	0.47		1.36		1.23
$ClNO_2$	0.53		1.80		1.20
$ONNO_2$		9.6			
O_2NNO_2		13	1.75		1.18
$HONO_2$			1.41		1.22

G-3 X$_2$Y$_2$ and X$_2$Y$_4$ Molecules

Molecule	μ	D_{X-N}	r_{X-N}	D_{N-O}	r_{N-O}
H_2O_2	2.2	51	1.47	90	0.97
O_2F_2	1.44		1.22		1.58
C_2N_2	0	145	—	—	—
cis-N_2F_2	0.16	—	1.25	—	1.44
gauche N_2F_4	0.26	19.9	(1.47)*	—	(1.37)*
N_2H_4	—	—	1.45	—	1.02
S_2H_2	—	68	2.05	—	1.33
S_2F_2	1.45	—	—	—	—
S_2Cl_2	—	—	1.97	—	2.07
S_2Br_2	—	—	1.98	—	2.24

* Parenthetical values are estimated.

Average Bond Energies (kcal/mole) and Average Bond Lengths (Å) in Polyatomic Molecules

Bond	Average Energy	Average Length	
H—C	98.8	1.10 (monosubstituted alkanes)	
H—N	93.4	1.00	
H—O	110.6	0.97	
C—C	82.6	1.54 } aromatic, 1.39	
C=C	145.8	1.34 } aromatic, 1.39	
C≡C	199.6	1.20	
C—N	72.8	1.47	
C=N	147		
C≡N	212.6	1.16	
C—O	85.5	1.43 (alcohols, ethers)	
		1.36 (carboxylic acids)	
C=O	178	1.22 (aldehydes and ketones)	
	192.1	1.16 (CO_2)	
C—F	116	1.38 (alkanes)	1.33 (others)
C—Cl	81	1.77 (alkanes)	1.72 (alkenes)
		1.70 (aromatics)	
C—Br	68	1.94 (alkanes)	1.89 (alkenes)
		1.85 (aromatics)	
C—I	51	2.21	
N—O	53	1.36	
N=O	145	1.21	
N—F	65	1.36	
N—Cl	46	1.75	
XeO	27	1.76	
XeF	32	1.97	

physical constants

Avogadro's number	N_0	6.023×10^{23} mole^{-1}
Planck constant	h	6.626×10^{-27} erg sec
		$9.537 \times 10^{-14} \left(\frac{\text{kcal}}{\text{mole}}\right)$ sec
Speed of light (vacuum)	c	2.998×10^{10} cm/sec
Elementary charge	e	1.602×10^{-19} coulomb
		4.803×10^{-10} e.s.u.
Electron rest mass	m_e	9.109×10^{-28} g
Proton rest mass	m_p	1.673×10^{-24} g
Bohr radius	a_0	0.5292×10^{-8} cm
H ionization energy	E_H	313.6 kcal/mole
Gas constant	R	1.987 cal/mole degK
		8.314 joule/mole degK
		62.36 liter torr/mole degK
		8.205×10^{-2} liter atm/mole degK
2.303 *RT* (25°C)		1.364 kcal/mole
Volume 1 mole ideal gas		
at STP (1 atm, 0°C)		22.41 liter
at 1 atm, 25°C		24.47 liter
Faraday constant	$\mathcal{F}$	96,487 coulombs per mole of electrons
		23,061 calories per volt per mole of electrons
Standard atmosphere	atm	760 torr = 760 mm of mercury
		1.013×10^6 dyne/cm^2
$\pi = 3.1416$	$e = 2.71828$	$\log_e 10 = 2.30259 = 1/0.43429$
		$\log_{10} x = 0.43429 \log_e x$

These values are those recommended by the National Bureau of Standards as listed in the October, 1963, NBS Technical News Bulletin and reprinted in the *Journal of Chemical Education*, **40**, 642 (1963). All values are based on the Unified atomic weight scale in which $C^{12} = 12$ exactly.

conversion factors

	TO CONVERT FROM →	TO	MULTIPLY BY ↓
6.947×10^{-14}	erg/molecule	kcal/mole	1.439×10^{13}
4.336×10^{-2}	$\frac{\text{electron volts}}{\text{molecule}}$	kcal/mole	23.061
3.498×10^{2}	$\frac{\text{cm}^{-1}}{\text{molecule}}$	kcal/mole	2.859×10^{-3}
5.034×10^{15}	$\frac{\text{cm}^{-1}}{\text{molecule}}$	erg/molecule	1.986×10^{-16}
2.390×10^{-4}	kcal/mole	joules/mole	4.1840×10^{3}
↑ MULTIPLY BY	TO ←	FROM TO CONVERT	

index

Formula Index

Subject Index